Chemistry and Physics of Carbon

VOLUME 28

Chemistry and Physics of Carbon

VOLUME 24

Chemistry and Physics of Carbon

Edited by
LJUBISA R. RADOVIC

Department of Energy and Geo-Environmental Engineering
The Pennsylvania State University
University Park, Pennsylvania

VOLUME 28

CRC Press
Taylor & Francis Group
Boca Raton London New York

CRC Press is an imprint of the
Taylor & Francis Group, an **informa** business

CRC Press
Taylor & Francis Group
6000 Broken Sound Parkway NW, Suite 300
Boca Raton, FL 33487-2742

First issued in paperback 2019

ISBN-13: 978-0-8247-0987-7 (hbk)
ISBN-13: 978-0-367-39527-8 (pbk)

**Visit the Taylor & Francis Web site at
http://www.taylorandfrancis.com**

**and the CRC Press Web site at
http://www.crcpress.com**

Library of Congress Cataloging-in-Publication Data
A catalog record for this book is available from the Library of Congress.

Preface

This volume contains four chapters written by distinguished scientists from four continents. The middle two deal with the chemistry and physics of carbon surfaces, while the first and the last are devoted to the structural properties of carbons and their precursors.

Five years ago, Peter Thrower lamented on these pages that we have not been able to publish reviews on the most novel forms of carbon. The contribution from Peter Harris, who recently wrote a notable monograph on carbon nanotubes, is thus timely and valuable, but it is inevitably controversial because much of the dust has not yet settled; the author takes the bull by the horns and makes a valiant attempt to bridge the gap between the flat and the curved sp^2-hybridized carbon structures. Its inclusion in this series is meant to stimulate additional research that will resolve two of the most intriguing fundamental questions in carbon science: "Where exactly do nanotubes end and vapor-grown carbon fibers begin?" and, more to the point, "Where exactly do fullerenes end and soot begin?" Indeed, ever since the discovery of buckminsterfullerene, almost two decades ago, its relationship to conventional or classic carbon materials (see *Carbon* 2002; 40: 2279–2282) has provoked extreme reactions: it has been both ignored and arguably exaggerated. The case of soot (or carbon black) has been highly publicized; there have been refusals to accept any connection by most prominent members of the combustion and carbon communities, and there have been reinterpretations of the structure by acknowledged experts under the influence of the beauty of C_{60}. Professor Harris boldly argues that even beyond soot—in chars and other flat sp^2-hybridized carbons—there is indeed evidence for the existence (and prevalence?) of five-member ring structures. He acknowledges that "[m]any questions . . . remain" and we look forward to publishing more definitive accounts in the not too distant future.

It is a privilege to welcome Professors Keith Gubbins and Katsumi Kaneko to this series. Their pioneering and highly productive research on adsorption phenomena had kept them very busy. In Chapter 2, together with an eclectic group of collaborators, they offer a comprehensive and authoritative insight into the molecular-level structure of carbons. Of particular interest is how they build upon their Figure 56, in a way that is both different from and complementary to what Professor Harris does with the same structural representation (see Figure 10 in Chapter 1), which has been the "state of the art" for half a century. They show how modern computational methods are increasingly allowing us to sacrifice less and less of the physical reality of microporous carbons, which has been revealed (or constructed!?) through many years of often painstaking experimentation. The authors optimistically conclude that, with such modeling approaches, we shall soon be able to predict with increasing confidence the ever-important adsorption, separation, and diffusion phenomena in carbon materials.

The group of Professor Duong Do, another acknowledged authority on adsorption in microporous carbons, joins Professor Kaneko in offering a brief but very important review of water/carbon interactions. I know of no other topic in which the chemistry and the physics of carbon surfaces *both* come to bear on the behavior of carbons in as many practical applications. This topic has hardly been discussed in this series (for a notable exception, see Chapter 2, by Turov and Leboda, in Vol. 27), despite much research. The reason for such an unsatisfactory state of affairs is reinforced by the authors themselves: they show that much progress has been made recently, but acknowledge at the same time that "much remains to be done to improve our understanding of the mechanisms" of water adsorption.

It is a pleasure to welcome Dr. Rosa Menéndez as a contributor to this volume. She and her colleagues at INCAR have been providing new insights into coal carbonization for over a decade, ever since she spent time with Professor Harry Marsh at the University of Newcastle upon Tyne. Pyrolysis and carbonization phenomena are of crucial importance in many applications of carbon materials. They have been reviewed in this series from many vantage points, but this is the first contribution from INCAR scientists, and the emphasis here is on an endangered, unique, and complex precursor: coal tar pitch. The authors show how much progress can be made when a battery of complementary techniques is applied to advance fundamental understanding, and they argue that such compositional information can be used to take maximum advantage of the virtues of this abundant raw material while minimizing its acknowledged environmental and toxicological liabilities.

As we continue with the publication of the Chemistry and Physics of Carbon series in its fifth decade, along the lines described in the preface to Volume 27, we want to stimulate our readers to submit authoritative, critical, and comprehensive reviews, by clearly spelling out our ambitious goals: 1) no issue shall be considered settled and no proposal shall be considered accepted unless and until consecrated in the series and 2) no researcher shall be considered an acknowledged expert unless and until his or her work has been analyzed and placed into historical context in the series. Enticing enough?

Ljubisa R. Radovic

Contributors to Volume 28

Teresa J. Bandosz   Department of Chemistry, City College of the City University of New York, New York, New York, U.S.A.

Mark J. Biggs   Department of Engineering, University of Edinburgh, Edinburgh, Scotland, United Kingdom

D. D. Do   Department of Chemical Engineering, University of Queensland, Brisbane, Queensland, Australia

Marcos Granda   Department of Chemistry of Materials, Instituto Nacional del Carbón (CSIC), Oviedo, Spain

Keith E. Gubbins   Department of Chemical Engineering, North Carolina State University, Raleigh, North Carolina, U.S.A.

Peter J. F. Harris   Department of Chemistry, University of Reading, Whiteknights, Reading, United Kingdom

Y. Hattori   Department of Chemistry, Shinshou University, Nagano, Japan

T. Iiyama   Center for Frontier Electronics and Photonics, Chiba University, Chiba, Japan

Katsumi Kaneko   Department of Chemistry, Chiba University, Chiba, Japan

Rosa Menéndez   Department of Chemistry of Materials, Instituto Nacional del Carbón (CSIC), Oviedo, Spain

D. Mowla*   Department of Chemical Engineering, University of Queensland, Brisbane, Queensland, Australia

Jorge Pikunic   Department of Chemical Engineering, North Carolina State University, Raleigh, North Carolina, U.S.A.

Ricardo Santamaría   Department of Chemistry of Materials, Instituto Nacional del Carbón (CSIC), Oviedo, Spain

Kendall T. Thomson   School of Chemical Engineering, Purdue University, West Lafayette, Indiana, U.S.A.

* *Current affiliation*: Department of Chemical Engineering, Shiraz University, Shiraz, Iran.

Contents of Volume 28

Contents of Other Volumes

Chemistry and Physics of Carbon

VOLUME 28

Chemistry and Physics of Carbon

VOLUME 19

1

Impact of the Discovery of Fullerenes on Carbon Science

Peter J. F. Harris

University of Reading, Whiteknights, Reading, United Kingdom

I. INTRODUCTION

The discovery of C_{60} in 1985, and subsequent identification of fullerene-related structures such as carbon nanotubes and nanoparticles, has stimulated a large volume of research. To date, most of this research has concentrated on the remarkable properties of the new carbons themselves. Carbon nanotubes in particular have been shown to have quite exceptional mechanical and electronic properties, which could have a significant technological impact [1]. However, the knowledge gained as a result of fullerene research may be equally important in providing us with a fresh perspective on forms of carbon that have been known for many years. Charcoal, or char, for example, has been used in various ways for millennia, and is currently of enormous importance in purifying water and air supplies [2–4]. Yet despite a huge amount of research, many questions remain about the atomic structure of char and the exact reasons for its remarkable adsorptive properties. Recently, detailed studies of chars both before and after high-temperature heat treatments have provided evidence that they might contain fullerene-related elements. There are also reasons to believe that the class of chars known as glassy carbons may have a fullerene-like structure. In this chapter we assess the evidence that well-known forms of carbon may have structures related to those of the fullerenes. A detailed discussion is given of the structure of chars (nongraphitizing carbons) and of the various models for their structure that have been put forward since the classic work of Rosalind Franklin. In subsequent sections, the structure of glassy carbon, carbon fibers, soot (carbon black), and graphitic particles in spherulitic graphite cast iron are considered. The evidence that fullerenes or fullerene-like structures may be present in naturally occurring forms of carbon is then summarized. To begin, an outline of the main structural features of fullerenes, carbon nanotubes, and carbon nanoparticles will be given, together with a brief discussion of their stability.

II. STRUCTURE AND STABILITY OF FULLERENES

The defining structural feature of the fullerenes is the pentagonal carbon ring. In buckminsterfullerene we have 12 pentagons distributed symmetrically in an

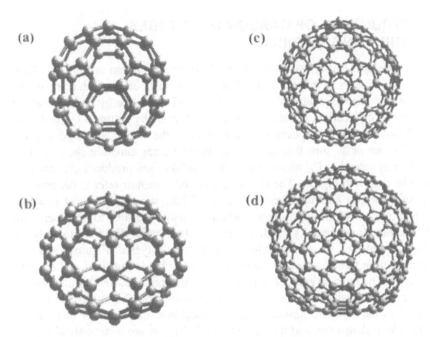

FIG. 1 The structure of (a) buckminsterfullerene, C_{60}, (b) C_{70} (D_{5h}), (c) icosahedral C_{180}, (d) icosahedral C_{240}.

icosahedral structure, as shown in Fig. 1a. It will be noted that in this icosahedral arrangement all the pentagons are isolated from each other. This is important because adjacent pentagonal rings form an unstable bonding arrangement. All other closed-cage isomers of C_{60}, and all smaller fullerenes, are less stable than buckminsterfullerene since they have adjacent pentagons. Above C_{60}, the next fullerene with isolated pentagons is the C_{70} structure shown in Fig. 1b. For higher fullerenes, the number of structures with isolated pentagonal rings increases rapidly with size. For example, C_{100} has 450 isolated-pentagon isomers [5]. Most of these higher fullerenes have low symmetry; only a very small number of them have the icosahedral symmetry of C_{60}. Examples of giant fullerenes that can have icosahedral symmetry are C_{180} and C_{240}, as shown in Fig. 1c and 1d.

    There have been many studies of the stability of fullerenes as a function of size [e.g., 6, 7]. These show that, in general, stability increases with size. Experimentally, there is evidence that C_{60} is unstable with respect to large, multiwalled fullerenes. This was demonstrated by Mochida and colleagues, who heated C_{60} and C_{70} in a sublimation-limiting furnace [8]. They showed that the cage structure broke down at 900–1000°C, whereas at 2400°C fullerene-like "hollow spheres" with diameters in the range 10–20 nm were formed.

III. STRUCTURE OF CARBON NANOTUBES AND NANOPARTICLES

Fullerene-related carbon nanotubes were discovered by Iijima in 1991 [9]. They consist of cylinders of graphite, closed at each end with caps which contain precisely six pentagonal rings. We can illustrate their structure by considering the two "archetypal" carbon nanotubes that can be formed by cutting a C_{60} molecule in half and placing a graphene cylinder between the two halves. Dividing C_{60} parallel to one of the threefold axes results in the zig-zag nanotube shown in Fig. 2a, whereas bisecting C_{60} along one of the fivefold axes produces the armchair nanotube shown in Fig. 2b. The terms *zig-zag* and *armchair* refer to the arrangement of hexagons around the circumference. There is a third class of structure in which the hexagons are arranged helically around the tube axis. Experimentally, the tubes are generally much less perfect than the idealized versions shown in Fig. 2, and may be either multilayered or singlelayered. Figure 3 shows a high-resolution transmission electron microscopy (TEM) image of multilayered nanotubes. The multilayered tubes range in length from a few tens of nanometers to several micrometers, and in outer diameter from about 2.5 nm to 30 nm. The end-caps of the tubes are sometimes symmetrical in shape, but more often asymmetric. Conical structures of the kind shown in Fig. 4a are commonly observed. This type of structure is believed to result from the presence of a single pentagon at the position indicated by the arrow, with five additional pentagons at the apex

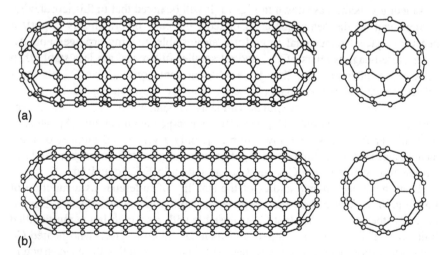

(a)

(b)

FIG. 2 Drawings of the two nanotubes that can be capped by one half of a C_{60} molecule: (a) zig-zag (9,0) structure, (b) armchair (5,5) structure [10].

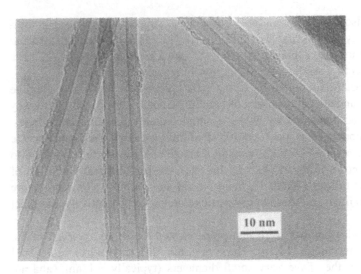

FIG. 3 TEM image of multiwalled nanotubes.

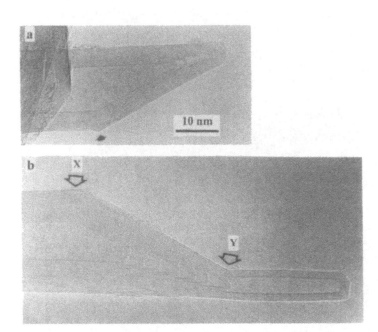

FIG. 4 Images of typical multiwall nanotube caps: (a) cap with asymmetric cone structure; (b) cap with bill-like structure [11].

of the cone. Also quite common are complex cap structures displaying a "bill-like" morphology such as that shown in Fig. 4b [11]. This structure results from the presence of a single pentagon at point "X" and a heptagon at point "Y." The heptagon results in a saddle point, or region of negative curvature.

The nanotubes first reported by Iijima were prepared by vaporising graphite in a carbon arc under an atmosphere of helium. Nanotubes produced in this way are invariably accompanied by other material, notably carbon nanoparticles. These can be thought of as giant, multilayered fullerenes, and they range in size from ~5 nm to ~15 nm. A high-resolution image of a nanoparticle attached to a nanotube is shown in Fig. 5 [12]. In this case, the particle clearly consists of three concentric fullerene shells. A more typical nanoparticle, with many more layers, is shown in Fig. 6. These larger particles are probably relatively imperfect in structure.

Single-layered nanotubes were first prepared in 1993 using a variant of the arc evaporation technique [13,14]. These are quite different from multilayered nanotubes in that they have very small diameters (typically ~1 nm), and are curled and looped rather than straight. They will not be considered further here

FIG. 5 Image of small carbon nanoparticle on nanotube surface [12]. The particle has three concentric layers.

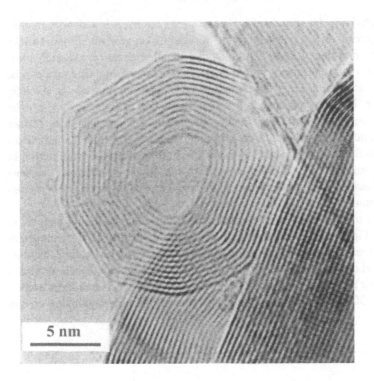

FIG. 6 Image of a typical faceted nanoparticle prepared by arc evaporation.

because they have no parallel among well-known forms of carbon discussed in this chapter.

The stability of multilayered carbon nanotubes and nanoparticles has not been studied in detail experimentally. However, we know that they are formed at the center of graphite electrodes during arcing, where temperatures probably approach 3000°C. It is reasonable to assume, therefore, that nanotubes and nanoparticles could withstand being reheated to such temperatures (in an inert atmosphere) without significant change.

IV. STRUCTURE OF CHARS (NONGRAPHITIZING CARBONS)

A. Background

Carbons formed by the pyrolysis of organic materials fall into two distinct classes, cokes and chars. The two types of carbon have quite different physical properties.

Cokes are relatively dense and soft, whereas chars are hard, low-density materials. Although cokes may be porous, this porosity is on a relatively large scale. Chars, on the other hand, have a high degree of microporosity, although some of this porosity is usually inaccessible to gases. The internal surface area can be enhanced by activation, i.e., mild oxidation with a gas such as carbon dioxide, steam, or air. In this way surface areas of the order of 2000 m^2 g^{-1} can be achieved. There is another key distinction between cokes and chars: the former can be converted to graphite by high-temperature annealing while the latter cannot. This was first demonstrated by Rosalind Franklin using X-ray diffraction in 1951 [15]. Franklin coined the terms *graphitizing* and *nongraphitizing carbons* to describe these two classes of carbon, and these terms will be used in the discussion that follows in place of the less descriptive "coke" and "char."

The effect of heat treatment on graphitizing and nongraphitizing carbons is illustrated in Fig. 7, taken from the work of Emmerich [16]. Here, the parameters L_a and L_c, defined as the length and thickness respectively of the graphite lamellae within the carbons, is plotted against temperature (note that the L_a and L_c scales are logarithmic). It can be seen that L_a for the graphitizing carbon reaches a value of 100 nm at 3000°C, while the maximum value for the nongraphitizing carbon

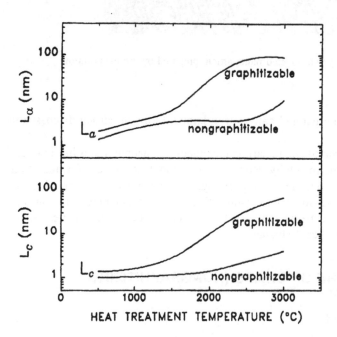

FIG. 7 Variation of L_a and L_c with heat treatment temperature for graphitizing and nongraphitizing carbons [16].

is only 10 nm. The L_c value for the graphitizing carbon also approaches 100 nm, while for the nongraphitizing carbons the maximal figure is ~4 nm. Extensive graphite crystallites are not formed in nongraphitizing carbons, even at the highest temperatures.

The distinction between graphitizing and nongraphitizing carbons can be illustrated most clearly using *transmission electron microscopy* TEM. Figure 8a shows a TEM image of a typical nongraphitizing carbon prepared by the pyrolysis of sucrose in an inert atmosphere at 1000°C. The inset shows a diffraction pattern recorded from an area approximately 0.25 μm in diameter. The image shows the structure to be disordered and isotropic, consisting of tightly curled single carbon layers, with no obvious graphitization. The diffraction pattern shows symmetrical rings, confirming the isotropic structure. The appearance of graphitizing carbons, on the other hand, approximates much more closely to that of graphite. This can be seen in the TEM micrograph of a carbon prepared from anthracene, which is shown in Fig. 8b. Here the structure contains small, approximately flat carbon layers, packed tightly together with a high degree of alignment. The fragments can be considered as somewhat imperfect graphene sheets. The diffraction pattern for the graphitizing carbon consists of arcs rather than symmetrical rings, confirming that the layers are preferentially aligned along a particular direction. The bright, narrow arcs in this pattern correspond to the interlayer {0002} spacings, while the other reflections appear as broader, less intense arcs.

Transmission electron micrographs showing the effect of high-temperature heat treatments on the structure of nongraphitizing and graphitizing carbons are

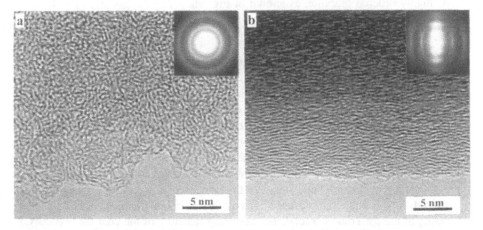

FIG. 8 (a) High resolution TEM image of carbon prepared by pyrolysis of sucrose in nitrogen at 1000°C. (b) Carbon prepared by pyrolysis of anthracene at 1000°C. Insets show selected area diffraction patterns [24].

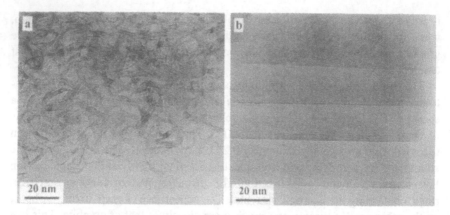

FIG. 9 Micrographs of (a) sucrose carbon and (b) anthracene carbon following heat treatment at 2300°C.

shown in Fig. 9 (note that the magnification here is much lower than for Fig. 8). In the case of the nongraphitizing carbon, heating at 2300°C in an inert atmosphere produces the disordered, porous material shown in Fig. 9a. This structure is made up of curved and faceted graphitic layer planes, typically 1–2 nm in thickness and 5–15 nm in length, enclosing randomly shaped pores. A few somewhat larger graphite crystallites are present, but there is no macroscopic graphitization. In contrast, heat treatment of the anthracene-derived carbon produces large crystals of highly ordered graphite, as shown in Fig. 9b.

B. Early Models

The first structural models of graphitizing and nongraphitizing carbons were put forward by Franklin in her 1951 paper. In these models, the basic units are small graphitic crystallites containing a few layer planes, which are joined together by cross-links. The precise nature of the cross-links is not specified. A schematic illustration of Franklin's models is shown in Fig. 10. Using these models, she put forward an explanation of why nongraphitizing carbons cannot be converted by heat treatment to graphite, and this is still generally accepted in the carbon community. Her theory will now be summarized. During carbonization the incipient stacking of the graphene sheets in the nongraphitizing carbon is largely prevented. At this stage the presence of cross-links, internal hydrogen, and the viscosity of the material is crucial. The resulting structure of the char (at ~1000°C) consists of randomly ordered crystallites, held together by residual cross-links and van der Waals forces. During high-temperature treatment, even though these cross-links may be broken, the activation energy for the motion of entire crystal-

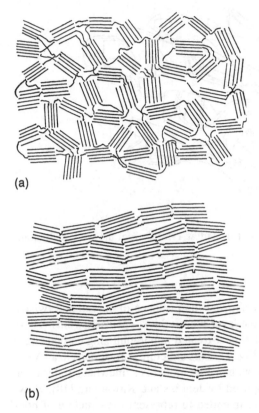

FIG. 10 Franklin's representations of (a) nongraphitizing and (b) graphitizing carbons [15].

lites, required for achieving the structure of graphite, is too high and graphite is not formed. On the other hand, the structural units in a graphitizing carbon are approximately parallel to each other, and the transformation of such a structure into crystalline graphite would be expected to be relatively facile. While Franklin's ideas on graphitizing and nongraphitizing carbons may be broadly correct, they are in some ways incomplete. The key point, as mentioned above, is that the nature of the cross-links between the graphitic fragments is not specified, so the causes of the sharply differing properties of graphitizing and nongraphitizing carbons are not explained.

The advent of high-resolution electron microscopy in the early 1970s made it possible to image the structure of nongraphitizing carbons directly. In a typical study, Ban et al. [17] examined carbons prepared from polyvinylidene chloride (PVDC) following heat treatments at temperatures in the range 530–2700°C.

FIG. 11 Model of a PVDC carbon heat treated at 1950°C from the work of Ban and colleagues [17].

Images of these carbons apparently showed the presence of curved and twisted graphite sheets, typically two or three layer planes thick, enclosing voids. These images led the authors to suggest that heat-treated nongraphitizing carbons have a ribbon-like structure, as shown in Fig. 11. This structure corresponds to a PVDC carbon heat treated at 1950°C. This ribbon-like model is somewhat similar to an earlier model of glassy carbon proposed by Jenkins and Kawamura [18]. However, models of this kind, which are intended to represent the structure of non-graphitizing carbons following high-temperature heat treatment, have serious weaknesses. Such models consist of curved and twisted graphene sheets enclosing irregularly shaped pores. However, graphene sheets are known to be highly flexible, and would therefore be expected to become ever more closely folded together at high temperatures in order to reduce surface energy. Indeed, tightly folded graphene sheets are frequently seen in carbons that have been exposed to extreme conditions [19]. Thus, structures like the one shown in Fig. 11 would be unlikely to be stable at very high temperatures.

It has also been pointed out by Oberlin [20] that the models put forward by Jenkins, Ban, and their colleagues were based on a questionable interpretation of the electron micrographs. In most micrographs of graphitized carbons, only the {0002} fringes are resolved, and these are only visible when they are approximately parallel to the electron beam. Therefore, such images tend to have a ribbon-like appearance. However, since only a part of the structure is being imaged, this appearance can be misleading, and the true three-dimensional structure may be more cage-like than ribbon-like. This is a very important point that must always be borne in mind when analyzing images of graphitic carbons.

The models of nongraphitizing carbons described so far have assumed that the carbon atoms are exclusively sp^2 and are bonded in hexagonal rings. Some authors have suggested that sp^3-bonded atoms may be present in these carbons [e.g., 21], basing their arguments on an analysis of X-ray diffraction patterns. The presence of diamond-like domains would be consistent with the hardness of nongraphitizing carbons and might also explain their extreme resistance to graphitization. A serious problem with these models is that sp^3 carbon is unstable at high temperatures. Diamond is converted to graphite at 1700°C, while tetrahedrally bonded carbon atoms in amorphous films are unstable above ~700°C [22]. Therefore, the presence of sp^3 atoms in a carbon cannot explain the resistance of the carbon to graphitization at high temperatures. It should also be noted that more recent diffraction studies of nongraphitizing carbons have suggested that sp^3-bonded atoms are not present (see next section).

C. Evidence for Fullerene-like Structures in Chars

In the mid-1990s this author and colleagues initiated a series of high-resolution TEM studies of some typical nongraphitizing microporous carbons, with the primary aim of establishing whether fullerene-like structures were present [23–25]. The first such study involved examining carbons prepared from PVDC and sucrose, after heat treatments at temperatures in the range 2100–2600°C [22]. The carbons subjected to very high temperatures had rather disordered structures similar to that shown in Fig. 9(a). Careful examination of the heated carbons showed that they often contained closed nanoparticles; examples can be seen in Fig. 12.

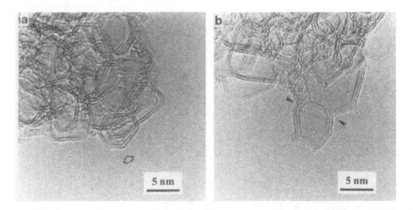

FIG. 12 (a) Micrograph showing closed structure in PVDC-derived carbon heated at 2600°C. (b) Another micrograph of same sample, with arrows showing regions of negative curvature [23].

The particles were usually faceted, and often hexagonal or pentagonal in shape. Sometimes, faceted layer planes enclosed two or more of the nanoparticles, as shown in Fig. 12b. Here, the arrows indicate two saddle-points, similar to that shown in Fig. 4b. Further examples of nanostructures in a heat-treated carbon are shown in Fig. 13. The structure in Fig. 13b strongly resembles the cap of a multilayered carbon nanotube. The closed nature of the nanoparticles, their hexagonal or pentagonal shapes, and other features such as the saddle-points strongly suggest that the particles have fullerene-like structures. Indeed, in many cases the particles resemble those produced by arc-evaporation in a fullerene generator (see Fig. 6) although in the latter case the particles usually contain many more layers.

The observation of fullerene-related nanoparticles in the heat-treated carbons suggested that the original, freshly prepared carbons may also have had fullerene-related structures (see next section). However, obtaining direct evidence for this is difficult. High-resolution electron micrographs of freshly prepared carbons, such as that shown in Fig. 8a, are usually rather featureless, and do not reveal the detailed structure. However, occasionally, very small closed particles can be found in the carbons [25]. The presence of such particles provides strong circumstantial evidence that the surrounding carbon may have a fullerene-related structure.

Some microporous carbons have somewhat less tightly packed structures than those shown in Fig. 8a. An example is the commercial activated carbon shown in Fig. 14. The "open" microstructure of this carbon, which may be partly a result of activation, allows the individual curved graphene sheets to be seen quite

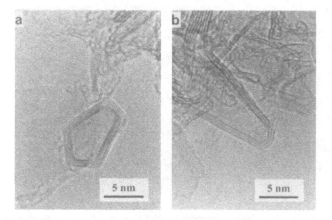

FIG. 13 Nanostructures in sucrose-derived carbon following heat treatment at 2600°C. (a) Discrete nanoparticle. (b) Structure resembling nanotube cap [23].

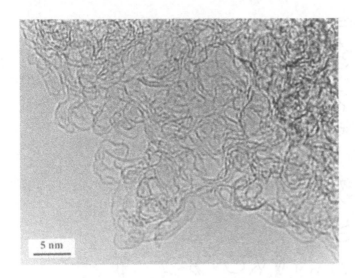

5 nm

FIG. 14 High-resolution TEM image of commercial activated carbon (Norit GSX).

clearly. It is notable that the micropores here tend to have random shapes, and are not always "slit-like," as assumed in many theoretical models of activated carbons [e.g., 26].

In addition to high-resolution TEM, diffraction methods have been widely applied to microporous and activated carbons [e.g., 26–28]. However, the interpretation of diffraction data from these highly disordered materials is not straightforward. As already mentioned, some early X-ray diffraction studies were interpreted as providing evidence for the presence of sp³-bonded atoms. More recent neutron diffraction studies have suggested that nongraphitizing carbons consist entirely of sp² atoms [27]. It is less clear whether diffraction methods can establish whether the atoms are bonded in pentagonal or hexagonal rings. Petkov *et al.* [28] have interpreted neutron diffraction data of carbons produced by pyrolysis of poly(furfuryl alcohol) in terms of a structure containing nonhexagonal rings, but other interpretations may also be possible.

D. A New Model for the Structure of Char

The observations described in the previous section suggested that nongraphitizing carbons may have fullerene-like microstructures. The present author and colleagues therefore proposed a model for the structure of nongraphitizing carbons consisting of discrete fragments of curved carbon sheets, in which pentagons and heptagons are dispersed randomly throughout networks of hexagons, as illus-

FIG. 15 Schematic illustration of a model for the structure of nongraphitizing carbons based on fullerene-like elements [24].

trated in Fig. 15 [23,24]. The size of the micropores in this model would be of the order of 0.5–1.0 nm, which is similar to the pore sizes observed in typical microporous carbons.

The model has some similarities to the "random schwarzite" structure put forward by Townsend and colleagues in 1992 [29]. This was a variation of the ordered schwarzite structures proposed earlier by Mackay and Terrones [30] and others. The key feature of schwarzite structures is the presence of negative curvature (i.e., saddle points), due to the presence of heptagonal rings. An example of a disordered schwarzite structure, is shown in Fig. 16 [29]. The fragment shown contains 38 five-membered rings, 394 six-membered rings, 155 seven-membered rings, 12 eight-membered rings, and 1 nine-membered ring. The structure is continuous, with no edges or unsatisfied valencies. Again the micropores in this structure are of the order of 0.5–1.0 nm in size, but there are aspects of the model that do not seem appropriate for nongraphitizing carbons. In particular, the random schwarzite structure consists of a single continuous sheet, while nongraphitizing carbons are believed to be made up of relatively small fragments. An unbroken sheet such as that illustrated in Fig. 16, with no edges or dangling bonds, would have a very low reactivity, unlike most nongraphitizing carbons, which can be readily oxidized at moderate temperatures. Also, our model envisages a higher proportion of pentagons and a smaller proportion of heptagons than in the random schwarzite structure. It is worth noting that a structure of the kind

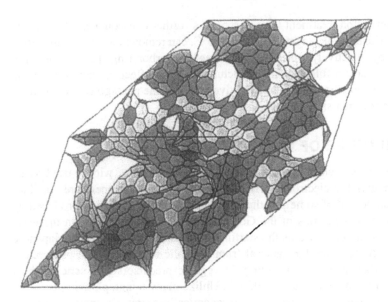

FIG. 16 Random schwarzite structure [29].

shown in Fig. 15 would be expected to be physically quite stable, since there is little strain associated with the curvature. Townsend and colleagues determined the energy per atom, ΔE, of their random schwarzite structure [29] and found a ΔE value of 0.23 eV, considerably lower than the value for C_{60} (0.42 eV).

If the model we are proposing for nongraphitizing carbons is correct, it suggests that these carbons are very similar in structure to fullerene soot, the low-density, disordered material that forms on walls of the arc evaporation vessel and from which C_{60} and other fullerenes may be extracted. Fullerene soot is known to be microporous, with a surface area, after activation with carbon dioxide, of approximately 700 m^2g^{-1} [31], and detailed analysis of high-resolution electron micrographs of fullerene soot has shown that these are consistent with a structure in which pentagons and heptagons are distributed randomly throughout a network of hexagons [32]. It is significant that high-temperature heat treatments can transform fullerene soot into nanoparticles very similar to those observed in heated microporous carbon [33].

The new model may also help in understanding the "activation" process, which is essential for developing a very high surface area in nongraphitizing carbons. Activation usually involves treatment with a mild oxidizing agent, such as CO_2 or water vapor, and it is generally believed that this has the effect of burning away carbon fragments inside micropores, thus enhancing surface area [2,3]. However, if our new model for the structure of microporous carbons is correct, then this activation treatment may also have a further consequence. It is known that mild oxidation, such as with CO_2 at 850°C, can remove the caps from carbon nanotubes by selectively attacking the pentagonal carbon rings [34]. If microporous carbons have a fullerene-like structure, then the effect of such a treatment would be to open closed pores by selective attack of the pentagons, thus increasing the surface area significantly.

V. STRUCTURE OF GLASSY CARBON

Glassy carbon is a class of nongraphitizing carbon that is widely used as an electrode material in electrochemistry, as well as for high temperature crucibles and as a component of some prosthetic devices [35,36]. It was first produced by workers at the laboratories of the General Electric Company, UK, in the early 1960s [37], using cellulose as the starting material. A short time later, Japanese workers produced a similar material from phenolic resin [38]. The preparation of glassy carbon involves subjecting the organic precursors to a series of heat treatments at temperatures up to 3000°C. Unlike many nongraphitizing carbons, they are impermeable to gases and are chemically extremely inert, especially those that have been prepared at very high temperatures. It has been demonstrated that the rates of oxidation of certain glassy carbons in oxygen, carbon dioxide, or water vapor are lower than those of any other carbon [36]. They are also highly resistant to attack by acids. Thus, while normal graphite is reduced to a powder by a mixture of concentrated sulfuric and nitric acids at room temperature, glassy carbon is unaffected by such treatment, even after several months.

High-resolution transmission electron micrographs of two commercial glassy carbons are shown in Fig. 17. Both were prepared from cross-linked aromatic polymers [39]. The carbon shown in Fig. 17a was heated to 1000°C and has a highly disordered, microporous structure made up of tightly curled graphene layers, very similar to that of the nongraphitizing carbon shown in Fig. 8a. The carbon shown in Fig. 17b was prepared at approximately 3000°C and has larger pores bounded by faceted or curved single- or multilayered graphitic fragments.

Some of the earliest structural models for glassy carbon assumed that both sp^2- and sp^3-bonded atoms were present [e.g., 40]. Graphitic domains were envisaged to be interspersed with tetrahedral domains, perhaps linked by short oxygen-containing bridges. These models were based primarily on an analysis of X-ray diffraction measurements and, as mentioned above, such measurements can be open to a number of interpretations. It should also be noted that neutron diffraction data have shown an absence of tetrahedrally bonded domains in glassy carbon heat treated at 2000°C [41].

A different model for the structure of glassy carbon was put forward by Jenkins and Kawamura in 1971 [18]. This model, illustrated in Fig. 18, is based on the

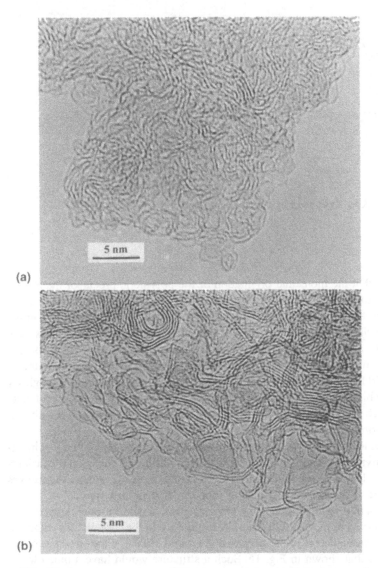

FIG. 17   (a) High-resolution TEM image of commercial glassy carbon, Sigradur K (HTW GmbH). (b) Image of glassy carbon, Sigradur G.

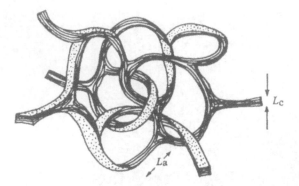

FIG. 18 Jenkins-Kawamura model of glassy carbon [18,35,36].

assumption that the molecular orientation of the polymeric precursor material is memorized to some extent after carbonization. Thus, the structure bears some resemblance to that of a polymer in which the "fibrils" are very narrow curved and twisted ribbons of graphitic carbon. The Jenkins-Kawamura model has been quite widely accepted but appears to be deficient in a number of aspects. For example, a structure such as that shown in Fig. 18, with many conjoined micropores, would be expected to be permeable to gases, whereas we know that glassy carbons are highly impermeable. The structure also has a high proportion of edge atoms, which are known to have a relatively high reactivity compared with "in-plane" carbon atoms.

Recently, Russian workers have proposed a model for glassy carbon that incorporates carbyne-like chains [42,43]. This model was based partly on a consideration of the electronic properties of glassy carbon but, as the authors themselves concur, there is no direct experimental support for their structure.

An alternative type of model for glassy carbon would be one composed of randomly curved or faceted fullerene-like fragments. In the case of "low-temperature" glassy carbons (i.e., those prepared at around 1000°C), the structure might be similar to that shown in Fig. 15. Such a structure would have a much lower reactivity and permeability than the Jenkins-Kawamura structure, particularly if there were a high proportion of completely closed particles. For the high-temperature glassy carbons, prepared at about 3000°C, the basic building blocks would be generally larger and more faceted, resembling incomplete giant fullerenes (together with some completely closed particles).

A model of heat-treated glassy carbon that involved closed cage–like components was proposed by Japanese workers in 1984 [44]. This model was put forward before the discovery of C_{60}, and the possibility that the closed particles might contain non-six-membered rings was not considered.

VI. STRUCTURE OF CARBON FIBERS

Most commercial carbon fibers are produced either from polyacrylonitrile (PAN) or from mesophase pitch [45]. Fibers derived from pitch are highly graphitic and have high elastic moduli, whereas those derived from PAN have a much more imperfect, lower density structure. The lack of extended structure in PAN-derived fibers makes them relatively insensitive to flaws, giving them higher strength but lower modulus than pitch-derived fibres. The properties of PAN-derived fibers result from the fact that PAN carbon is nongraphitizing.

X-ray diffraction of PAN-derived fibers produces crystallite lengths in the a direction of approximately 4–10 nm depending on the annealing temperature [46]. High-resolution electron microscopy shows that the fibres have an imperfect structure, containing many elongated voids. Several models have been put forward for the structure of PAN-derived carbon fibers, all based on the assumption that the basic structural units are graphite sheets or ribbons. A model suggested by Crawford and Johnson [47] is shown in Fig. 19. Here the structure consists of a random arrangement of flat or crumpled graphite sheets, with all the a-b planes running parallel to the fiber axis. However, given the flexibility of graphite sheets, it is difficult to see how the voids in such a structure could survive high-temperature heat treatment. Therefore, the possibility that the voids in fact result from the presence of fullerene-like elements is worthy of consideration. The elongated shapes of the voids suggests that they may have structures related to those

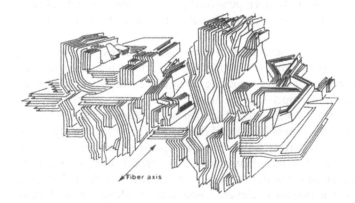

FIG. 19 Crawford-Johnson model of PAN-derived carbon fibers [47].

of carbon nanotubes, but additional high-resolution TEM studies will be needed to confirm this possibility.

VII. SOOT AND CARBON BLACK

A. Background

Like char, soot is a carbon material that has been intensively researched over a long period but whose structure is poorly understood. It usually consists of quasi-spherical particles ranging from about 10 nm to about 500 nm in size, which are often joined together in clusters or "necklace" chains. In most soots, these particles are accompanied by a poorly defined mixture of ash, polyaromatic hydrocarbons, and inorganic material. However, when formed under carefully controlled conditions, samples of soot consisting almost entirely of carbon particles can be produced, and this material, known as carbon black, is manufactured industrially on a large scale, primarily for use as a filler in rubber products. It is also used as a pigment and as a component of xerographic toners. A variety of industrial processes are used, the most important of which is the furnace black process, which involves the partial combustion of petrochemical or coal tar oils. Other methods for producing carbon black involve gas phase pyrolysis rather than combustion. A TEM micrograph of typical carbon black particles is shown in Fig. 20.

Research into soot formation can be traced back to the work of Humphrey Davy and Michael Faraday at the Royal Institution in the 19th century. Despite the vast amount of work that has been carried out on the subject since that time, the mechanism of soot formation is still a matter for debate. A detailed discussion of the various theories would be beyond the scope of the present work; a number of comprehensive reviews have been published [48–50]. However, a brief summary can be given of the basic aspects of soot formation as they are currently understood.

It is believed that soot particles form in three distinct phases. The first stage, known as particle inception, involves homogeneous reactions between hydrocarbon species that combine into larger aromatic layers and eventually condense out of the vapor phase to form nuclei. In the second stage, called growth, two processes occur: nuclei coalescence and surface deposition, the latter process being responsible for most of the mass increase of the primary particles. Finally, in the chain formation stage, relatively large spheroidal particles become joined together, without coalescing, to form long chains. Many aspects of this mechanism are poorly understood, in particular the nature of the initial nucleus. The most widely accepted view is that the nuclei are liquid microdroplets, thus explaining the particles' sphericity. However, there is much uncertainty about the mechanism whereby the droplets transform into solid graphite-like particles.

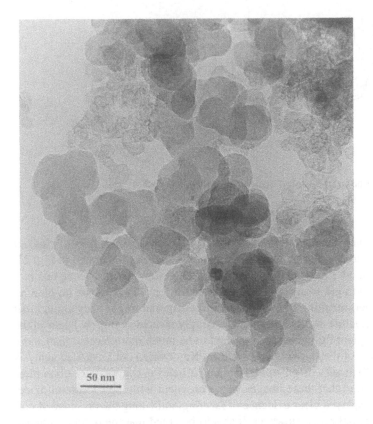

FIG. 20   Micrograph of commercial carbon black particles (Vulcan XC-72 Cabot Corp.).

Kroto, Smalley, and their coworkers became interested in the subject of soot formation in the aftermath of their discovery of buckminsterfullerene in 1985. They were well aware of the uncertainty in the carbon community concerning the initial stages of soot formation and wondered whether fullerene-like structures might be involved. After all, the carbon particles in soot, like fullerenes, were spheroidal in form, and both result from the condensation of carbon species from the gas phase. In 1986 they put forward a new mechanism for soot formation, the icospiral growth mechanism [51], which was later refined by Kroto and McKay [52]. The mechanism was based on the "pentagon-road" model of fullerene formation. The essential element of the pentagon-road model is the incorporation of pentagonal rings into a growing carbon network, driven by the need to eliminate dangling bonds. If the pentagons occur in the "correct" positions then C_{60} and other fullerenes will result, but in general closed structures will not be

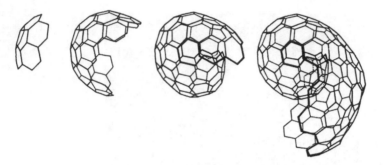

FIG. 21   Illustration of icospiral nucleation model [52].

formed. Kroto and Smalley suggested that if the growing shell fails to close, it would tend to curl around on itself like a nautilus shell, as illustrated in Fig. 21. In the refined model [52], Kroto and McKay suggested that the pentagons in a new layer would tend to form directly above those in the previous layer in an epitaxial manner, resulting in 12 columns of pentagonal rings. They argued that the spiraling structure would become increasingly faceted as it grew larger in the same way that giant fullerenes are expected to be much more faceted than small ones. In support of these ideas, Kroto and McKay carried out a detailed analysis of Iijima's 1980 images of spheroidal carbon particles in evaporated carbon films [53]. They showed that these images were consistent with their icospiral model particles and displayed clear evidence of faceting, indicating the presence of columns of pentagonal rings. However, it should be noted that the particles imaged by Iijima have a relatively ordered structure and resemble the carbon "onions" later studied by Ugarte [54] rather more than soot particles.

B.   Structure of Soot Particles

Structural studies of carbon black and soot particles would be expected to yield a better understanding of the growth mechanism, and this has been a primary motivation for the large number of such studies that have been carried out during the past 70 years or so. The first X-ray diffraction studies were carried out by Warren in the 1930s and 1940s [55,56], who demonstrated the presence of layer planes of graphite-like carbon. However, no evidence of the three-dimensional graphite structure was detected. Many subsequent X-ray diffraction studies have been carried out [50], making it possible to determine the dimensions of the individual crystallites. As noted earlier, the two parameters used to characterize the structure of disordered carbons are L_a, the length of the fragment in the in-plane direction, and L_c, the length in the c direction. For carbon blacks, L_a usually falls in the range 1–3 nm, whereas L_c is usually of the order of 1.5 nm. However, these values probably underestimate the size of the sheets that make up carbon black, as discussed below.

X-ray diffraction reveals the nature of the basic structural units, but detailed information on the internal organization of the carbon black particles can only be achieved through high-resolution electron microscopy. It is important to point out, however, that high resolution electron microscope images can be extremely difficult to interpret, particularly with disordered materials such as carbon blacks. Thus, the contrast displayed by a graphene sheet depends critically on its orientation with respect to the beam, so that the individual sheets making up the carbon black particle may be much longer than they appear to be. Therefore, it should not always be assumed that a one-to-one relationship exists between the TEM image and the true structure of the carbon.

A review of early TEM studies of carbon black was given by Heidenreich et al. in 1968 [57]. These authors interpreted the TEM results in terms of the model shown in Fig. 22. In this model, the basic structural units are somewhat flat graphene planes, arranged in a concentric manner around a hollow center. More recent high-resolution TEM studies have suggested that the structural units making up carbon black particles may be less perfect, and somewhat larger than those envisaged by Heidenreich *et al.* Figure 23 shows a high-resolution TEM image of a typical carbon black particle, with a region enlarged to show the detailed structure. It can be seen that the individual lattice fringes are not flat but rather wavy and curved. Importantly, the fringes are not preferentially curved about the center of the particle, as might be expected if the "icospiral" model were correct. A detailed analysis of high-resolution TEM images of carbon black particles has

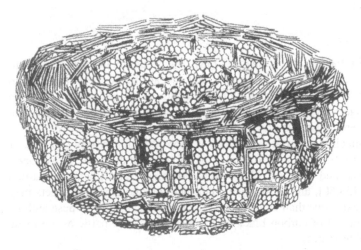

FIG. 22   Heidenreich-Hess-Ban model of carbon black structure [57].

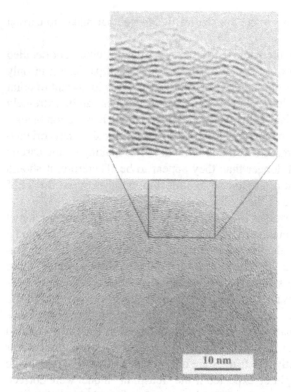

FIG. 23 High-resolution TEM image of commercial carbon black particle (Raven 430 Columbian Chemicals Co.). Enlarged area shows individual graphene layers. (Courtesy of Adrian Burden.)

been reported by Palotás and colleagues [58]. These authors used filtering techniques to remove noise from the images and reveal clearly the contrast from individual carbon planes. An example of their filtered images is reproduced in Fig. 24. This again shows the planes to be wrinkled rather than flat, and that there is no preferential curvature about the center.

Scanning tunneling microscopy (STM) is more difficult to apply to soot or carbon black particles than TEM, owing to the curvature of the particles' surfaces. However, Donnet and Custodéro have obtained some useful images of carbon black particles [59,60], and these have confirmed that the graphene sheets may be much more extensive than assumed by Heidenreich *et al.* [57]. These authors put forward a model of carbon black structure in which the particles are made up of overlapping scales.

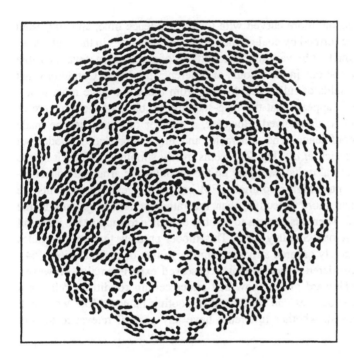

FIG. 24 Filtered pattern of soot particle structure [58].

Another technique that has been used to study carbon black and soot structure is ^{13}C nuclear magnetic resonance (NMR) spectroscopy. Frenklach and Ebert have used evidence from ^{13}C NMR spectroscopy of soot to argue against the icospiral model [61,62]. They point out that ^{13}C NMR spectra of soot resemble those of aromatic molecules much more closely than those of fullerenes. These authors have also stated that the kinetics of soot formation are inconsistent with icospiral model of the structure. They have used computer simulations to show that the growth of shell structures would be much slower than that of planar fragments. Soot formation is known to be extremely rapid, so it seemed unlikely that shell growth of the kind described by Kroto and colleagues could be involved.

While Frenklach and Ebert, and other investigators, have argued strongly against the idea that fullerenes have any relevance to soot formation, Howard and colleagues have shown that C_{60} can be formed in sooting acetylene-oxygen and benzene-oxygen flames [63]. This group has also obtained high-resolution TEM images of closed, fullerene-like structures in soot [64,65]. Similar images have also been obtained by Donnet *et al.* [66]. The Howard group carried out a detailed analysis of high-resolution TEM images of the soot and found that the structural units become more curved with longer residence time, and suggested that fullerenes were formed by coalescence of curved structures in the soot. However, they also found that by varying the conditions of combustion it was possible to produce wide variations in the amount of fullerenes produced. Conditions that led to the largest yields of fullerenes differed from those that gave most soot.

To summarize, it appears that the long-standing problem of the structure of soot particles remains unresolved. However, a number of points can be made with reasonable confidence. The venerable, and much reproduced, model put forward by Heidenreich *et al.* (Fig. 22) seems overly simplistic. While it is true that the basic structural units that make up the particles are arranged concentrically around the center, high-resolution TEM images suggest that the units are more extensive, and more defective, than the rather perfect fragments in the Heidenreich model. There also seems to be little evidence to support the icospiral growth model of Kroto, Smalley *et al.* This model predicts that soot particles should have a relatively ordered structure, similar to that of carbon onions [54], rather than the disordered structure that is observed experimentally. However, the results of the Howard group show that under certain conditions fullerenes and fullerene-like structures *can* be formed in sooting flames. More work will be needed to establish whether fullerenes have any general relevance to soot formation.

VIII. SPHERULITIC GRAPHITE IN CAST IRON

A. Background

In cast iron, the carbon content ranges from around 2% to 5% by weight. Other elements are also present, principally silicon. The nature of the carbon in the alloy depends on how the carbon comes out of solution, which in turn depends on the cooling rate and silicon content. Slow cooling and high silicon content leads to the formation of "gray iron" in which the precipitated carbon is in the form of flakes of graphite typically 10–100 μm in length. These flakes have very little strength, and with their sharp ends they act like numerous cracks within the microstructure. One result of this is that gray iron has very little tensile strength, but it can be utilized usefully in compression. Indeed, during the early part of the 19th century cast iron beams were used widely for bearing compressive loads in buildings.

If carbon could be persuaded to precipitate as compact nodules in place of flakes then this would enormously increase the range of application of cast iron. A method for achieving this, known as "malleablizing", was introduced in the 19th century. For this process the composition and casting conditions are chosen such that during the solidification of cast iron the carbon comes out of solution not as graphite but as cementite, the hard white iron-carbide Fe_3C. The "white iron" casting is then malleablized, i.e., given a prolonged anneal at subeutectic temperatures, which results in the cementite decomposing and, depending on the annealing temperature, compact graphite nodules are formed. This malleablized cast iron is much tougher than conventional gray iron, and the product found extensive application throughout the 19th century and the first half of the 20th century.

However, malleablizing is an expensive process, and clearly it was desirable to develop a composition or procedure that would produce a cast iron in which in the as-cast condition the graphite consisted of compact near-spherical nodules. This breakthrough was achieved in the late 1940s when Morrogh and Williams [67] and others discovered that the addition of rare-earth metals or magnesium immediately before pouring the casting did indeed produce the desired result— the graphite phase appearing in the microstructure not as flakes but as compact spherulites. This was one of the most significant discoveries in the field of casting in the 20th century. Known as spherulitic graphite (SG) cast iron, this alloy can be used in engineering, particularly in the automobile industry, and in construction.

B. Structure of Spherulitic Graphite

Typical graphite spherulites range from about 5 to 25 µm in diameter, although they can grow as large as 100 µm. They are thus far larger than typical soot particles. An optical micrograph of an individual spherulite, recorded with polarized light, is shown in Fig. 25. This shows that the structure of the spherulite is not onion-like but rather consists of closely packed and intertwined filamentary segments growing from a central nucleus. Electron microscopy and diffraction shows that the radial segments grow primarily in the c direction, so that the basal planes (graphene sheets) remain approximately parallel to the surface of the growing particle. However, the detailed structure of the segments is not known; this will be discussed further below.

The nature of the initial nucleus for spherulitic graphite growth is also unknown. The possibility that fullerene-like structures might be involved in the nucleation of spherulitic graphite has been discussed by a number of groups [69–72]. The most detailed discussion has been given by Double and Hellawell [72] who propose a model for spherulitic graphite growth similar to the McKay-Kroto icospiral nucleation model (although they are vague about the precise details). They point out that spheroidal growth is the only mechanism that allows large graphitic particles to form by continuous addition to a growing sheet. The formation of flake-like graphite, made up of stacks of layer planes, would require repeated nucleation events (unless the graphene sheets folded back on themselves), and therefore a greater activation energy.

It has been suggested that the involvement of fullerenes at the nucleation stage might explain the role of foreign elements in promoting spherulitic graphite

FIG. 25 Optical micrograph showing filamentary structure of individual graphite spherulite in spherulite graphite iron [68].

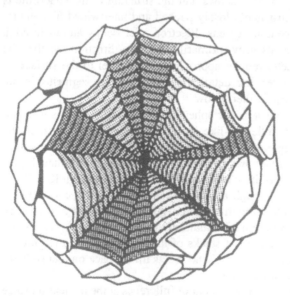

FIG. 26 Sketch of cone-helix model of the structure of spherulitic graphite [74].

growth. The traditional view is that these elements increase the interfacial energy between the carbon and the matrix, favoring spheroidal precipitates, but this is by no means certain. Lanthanum, which is one of the rare-earth metals used in promoting spherulitic graphite growth, is readily incorporated into fullerene cages [73]. This prompted the idea that a metallofullerene might be the seed for spheru-lite growth [72]. However, it should be borne in mind that spherulitic graphite does not necessarily require the presence of additives but can also form in a "clean" melt from which most impurities have been removed.

The structure of the filamentary segments has been discussed by a number of groups. Double and Hellawell (hereafter DH) suggest that the filamentary seg-ments consist of helical cones, as illustrated in Fig. 26 [74]. In support of this idea, DH note that graphitic whiskers with a rather similar structure can be produced by pyrolysis of carbon monoxide on a silicon carbide substrate [75]. They point out that the lowest energy configuration of such a cone helix will occur when there is a favorable coincidence configuration between the growing sheet and the under-

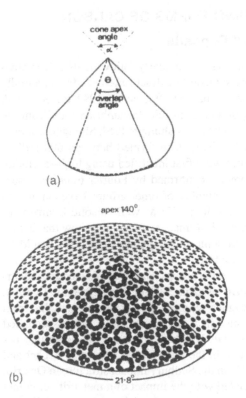

FIG. 27 Formation of a cone from a graphene sheet [74].

lying one, which in turn implies that there will be certain favored cone apex angles. One of these favored angles will be 140°, which happens to be precisely the angle observed in the graphitic whiskers referred to above. This case is illustrated in Fig. 27. DH describe the defect at the apex as a screw dislocation but do not describe the atomic structure of this dislocation in detail. One possibility that DH do not seem to have considered is that there might be an individual pentagon at the apex. This would result in a cone angle of 112.9°. Of course, in the absence of any other defects, a single pentagon would simply result in a seamless cone, but the introduction of an imperfection such as a foreign atom from the matrix might produce a discontinuity in the growing sheet, resulting in overgrowth, and a spiraling filament of the kind envisaged by DH.

Very little experimental work has been carried out on the possible role of fullerenes in spherulitic graphite growth. DH have looked for evidence of fullerenes in the graphitic residues left after chemical and electrolytic dissolution of spherulitic graphite iron, but with no success.

IX. NATURALLY OCCURRING FORMS OF CARBON

A. Carbon from Geological Deposits

Following the discovery of C_{60} in 1985, and particularly following its bulk synthesis in 1991, a number of attempts were made to extract fullerenes from naturally occurring forms of carbon. Most were unsuccessful, but in 1992 Buseck et al. claimed to have found C_{60} and C_{70} in a carbon-rich Precambrian rock from the Karelia region of northeastern Russia, known as shungite [76]. Shungite is a rare, carbon-rich variety of rock believed to have been formed between 600 million and 4 billion years ago. The fullerenes were first identified using high-resolution TEM, and their presence was apparently confirmed by Fourier transform mass spectrometry. Since this initial work, a number of other groups have claimed to have extracted fullerenes from shungite [e.g., 77] as well as solid bitumens of Precambrian age [78]. High-resolution TEM images of shungite show that it consists of curved graphitic layers, with some closed shells [79]. This would be consistent with a fullerene-related structure.

As well as shungite and related materials, it has also been claimed that fullerenes are present in geological deposits associated with impact events. In 1994, Heymann, Smalley, and their colleagues reported finding C_{60} in samples from two Cretaceous-Tertiary boundary sites in New Zealand [80]. They speculated that the C_{60} formed in wildfires associated with the catastrophic event that occurred at the end of the Cretaceous era. In the same issue of *Science*, Becker and colleagues reported finding C_{60} and C_{70} in the Sudbury impact structure in Ontario, Canada [81]. This structure is associated with the impact of a meteorite or comet 1.85 billion years, and the authors suggested that the fullerenes were delivered to Earth by the impact. This was apparently confirmed when Becker *et al.* subsequently showed that the fullerenes contained helium with $^3He/$ ^4He ratios that were consistent with an extraterrestrial origin [82]. Recently, the same group has claimed to show that fullerenes from the Permian/Triassic boundary layer contain trapped helium with $^3He/^4He$ ratios approximately 100 times that of atmospheric helium [83].

It should be noted that a number of authors have found the reports of naturally occurring fullerenes hard to accept [84–92]. It has been suggested that the fullerenes found in these samples could be contaminants arising from previous work on fullerenes. Alternatively, the fullerenes could have been created during the preparation of samples for analysis, e.g., during laser vaporization prior to analysis by mass spectroscopy. Taylor and colleagues have also pointed out that C_{60} fairly rapidly degrades to $C_{120}O$ in air [86], as well as being sensitive to ultraviolet light [88], which makes it hard to believe that fullerenes could survive in the natural environment for billions of years. On the other hand, the results of Becker *et al.* showing the presence of extraterrestrial gases within the fullerenes appears to confirm the extreme age of the fullerenes. There is currently no sign of a resolution to these arguments.

B. Carbon in Meteorites

The discovery that fullerenes are apparently present in geological features associated with impacts suggests that they should also be present in meteorites. In fact, this possibility was first put forward by Heymann in 1986 [93], shortly after the discovery of C_{60}. However, most searches for C_{60} and other fullerenes in meteorites have been unsuccessful. Once again, it is only the group of Becker and colleagues who have successfully extracted fullerenes from a meteorite. These workers reported in 1994 that C_{60} was present at a level of approximately 0.1 ppm in the Allende meteorite [94], and subsequently found a distribution of higher fullerenes, from C_{100} to C_{400} in the same meteorite [95]. This group showed in 2000 that fullerenes from the Allende and Murchison meteorites apparently contained trapped noble gases with isotopic ratios indicative of an extraterrestrial origin [96]. However this work, like the studies described in the previous section, has been extremely controversial

This author and colleagues have used TEM to examine carbon extracted from a sample of Allende, with the primary aim of seeking evidence for the presence of fullerene-related structures [97]. The carbon was found to be somewhat disordered, consisting of curved and faceted graphene sheets with little large-scale graphitization, resembling a microporous carbon following high-temperature heat treatment. Closed fullerene-like carbon nanoparticles, typically ~2–10 nm in diameter, were indeed commonly observed. A typical micrograph showing closed

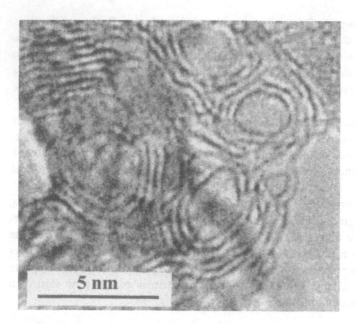

FIG. 28 TEM image of carbon extracted from the Allende meteorite, showing closed nanostructures.

structures is shown in Fig. 28. Additional TEM studies of the Allende carbon and of carbon from other meteorites are underway.

X. DISCUSSION

It is now 18 years since the first synthesis of C_{60} and 12 years since the publication of Iijima's paper on carbon nanotubes. However, the importance of these discoveries for our understanding of well-known forms of carbon is still not widely recognized. Thus, review papers on conventional carbons still tend to discuss their structure in terms of models based on "flat graphite." In this chapter we have attempted to outline the evidence that fullerene-related structures may be present in well-known carbon materials and to explain the ways in which this may help in understanding the properties of these carbons.

A detailed discussion has been given of the structure of nongraphitizing carbons. There are a number of reasons for believing that these carbons may be fullerene-like. Such a structure would explain their low density, microporosity, and hardness. The presence of pentagonal rings might also explain why high-temperature heat treatment of these carbons results in the formation of partially and completely closed fullerene-like carbon cages. Of course, it could be argued that the pentagons form during the high-temperature heat treatment and are not present in the original carbon. However, this raises the question of why pentagons are *not* formed during the heat treatment of graphitizing carbons such as polyvinyl chloride-derived carbon and petroleum coke. It should be stressed that there is no *direct* proof that nongraphitizing carbons contain pentagonal rings, and achieving such evidence will not be easy. As already noted, diffraction methods do not provide unequivocal evidence that pentagons or other nonhexagonal rings are present. In principle, modern high resolution transmission electron microscopes should be capable of imaging individual pentagonal rings, but this would be pushing their capability to the limits.

The nature of the hypothetical transformation from microporous nongraphitizing carbon to a structure containing closed carbon cages is not well understood, and further work on this problem would be welcome. The transformation may involve ring migration mechanisms such as the Stone-Wales rearrangement [98], which has been invoked to explain fullerene isomerization. It is notable that the cage structures observed in heat-treated nongraphitizing carbons fall into quite a narrow size range—typically 5–15 nm—suggesting that structures in this size range have a special stability.

There are other aspects of nongraphitizing carbons that remain inadequately understood. In particular, there is the very basic question of why some organic materials produce graphitizing carbons and others yield nongraphitizing carbons. If the ideas put forward in this review are correct, this question becomes, why do some precursors form five-membered rings when carbonized, and others only hexagonal? It is unlikely that the answer to this question is simple because the carbonization of organic materials is an immensely complex process, and there is rarely a simple relationship between the structure of the original precursor and the nature of the carbon produced by pyrolysis. For example, structures that contain five-membered rings can yield either graphitizing or nongraphitizing carbon [99]. In fact, it is generally believed that the physical properties of the precursors, and the conditions under which pyrolysis is carried out, are more important than chemical structure in determining whether the final carbon is graphitizing or non-graphitizing. Thus, graphitizing carbons usually form a liquid on heating to temperatures around 400–500°C, while nongraphitizing carbons generally form solid chars without melting. The liquid phase produced on heating graphitizing carbons is believed to provide the mobility necessary to form oriented regions. However, this may not be a complete explanation because some precursors, such as sucrose, form nongraphitizing carbons despite passing through a liquid phase. There is clearly a need for more work in this area.

If the idea that microporous carbons are fullerene-like is correct, this has important implications. As already noted, it may suggest that models of activated carbons that assume that the pores are slit-like in shape may have to be modified;

in a fullerene-like structure, many pores would be expected to have more rounded shapes. The chemistry of microporous carbons has also traditionally been understood in terms of a graphite-like structure. The presence of five-membered rings in the structure would significantly affect the chemical properties of the carbon.

As far as soot and carbon black particles are concerned, the evidence that these might have fullerene-related structures is less clear, despite their spheroidal shapes. Certainly the work of the Howard group shows that fullerenes and fullerenic nanostructures can be formed in sooting flames, but the highest yields are formed under rather specific conditions, namely, in low-pressure premixed benzene/oxygen flames with or without an argon diluent. It is important that the question of whether fullerene-like structures have a wider importance in soot is resolved since, as in the case of microporous carbons, the presence of such structures would have a major influence on their properties [100].

Many questions also remain about the other materials discussed in this chapter. There seem to be good reasons for believing that glassy carbon and PAN-derived carbon fibers may be fullerene related, but there is no direct proof of this. It seems less likely that fullerene-related elements are important in spherulitic graphite cast iron, but the detailed structure of the spherules remain unknown. The evidence for fullerenes and fullerene-related structures in naturally occurring forms of carbon remains highly controversial. More work is needed on all of these forms of carbon in light of the new knowledge that has been gained since the discovery of fullerenes.

ACKNOWLEDGMENTS

The author is grateful to Rainer Dübgen for supplying samples of glassy carbon. Thanks are also due to Ljubisa Radovic for helpful comments.

REFERENCES

1. PJF Harris. Carbon Nanotubes and Related Structures: New Materials for the Twenty-first Century. Cambridge: Cambridge University Press, 1999.
2. H Jankowska, A Swiatkowski, J Choma. Active Carbon. New York: Ellis Horwood, New York, 1991.
3. JW Patrick (ed.). Porosity in Carbons: Characterisation and Applications. London: Arnold, London 1994.
4. PJF Harris. Interdisc Sci Rev 24, 301 (1999).
5. PW Fowler, DE Manolopoulos. An Atlas of Fullerenes. Oxford: Oxford University Press, 1995.
6. BI Dunlap, DW Brenner, JW Mintmire, RC Mowrey, CT White. J Phys Chem 95, 8737 (1991).
7. E Hernández, P Ordejón, H Terrones. Phys Rev B. 63, 193403 (2001).

8. I Mochida, M Egashira, Y Korai, K Yokogawa. Carbon 35, 1707 (1997).
9. S Iijima. Nature 354, 56 (1991).
10. M Ge, K Sattler. Appl Phys Lett 65, 2284 (1994).
11. S Iijima, T Ichihashi, Y Ando. Nature 356, 776 (1992).
12. Y Ando, S Iijima. Jap. J Appl Phys 32, L107 (1993).
13. S Iijima, T Ichihashi. Nature 363, 603 (1993).
14. DS Bethune, CH Kiang, MS de Vries, G Gorman, R Savoy, J Vasquez, R Beyers. Nature 363, 605 (1993).
15. RE Franklin. Proc R Soc A 209, 196 (1951).
16. FG Emmerich. Carbon 33, 1709 (1995).
17. LL Ban, D Crawford, H Marsh. J Appl Cryst 8, 415 (1975).
18. GM Jenkins, K Kawamura. Nature 231, 175 (1971).
19. PR Buseck, H Bojun, LP Keller. Energy Fuels 1, 105 (1987).
20. A Oberlin. Chemistry and Physics of Carbon, Vol. 22 (P.A. Thrower, ed.). New York: Marcel Dekker, 1989, p.1.
21. S Ergun VH Tiensuu. Acta Cryst 12, 1050 (1959).
22. DR McKenzie, Rep Prog Phys 59, 1611 (1996).
23. PJF Harris, SC Tsang. Phil Mag A 76, 667 (1997).
24. PJF Harris. Int Mater Rev 42, 206 (1997).
25. PJF Harris, A Burian, S Duber. Phil Mag Lett 80, 381 (2000).
26. MB Sweatman, N Quirke. Mol Sim 27, 295 (2001).
27. A Burian, A Ratuszna, JC Dore, SW Howells. Carbon 36, 1613 (1998).
28. V Petkov, RG Difrancesco, SJL Billinge, M Acharya, HC Foley. Phil Mag. B 79, 1519 (1999).
29. SJ Townsend, TJ Lenosky, DA Muller, CS Nichols, V Elser. Phys Rev Lett 69, 921 (1992).
30. AL Mackay, H Terrones. Nature 352, 762 (1991).
31. SC Tsang, PJF Harris, JB Claridge, MLH Green. J Chem Soc. Chem Commun 1519 (1993).
32. LA Bursill, LN Bourgeois. Mod Phys Lett B 9, 1461 (1995).
33. WA De Heer, D Ugarte. Chem Phys Lett 207, 480 (1993).
34. SC Tsang, PJF Harris, MLH Green. Nature 362, 520 (1993).
35. GM Jenkins, K Kawamura, LL Ban. Proc R Soc A 327, 501 (1972).
36. GM Jenkins, K Kawamura. Polymeric Carbons: Carbon Fibre, Glass and Char. Cambridge: Cambridge University Press, 1976.
37. HW Davidson. Nucl Eng 7, 159 (1962).
38. S Yamada, H Sato. Nature 193, 261 (1962).
39. R Dübgen, personal communication.
40. J Kakinoki. Acta Cryst 18, 578 (1965).
41. DFR Mildner, JM Carpenter. J Non-cryst Solids 47, 391 (1982).
42. LA Pesin. J Mater Sci 37, 1 (2002).
43. LA Pesin, EM Baitinger. Carbon 40, 295 (2002).
44. M Shiriashi. Introduction to Carbon Materials (in Japanese). Carbon Society of Japan, Tokyo, 1984.
45. MS Dresselhaus, G Dresselhaus, K Sugihara, IL Spain, HA Goldberg. Graphite Fibers and Filaments. Berlin: Springer-Verlag, 1988.

46. DJ Johnson. Phil Trans R Soc A294, 443 (1980).
47. D Crawford, DJ Johnson. J Microsc 94, 51 (1971).
48. HB Palmer, CF Cullis. In: Chemistry and Physics of Carbon, Vol. 1 (PL Walker Jr, ed.). New York: Marcel Dekker, 1965, p. 265.
49. SJ Harris, AM Weiner. Annu Rev Phys Chem 36, 31 (1985).
50. J-B Donnet, RC Bansal, M-J Wang. Carbon Black, 2nd ed. New York: Marcel Dekker, 1993.
51. QL Zhang, SC O'Brien, JR Heath, Y Liu, RF Curl, HW Kroto, RE Smalley. J Phys Chem 90, 525 (1986).
52. HW Kroto, K McKay. Nature 331, 328 (1988).
53. S Iijima. J. Microsc 119, 99, (1980).
54. D Ugarte. Nature 359, 707 (1992).
55. BE Warren. J Chem Phys 2, 551 (1934).
56. J Biscoe, BE Warren. J Appl Phys. 13, 364 (1942).
57. RD Heidenreich, WM Hess, and LL Ban. J Appl Cryst 1, 1 (1968).
58. AB Palotás, LC Rainey, CJ Feldermann, AF Sarofim, JB Vander Sande. Micros Res Tech 33, 266 (1996).
59. J-B Donnet, E Custodéro. Bull Soc Chim Fr 131, 115 (1994).
60. J-B Donnet, E Custodéro. In: Carbon black, 2nd ed. (J-B Donnet, RC Bansal, M-J Wang, eds.). New York: Marcel Dekker, 1993, p. 221.
61. M Frenklach, LB Ebert. J Phys Chem 92, 561 (1988).
62. LB Ebert. Carbon 31, 999 (1993).
63. JB Howard, JT McKinnon, Y Makarovsky, AL Lafleur, and ME Johnson. Nature 352, 139, (1991).
64. WJ Grieco, JB Howard, LC Rainey, JB Vander Sande. Carbon 38, 597 (2000).
65. A Goel, P Hebgen, JB Vander Sande, JB Howard. Carbon 40, 177 (2002).
66. J-B Donnet, MP Johnson, DT Norman, TK Wang. Carbon 38, 1885 (2000).
67. H Morrogh, WJ Williams. J Iron Steel Inst 155, 321 (1947).
68. JE Harris. The Formation of Graphite in Cast Iron, PhD thesis, University of Birmingham, 1956.
69. B Miao, K Fang, W Bian, G Liu. Acta Metall Mater 38, 2167 (1990).
70. B Miao, DO Northwood, W Bian, K Fang, MH Fan. J Mater Sci 29, 255 (1994).
71. HW Kroto, JP Hare, A Sarkar, K Hsu, M Terrones, R Abeysinghe. MRS Bull 19(11), 51 (1994).
72. DD Double, A Hellawell. Acta Metall Mater 43, 2435 (1995).
73. DS Bethune, RD Johnson, JR Salem, MS de Vries, CS Yannoni. Nature 366, 123 (1993).
74. DD Double, A Hellawell. Acta Metall 22, 481 (1974).
75. HB Haanstra, WF Knippenberg, G Verspui. In: Proceedings of the 5th European Conference on Electron Microscopy, Institute of Physics, Manchester, 1972, p. 214.
76. PR Buseck, SJ Tsipursky, R Hettich. Science 257, 215 (1992).
77. G Parthasarathy, R Srinivasan, M Vairamani, K Ravikumar, AC Kunwar. Geochim Cosmochim Acta 62, 3541 (1998).
78. J Jehlicka, M Ozawa, Z Slanina, E Osawa. Fullerene Sci Technol 8, 449 (2000).
79. VV Kovalevski. Russ J Inorg Chem 39, 28 (1994).

80. D Heymann, LPF Chibante, RR Brooks, WS Wolbach, RE Smalley. Science 265, 645 (1994).
81. L Becker, JL Bada, RE Winans, JE Hunt, TE Bunch, BM French. Science 265, 642 (1994).
82. L Becker, RJ Poreda, JL Bada. Science 272, 249 (1996).
83. L Becker, RJ Poreda, AG Hunt, TE Bunch, M Rampino. Science 291, 1530 (2001).
84. Y Gu, MA Wilson, KJ Fisher, IG Dance, GD Willett, D Ren, IB Volkova. Carbon 33, 862 (1995).
85. TW Ebbesen, H Hiura, JW Hedenquist, CEJ Deronde, A Andersen, M Often, VA Melezhik, PR Buseck, S Tsipursky. Science 268, 1634 (1995).
86. R Taylor, MP Barrow, T Drewello. J Chem Soc Chem Commun 2497 (1998).
87. R Taylor AK Abdul-Sada. Fullerene Sci Technol 8, 47 (2000).
88. R Taylor, JP Parsons, AG Avent, SP Rannard, TJ Dennis, JP Hare, HW Kroto, DRM Walton. Nature 351, 277 (1991).
89. RA Kerr, Science 291, 1469 (2001).
90. KA Farley, S Mukhopadhyay. Science 293 (5539), U1 (2001). [published online: http://www.sciencemag.org/cgi/content/full/293/5539/2343a].
91. Y Isozaki. Science 293 (5539), U3 (2001) [published online: http://www.sciencemag.org/cgi/content/full/293/5539/2343a].
92. T Braun, E Osawa, C Detre, I Tóth. Chem Phys Lett 348, 361 (2001).
93. D Heymann. J Geophys Res 91 E135 (1986).
94. L Becker, JL Bada, RE Winans, TE Bunch, Nature 372, 507 (1994).
95. L Becker, TE Bunch, LJ Allamandola. Nature 400, 227 (1999).
96. L Becker, RJ Poreda, TE Bunch. Proc Natl Acad Sci USA 97, 2979 (2000).
97. PJF Harris, RD Vis, D Heymann. Earth Planet Sci Lett 183, 355 (2000).
98. AJ Stone, DJ Wales. Chem Phys Lett 128, 501 (1986).
99. IC Lewis. Carbon 20, 519 (1982).
100. F Cataldo. Carbon 40, 157 (2002).

2

Molecular Models of Porous Carbons

Teresa J. Bandosz

*City College of the City University of New York,
New York, New York, U.S.A.*

Mark J. Biggs

University of Edinburgh, Edinburgh, Scotland, United Kingdom

Keith E. Gubbins

North Carolina State University, Raleigh, North Carolina, U.S.A.

Y. Hattori

Shinshou University, Nagano, Japan

T. Iiyama

Chiba University, Chiba, Japan

Katsumi Kaneko

Chiba University, Chiba, Japan

Jorge Pikunic

North Carolina State University, Raleigh, North Carolina, U.S.A.

Kendall T. Thomson

Purdue University, West Lafayette, Indiana, U.S.A.

I. INTRODUCTION

The widespread interest in porous carbons stems from their high surface activity, arising from strong adsorbate-carbon forces and large surface area. This activity leads to high adsorption capacity and selectivity in mixture separation. In addition, such carbons are relatively cheap to produce, and can be prepared with a range of pore sizes. The strong surface forces in carbons are due to the high surface density of carbon atoms. In graphite, the C-C distance is only 0.142 nm, much less than the van der Waals radius of carbon, which is about 0.335 nm. The short C-C distance, a result of the strong covalent sp^2 bonding in graphite, leads to a surface density of about 38.2 carbon atoms/nm^2.

The need for more convenient forms of porous carbon has in more recent years motivated the development of activated carbon fibers, porous carbon membranes, and carbon aerogels, while other completely new carbon materials, such as fullerite and the carbon nanotube, have also been discovered in the last two decades. In this chapter, our main focus is on carbons that contain micropores (pore size up to 2 nm) and mesopores (pore size of 2–50 nm), which are the most useful pore sizes for adsorption, kinetic sieving, and catalysis. Materials specifically considered include the conventional porous carbons such as the various particulate activated carbons, activated carbon fibers, and porous carbon membranes. More novel carbon forms are also considered, including carbon nanotubes and fullerites.

Activated carbons and carbon nanotubes find widespread use in fundamental scientific studies of the effects of confinement on phase changes, selective adsorption, reaction equilibrium, etc. The pores in activated carbons are usually approximately slit shaped; the slit geometry, coupled with the unusually strong intermolecular forces between the walls and the adsorbate molecules, provides a contrast with the roughly cylindrical pore geometry and much weaker wall forces found in porous glasses and silicas of various kinds.

Conventional porous carbons are the most widely used of all general-purpose adsorbents. Reviews of the more common uses of particulate activated carbons may be found in the monographs of Smísek and Cerný [1] and others [2,3], while some examples for activated carbon fibers may be found in Mays [4]. Conventional porous carbons have been used as adsorbents across the food, biotechnology, and agricultural sectors, and in the medical and health settings—many examples may be found in the recent review of Roy [5]. Specific examples include treatment of ingested poisons [6] and extensive use in respiratory protection [7]. More recent examples of use in the biotechnology field include separation of proteins and amino acids [8]. Conventional carbons are also used to effect separations via sieving, the best known example being the separation of air to generate O_2 and N_2 [9]. Use of more novel forms of carbon in the separation role have also been recently investigated [10]. Conventional and more novel forms of carbon have been investigated as a means of storing methane and hydrogen [11–13]. Conventional carbons have long been used as catalyst supports [14–18], while fullerenes have also been more recently considered in this role [19]. Conventional and more novel forms of carbon are also now being used in electrochemical applications, including as electrode materials in the Li-ion battery [20–22], supercapacitors [22–24], and fuel cells [25]. Critoph [26] reviews the application of conventional carbons in adsorption refrigerators and heat pumps, while Tanahashi et al. [27] investigate the use of activated carbon in controlled-atmosphere storage applications. This is just a partial list of applications, and new applications are constantly being investigated and developed.

With the exception of carbon nanotubes and nanocones (also referred to as nanohorns), porous carbons are in general disordered materials and as yet cannot be fully characterized from experiment. Techniques such as X-ray and neutron scattering, and high-resolution transmission electron microscopy (HRTEM), give useful partial information about the molecular structure but are not yet able to provide a complete picture at the atomic level. In order to interpret experimental data on the carbons themselves, and on the behavior of adsorbates in carbons, we must resort to structural models of the pore morphology and topology, in addition to models for the intermolecular forces involved. We define the *pore morphology* to be a description of the geometrical shape and structure of the pores, including the pore width and volume, and the detailed nature of the surfaces of the pore walls. We define *pore topology* to be the description of the arrangement of the pores relative to each other; pore topology describes the connectivity of the pores and the overall macroscopic environment seen by adsorbed molecules within the carbon structure.

For porous carbons, the simplest structural models are *single-pore* models, which include the effects of the pore morphology but not pore topology; the most widely used model of this type is the *slit-pore model*. In this model the pore structure is represented as a collection of nonconnected pores of slit shape and various pore widths, H. These pores have parallel walls, which are composed of graphene sheets; the walls are often treated as being structureless and smooth, and extend to infinity in the x and y directions (z is the direction normal to the walls). The advantage of such a simple model is that it is purely geometrical and it does not depend on the specifics of the carbon structure. The slit-pore model has been extensively used to estimate pore size distributions for carbons, $f(H)$, by assuming that a given measured property (usually the adsorption isotherm for a simple gas such as argon or nitrogen) is a linear combination of isotherms calculated for pores of different sizes, H. The slit-pore model is convenient to use, but it is a crude model that omits many important effects present in real carbons; connectivity, edge effects, curved and defective graphene sheets, chemical groups attached to the carbon surfaces, wedge- and other shaped pores, etc. More recently, molecular models have been developed that also consider the pore topology, which accounts for many of the phenomena observed experimentally that are not obtained from the more simple models.

Harris [28] has reviewed structural models of nongraphitizing carbons, and covers the literature up to 1996; his emphasis is mainly on qualitative models. Here we review the complete range of quantitative molecular models for the conventional and more novel forms of porous carbon. The simpler of these molecular models are now used routinely to interpret experimental data such as adsorption isotherms, while the building of the more complex models requires considerable interaction with experimental data, such as that from X-ray and neutron scattering, high-resolution transmission electron microscopy, and Fourier transform infrared spectroscopy (FTIR). Although there have been a number of reviews of the experimental studies of carbon structures [e.g., 1,29–36], none has really focused on the connection with quantitative molecular models such as those reviewed here. Therefore, we also review the relevant experimental methods in a way that will be useful to those who wish to build quantitative molecular models.

Following a discussion of the types of carbon to be considered and their synthesis (Section II), we survey useful experimental methods for investigating the structure of porous carbons (Section III). In this section our emphasis is on the molecular level information on the pore structure, surface chemistry, and other properties that can be obtained from each technique. Methods for modeling porous carbons are described in Section IV. We consider regular porous carbons such as nanotubes and nanocones, and disordered carbons that are microporous or mesoporous. The latter include activated carbons, activated carbon fibers, carbon membranes, and carbon aerogels. We do not explicitly consider macroporous carbons.

II. CARBONS: TYPES, CLASSIFICATION, PRECURSORS, AND PREPARATION

A. Types of Carbon Solids

Solid carbons include all solids that consist primarily of carbon atoms [37]. They may be pure carbon or include noncarbon atoms of any type and in any manner (e.g., as an integral part of the molecular structure or as an intercalate). A brief overview is given here of the various carbon solids that exist or that have been hypothesized and are yet to be disproved. In line with the aim of this review, attention is focused on the molecular structure; relevant reviews are given in case other details are desired.

1. Graphite and Diamond

Graphite and diamond are the two most common allotropes of carbon. Both allotropes occur naturally and are produced on an industrial scale. The molecular structures of many of the more complex carbon solids are often described in terms of their deviation from those of graphite or diamond. For example, so-called *turbostratic carbon* is essentially graphite without significant order in the direction perpendicular to the graphite planes. The graphite and diamond structures also underpin the simpler molecular models of materials such as activated carbon. The structures of graphite and diamond are, therefore, briefly reviewed here.

As illustrated in Fig. 1, graphite is assembled from parallel sheets of sp^2 hybridized carbon atoms, each being linked to three other sp^2 hybridized carbon atoms by sigma bonds of length 0.1415 nm. The in-plane lattice constant is $a = 0.2456$ nm. The parallel sheets, termed *graphene layers* [37] or *basal planes*, are *nominally* 0.3354 nm apart and are held in place by the delocalized π electrons of the carbon atoms. The two most common *polytypes* [39] of graphite are differentiated by the stacking of the graphene layers. The layers alternate ABABAB . . . in *hexagonal graphite* giving a lattice constant of $c = 0.6708$ nm, and ABCABC . . . in *rhombohedral graphite* where $c = 1.0062$ nm. Rhombohedral graphite is thermodynamically less favored than hexagonal graphite but can account for anything up to 30% of a graphite [40]; it has never been found in isolation [41]. The fraction of each polytype and the mean graphene layer extent for a particular graphite are dictated by the conditions of formation and processing [41]. Other less common polytypes of graphite also exist [38]. Although graphite is commonly considered in terms of the structures outlined

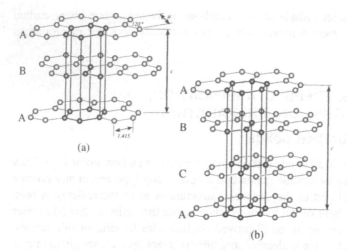

FIG. 1 The two most common polymorphs of graphite: (a) *hexagonal graphite* (also termed *Bernal graphite, 2H graphite, CI*, and *α-graphite* [38]); and (b) *rhombohedral graphite* (also termed *3R graphite, CII*, and *β-graphite* [38]).

here, they are invariably disturbed by many defects [42,43], including *stacking faults* where the graphene layer alternations differ from those of the main polytypes [42].

Diamond is assembled from sp^3 hybridized carbon atoms linked to four other sp^3 hybridized carbon atoms to form a three-dimensional tetragonal network. As with graphite, the two most common polytypes of diamond are differentiated by the stacking of the tetrahedra. In the case of the more common *face-centered cubic diamond* (Fig. 2a), the sheets of tetrahedra alternate ABCABC . . . , with a lattice constant of $a = 0.3567$ nm and C-C bond length 0.154 nm. As illustrated in Fig. 2b, the sheets of *hexagonal diamond* alternate ABAB . . . , with lattice constants $a = 0.252$ nm and $c = 0.412$ nm, and a C-C bond length of 0.152 nm. The hexagonal form is much rarer in nature, normally occurring as particles no larger than 30 μm, and is produced under conditions different from those of cubic diamond [44]. Other polytypes of diamond have been hypothesized (e.g. [45,46]), with a few of these being observed in recent years ([47] and references therein). As with graphite, the molecular structure of real diamonds is normally disrupted by the presence of defects and heteroatoms such as nitrogen, boron, hydrogen, and oxygen [48].

2.   Carbyne: An sp^1 Hybridized Carbon Allotrope?

Workers in the carbon field have for many years now been debating the existence of so-called *carbyne* [49–51], a hypothesized allotrope based on sp^1 hybridized

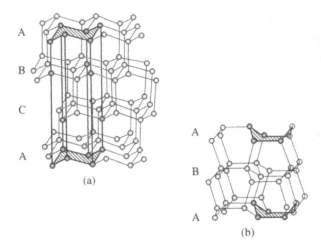

FIG. 2 The two most common polymorphs of diamond: (a) *face-centered cubic diamond* (also known as *3C diamond* and *CIII* [38]); and (b) *hexagonal diamond* (also termed *Lonsdaleite, 2H diamond, δ-phase,* and *CIV* [38]).

carbon that was first reported in 1961 by Kudryavtsev [52]. The essence of this debate has been recently summarized by Heimann [53] who concludes that the literature, while not unequivocal, strongly supports the existence of this controversial allotrope. The structural models of carbyne, which are to be considered further below, suggest that many different polymorphs may exist. Approximately 10 polymorphs have been reported to date [50], including the much studied α- and β-*carbynes* [54] and the recently discovered *carbolite* [55]. Naturally occurring carbyne was termed *chaoite*, but it is now thought to be a mixture of α- and β-carbyne [49]. The main obstacle to establishing the existence of carbyne beyond doubt is the current inability to isolate macroscopic quantities of the allotrope. Although the increased availability of more sophisticated experimental and computational methods has offset this problem to a certain extent, definitive statements on the structure of carbyne still cannot be made. The following outline of the structure should, therefore, be read in this context.

The various hypothesized models for the carbyne structure have been recently reviewed in a number of places [49,50,56]. Carbyne, if it exists at all, is built from *polyynes*, $(-C \equiv C-)_{n/2}$, and *cumulenes*, $(=C=)_n$, the same molecules strongly implicated in the formation of fullerenes [57]. It is now generally believed [58] that the carbyne crystal consists of hexagonal-packed *kinked* polyyne and/or cumulene chains stabilized by van der Waals forces, with more substantive cross-linking at the kinks, and bulky nonreactive end-groups; an example structure (without end groups) is illustrated in Fig. 3, with the different types of kinked

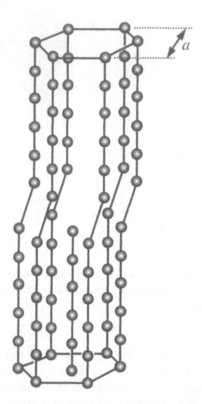

FIG. 3   An example of carbyne structure showing the hexagonal packing and kinks [59].

chains being shown in Fig. 4b–d. This model was first proposed by Heimann et al. [60] in 1984 in order to reconcile the various carbyne polytropes that had been reported up to that time. Subsequent computational studies [61] have confirmed that both isolated and grouped polyyne chains of any reasonable length naturally take up kinked conformations. These studies have also shown that interchain van der Waals interactions and cross-linking between kinks in adjacent chains can yield stable crystal structures provided the chains are prevented from coming too close by, for example, large end-groups or metal atoms. Although the a-direction lattice constant of the hexagonal carbyne crystal is dictated by the method of stabilization, it varies between 0.8 and 1.0 nm for almost all the known allotropes [60], the exception being carbolite, which is far less dense than the other carbynes [55]. The c-dimension lattice constant is defined by the number of carbons between the kinks, n, and the chain type. The number of carbons is believed to vary from about 6 to 12 [60], while computational studies suggest that both polyyne- and cumulene-based carbynes are equally likely [58].

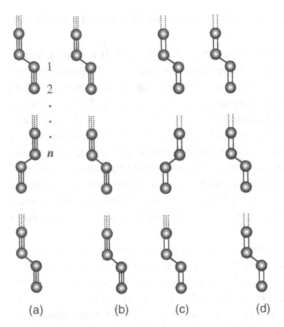

FIG. 4 The building blocks of carbynes as hypothesized by Heimann et al. [60]: (a) *cis*-transoid polyyne; (b) all-transoid polyyne; (c) *cis*-transoid cumulene; and (d) all-transoid cumulene.

The difficulty faced in isolating macroscopic quantities of carbyne and computational studies all suggest that carbyne structures are highly defective. Unfortunately, the lack of macroscopic quantities of carbyne makes it very difficult to comment in any concrete way on the nature of any defects. However, this current inability to produce three-dimensional crystals of carbyne despite immense efforts has prompted Heimann to suggest [62] that the structure of a specific carbyne film may in fact be best described by *paracrystal theory*, a theory that was developed many years ago to describe the structure of polyethylene. This theory envisages that the number of carbon atoms between successive kinks, the kink angles, and the a-direction lattice constant for a particular film are distributed about mean values.

3. Fullerenes and Their Phase Diagram

The *buckminsterfullerene* molecule, C_{60}, was first reported in 1985 by Kroto et al. [63], and isolated in a pure solid form some 5 years later [64]. Although the absence of any industrial scale production method currently limits the use of fullerenes, a wide range of potential applications have been suggested and explored. It is, therefore, appropriate to briefly review the structure of the fullerenes and their solid forms.

(a) Fullerene Molecules. Fullerenes, denoted by C_n, are shell-like *molecules* consisting of n three-coordinated carbon atoms arranged as 12 pentagons and $\frac{1}{2}(n - 20)$ hexagons where n is even, and greater than 20 [65] except 22 [66] which cannot be built from 12 pentagons and 1 hexagon [67]. Those fullerene-like molecules that are not of this configuration (e.g., contain other ring sizes) are officially termed *quasi-fullerenes* [65], although this term is not yet widely used in the literature [68]. Various other nonofficial terms are used in the literature to classify fullerenes. Fullerenes of between 72 and 100 carbon atoms are termed *higher fullerenes* [69], while the term *giant fullerene* is used when n is greater than 100 [70]. Fowler [67] has described fullerenes of $n \leq 60$ as *lower fullerenes*. Various IUPAC suffixes [71] are also used in conjunction with the word *fuller* (e.g., *fulleronium* [72]).

At this stage high-purity macroscopic quantities of fullerenes can only be produced for a very small number of fullerenes. By far the most common fullerenes are C_{60} and C_{70}, which can now be produced in macroscopic quantities (i.e., grams per day) with relative ease [73,74]. Of the higher fullerenes, milligram levels per day of high-purity C_{76}, C_{78}, and C_{84} can be isolated [75] with some effort [69]. Apart from C_{60}, the only lower fullerene to be isolated in significant quantities is C_{36} [76]. Vast numbers of other fullerene molecules have been *detected* experimentally. Maruyama et al. [77], for example, have identified fullerenes as large as 700 carbon atoms, while giant fullerenes up to $n \sim 400$ have been routinely reported [73,78–80]. Fullerenes smaller than C_{60} have also been regularly observed, especially those of 30 or more atoms [81]. The smallest possible fullerene, C_{20}, has been specifically produced for the first time recently by Prinzbach et al. [82].

Analysis of the energetics of *isolated* fullerenes indicates that they are most stable when no two pentagons are adjacent [83]. Experimentally observed fullerenes by and large conform to this so-called *(isolated) pentagon rule (IPR)*. However, it need not be true for all fullerenes, and it clearly cannot be true for $n <$ 60 where it is impossible to avoid adjacent pentagons. Although there has been considerable debate over the exact shape of fullerenes ([84–86] vs. [87,88]), modeling studies indicate that fullerene molecules are essentially polyhedral rather than spherical. It is worthwhile noting, however, that the sphericity of a (strictly speaking, quasi-)fullerene can be increased while lowering the net energy per atom by allowing seven- and higher membered rings [89]. As with other large molecules, each fullerene can exist as one of several, often many, different isomers [67,90,91]. By way of example, isomers of a variety of observed and hypothesized fullerenes are shown here in Fig. 5. As illustrated here for C_{240}, fullerenes can appear spherical when viewed from certain directions even though they

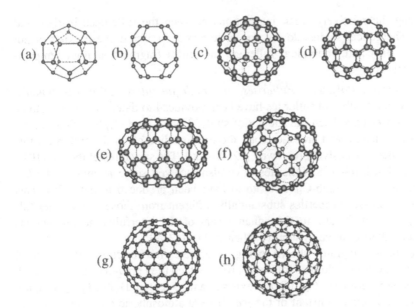

FIG. 5 A variety of observed and hypothesized fullerenes: (a) C_{20} [66]; (b) C_{36} [76]; (c) C_{60} [75]; (d) C_{70} [75]; (e) ellipsoidal, and (f) icosahedral isomers of C_{80} [75]; (g) and (h) a C_{240} isomer viewed from two directions [85].

are not so in general [84]. This has led to confusion in the past over the shape of fullerenes. Fowler [67] very nicely reviews practical methods for enumerating and generating fullerene structures, and for assessing their stability.

The phase diagrams of C_{60} and, to a lesser extent, C_{70} have been the subject of study since the first fullerite was produced in 1990 by Krätschmer et al. [64]; much of this work was reviewed very recently by Sundqvist [92]. At low to moderate temperatures and pressures, fullerenes aggregate to form a *molecular solid* termed *fullerite*. Fullerites can take on a variety of crystal structures that are dictated by the substrate, composition, temperature, and pressure; additional details follow in the next section. At high pressures and temperatures, and under other conditions such as exposure to UV radiation, the fullerenes of a fullerite may chemically bond to form *fullerene polymers*; once again, brief details of these materials will be given in the next section. At very high pressures, the fullerene molecules of a fullerite collapse reversibly or irreversibly to yield amorphous carbon [93], nongraphitic carbon (see later for definition) [94], graphite [95], diamond [96], or new carbon phases [97]. Laser irradiation can also lead to similar phase changes [96]. Only fullerite has ever been observed to sublime, reflecting the relatively weak interactions between the fullerenes in the solid phase. Theoretical studies do not appear to agree if a liquid phase exists or not [92]. As expected, individual fullerene molecules fragment at high temperatures [92].

(b)   Fullerites, Polymeric Fullerites, and Fullerite Intercalation Compounds. Only a small number of fullerites have been produced to date, and the structures of even fewer have been the subject of study. The C_{60} and C_{70} fullerites are the easiest to generate and are, therefore, the most studied of all the fullerites; this work has been extensively reviewed [92,98–100]. The structures of the C_{84} [101–103] and C_{76} [104–108] fullerites have also received some attention. The C_{36} fullerite is the only sub-C_{60} fullerite to have been produced to date [76]. This material possesses properties substantially different from those of the other fullerites, most likely due to the enhancement of the interfullerene interactions caused by their relative instability when isolated [109].

The "phase diagram" shown in Fig. 6 indicates the structures of pure solid C_{60} under the equilibrium conditions [110] at which they prevail or are produced. At higher temperatures, C_{60} fullerite exists as a face-centered cubic (*fcc*) structure (Fig. 7). The lattice constant of the *fcc* phase at atmospheric pressure decreases from $a \sim 1.4165$ nm at 320 K to $a \sim 1.415$ nm at the *fcc-sc* transition point (to be defined below) [111]. As the integrity of the C_{60} fullerite lattice is primarily maintained by short-range van der Waals forces, the C_{60} molecules in the *fcc* phase rotate freely about their three rotational degrees of freedom [111], as indicated by the arrows in Fig. 7. The free rotation of the molecules means they are essentially equivalent and the fullerite is considered an *orientational disordered solid* (i.e., there is no correlation of the orientations of the fullerene molecules [112]). The strength of the fullerene molecules relative to the interfullerene forces also means the lattice constant decreases with increasing pressure [113], while it is expected that rotational freedom will also diminish as pressure rises [92].

A weak first-order phase transition from the *fcc* structure to a simple cubic (*sc*) structure occurs at a temperature T_{fcc-sc}, which increases with pressure from ~260 K at room pressure [114]. This phase change is subtle, as the molecules do not move relative to each other, leaving the lattice constant essentially unchanged ($\Delta a \sim 0.3$–0.4% [111]). Instead, the C_{60} rotations cease about two axes and becomes activated and ratchet-like about the third [111]. This loss of rotational freedom results in an *orientational correlated solid* where the molecules are no longer equivalent; the *fcc* → *sc* phase change is attributable to this loss of equivalence. The activated rotation occurs between the two lowest energy configurations where the electron-rich "double bonds" of a fullerene align with the electron deficient centers of its neighbors: (1) hexagons (denoted H) and (2) pentagons (denoted P) [111]. The P configuration dominates at lower pressures [115] and is favored by lower temperatures at these pressures [111]. The hexagonal configuration becomes dominant at pressures beyond ~200 MPa and is

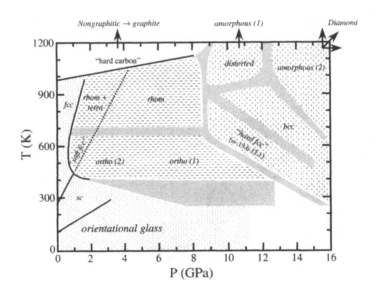

FIG. 6 The tentative "phase diagram" of Sundqvist for C_{60} [92], where the lines and light gray areas indicate the boundaries between the various phases. Structures above about 8 GPa are not guaranteed to be equilibrium structures but, rather, those which are observed at room temperature and pressure after treatment at the indicated conditions for a minute or less. This restriction on the high-pressure states is necessary because of the difficulties faced in maintaining high temperatures/pressures for extended periods, and the inability to observe structures at these extreme conditions. Regions: fullerite ▭, and polymeric fullerite ▬. All lattice parameters are in angstroms. Refer to text for overview of fullerite and polymeric fullerite structures. The high-pressure/high-temperature structures, ▭, demonstrate increased cross-linking between the fullerene molecules compared with the polymeric fullerite states. The cross-linking multiplies and becomes more random as the pressure/temperature rises causing increasing distortion of the lattice leading eventually to amorphous structures. Crystalline graphite and diamond are thought to be recovered at sufficiently high pressures and/or temperatures via nongraphitic carbon and amorphous diamond structures, respectively. (From Ref. 92.)

virtually the only configuration at pressures approaching 1 GPa [115]. The degree of rotational motion decreases as the temperature descends from T_{fcc-sc} to the *orientational glass transition* temperature, T_{og}, where it essentially ceases. The particular symmetry of the C_{60} molecule means that fullerites can never achieve full orientational order. The fullerites at temperatures below T_{og} are, therefore, termed *orientational fullerene glasses*, while the residual orientational disorder is termed *merohedral disorder*. The *orientational glass transition* temperature, which is ~85 K at room pressure, increases with pressure [115].

Fullerite crystals, whether singular or part of a polycrystalline solid or film,

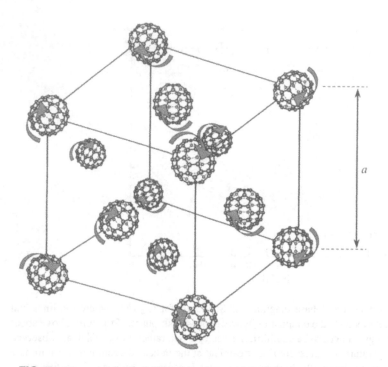

FIG. 7 Face-centered cubic (*fcc*) structure of C_{60} fullerite. Note that the fullerene molecules of this structure rotate freely and rapidly, and orientational order is essentially nonexistent.

can vary in size from nanometers to micrometers depending on the method and conditions of manufacture. Fullerites are disturbed by defects that can be both natural (e.g., stacking faults [113,116]) and induced (e.g., from substrate [117, 118]). Amorphous fullerite films can also be generated under appropriate conditions [119]. It has been shown that fullerites can contain pores ranging in size from 3 to 300 nm [120,121]. The presence of pores and cavities between the fullerene molecules means fullerites are excellent traps for guest molecules such as residual solvent [122,123] and oxygen [124,125]. The presence of such guest molecules and of other fullerenes is known to strongly affect the structure of a fullerite, the rotational dynamics of its fullerene molecules, and its properties [122–127]. In the case of films, the nature of the substrate also affects the fullerite structure (e.g., formation of superlattices [128]) and the dynamics of the fullerene molecules (e.g., retardation of rotation [100]).

The structures of other fullerites are far less well understood. For example, the nonspherical shape of C_{70} and the likely coexistence of both *fcc* and *hcp* structures at higher temperatures means there is still no definitive understanding of the C_{70} fullerite structure under almost any condition [92]. This uncertainty means it is not appropriate to review the structure of C_{70} here; the reader is referred to Sundqvist [92] for a review of the current understanding. The state of play for the higher fullerenes is probably only less uncertain due to the dearth of studies. A few experimental studies [101–107] suggest the C_{76} and C_{84} fullerites are *fcc* at room temperature, although recent experiments [108] indicate that, like the similarly nonspherical C_{70}, multiple phases may coexist at room temperature for the C_{76} fullerite. The data of Saito and coworkers [104] indicate that the lattice constant of the C_{60}, C_{70}, C_{76}, and C_{84} *fcc* phases at room temperature is given by $a \sim 0.097685\bar{d} + 0.78$ nm, where $\bar{d} = 7.1\sqrt{n/60}$ is a measure of the average diameter of the fullerene, C_n, in angstroms. While there is currently a great deal of uncertainty surrounding the C_{70} and higher fullerites, a brief survey of the literature indicates that the variation of their structure with temperature depends on, probably among other things, the number of fullerene isomers present in the fullerite, and their shape and symmetry. For example, the fact that C_{84} fullerite is made up of at least two isomers appears to forbid any orientational correlations setting in with cooling despite C_{84} being near-spherical [101]. This fact is reflected in a complete lack of evidence for any phase change even at temperatures as low as 20 K [103]. The nonspherical nature of C_{70} and C_{76} appears to provide opportunities for multiple states to coexist [92,108], while the low symmetry and nonspherical shape of C_{76} promotes greater orientational and spatial disorder at low temperatures [105].

Various methods have been used to promote covalent sp^3 bonding between adjacent fullerenes of a fullerite to form insoluble *polymerized fullerite*. Much of the work on C_{60} and C_{70} polymerized fullerites has been reviewed a number of times in recent years [92,129–131]. A variety of polymerized fullerite structures can be formed, some of which are illustrated in Fig. 8 along with their lattice constants at room temperature and pressure. The structures of some polymerized fullerites are not completely understood despite extensive research. For example, the polymer structure formed at low pressures and temperatures or via light irradiation is thought to be a disordered material consisting of either a mixture of orthorhombic, rhombohedral, and tetragonal phases, or a collection of linear chains of fullerenes of varying length and direction [92]. Various names are used to describe this disordered structure, including *soft-fcc* and *fcc(pC_{60})*. The structure of polymerized fullerite formed by application of very high pressures (greater than 8 GPa) is also not well understood [92].

The spaces within the lattice of fullerites are sufficiently large to accommodate a number of different noncarbon atoms or clusters as *intercalates*. Indeed, it is the ability for fullerites to take up and hold foreign species such as oxygen and solvents as intercalates that makes experimentation with them so fraught. These so-called *fullerides* have been the subject of much study since first reported in 1991 [133]; much of it has been reviewed by Rosseinsky [134,135].

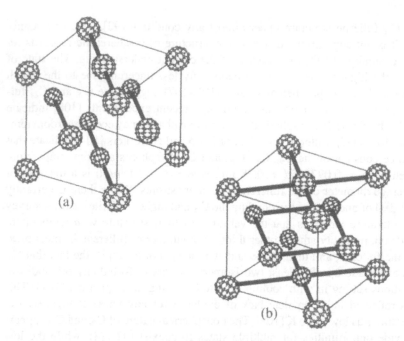

FIG. 8 Schematic of some of the possible polymeric fullerite structures (lattice parameters as measured at room temperature and pressure [132]): (a) orthorhombic ($a = 0.926$ nm, $b = 0.988$ nm, $c = 1.422$ nm); and (b) tetragonal ($a = b = 0.909$ nm, $c = 1.495$ nm). The chemical bonds between the C_{60} molecules are shown as heavy lines.

(c) Endohedral Fullerenes, Exohedral Fullerenes, and Heterofullerenes. A variety of fullerenes have been produced that include noncarbon atoms. One of the first to be reported [136] is the so-called *endohedral fullerene* which, as illustrated in Fig. 9a, has a noncarbon species trapped within it. Endohedral fullerenes are commonly denoted by $E_m@C_n$, where E_m is the guest species [136]. A wide variety of endohedral fullerenes have now been produced both as isolated fullerenes in solution and as fullerites including those that contain noble gases [137], complex clusters or molecules [138], and metal atoms or clusters. Metal-containing fullerenes have been the most studied of the endohedral fullerenes and have, therefore, been reviewed a number of times [139–141]. The presence of the species within the cage promotes the production of larger fullerenes—C_{82} endohedral fullerenes were the first to be produced [136]—in favor of C_{60}. Depending on its electronic structure, the trapped species may move and/or rotate within the cavity, or it may be intimately associated with the cage [141].

The rich chemistry [142] of fullerenes and their ions (*fulleride anions* and *fullerenium cations* [143]) permits the synthesis of a vast range of so-called *exohedral fullerenes* where atoms or functional groups are chemically attached to

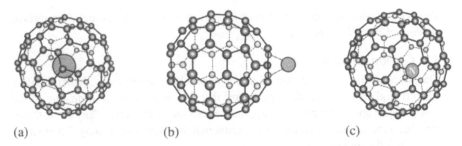

(a)   (b)   (c)

FIG. 9 Classes of fullerenes involving noncarbon species: (a) an *endohedral fullerene* where the noncarbon species is located within the shell; (b) an *exohedral fullerene* where the noncarbon species is chemically bound to the external surface of the shell; (c) a hetero-fullerene where the noncarbon species replace an equivalent number of carbon atoms in the shell itself.

the outside of the fullerene cage. One of the simplest classes of exohedral fuller-enes consist of those which are coated in noncarbon atoms. Martin, Malinowski and coworkers have produced fullerenes in the vapor phase coated in one or more layers of alkali [144], alkaline [145], and transition metal [146] atoms, and nonmetallic atoms [147]. These workers found that both the geometrical and the electronic configuration of the fullerene-heteroatom system determined the number of atoms within a layer [144]. Ohno et al. [148] observed that titanium and lanthanum at low loading preferentially bond with the outer surface of fuller-enes in the solid phase. The second and by far the largest [149] class of exohedral fullerenes includes those that are functionalized via synthetic chemistry routes; an example is shown in Fig. 9b. There are a vast array of functionalized fullerenes that may be classified on the basis of the nature of the functional groups and how they are attached to the fullerenes. The major groups of functionalized fullerenes include the *fulleroids* [150], *methanofullerenes* [151], *fullerene-metal complexes* [152,153], and *organometallofullerenes* [154,155]. Among many other things, the functionalized fullerene class of fullerenes offers the hope of functional films [156,157], including those used in electrochemical devices, and new carbon archi-tectures through self-assembly [157–159], and fullerene polymers [160, 161].

Synthetic chemistry appears [162,163] to also offer the best means to date of producing bulk quantities of so-called *heterofullerenes* where one or more heteroatoms replace an equivalent number of carbon atoms *within* the carbon shell, as illustrated in Fig. 9c. To date only the aza[60]fullerene, $C_{59}N$, and aza-[70]fullerene, $C_{69}N$, have been produced by this means; much of the work sur-rounding these two heterofullerenes has been reviewed recently by the two main groups in this area of endeavor [164,165]. Submacroscopic quantities of boron [166], lanthanum [167], niobium [168], and silicon [169] heterofullerenes have also been produced. Heterofullerenes involving metallic atoms are often termed *networked metallofullerenes*.

4. Carbon Nanotubes

The novel physics and structure of carbon nanotubes, combined with their relative stability and ease of manufacture, means that they are the most promising of the new carbon forms with regard to applications in the near future. They have, therefore, been the subject of extensive experimental and theoretical study. Their structure is briefly reviewed here.

(a) Carbon Nanotube Unit Cell. The unit cell of a single carbon nanotube (Fig. 10), is constructed from a portion of a graphene sheet defined by the *chiral vector* and the *translation vector* [170]. The chiral vector connects two crystallographically equivalent sites on the graphene sheet at an angle, θ, to the *zig-zag axis*, which is uniquely defined by the so-called *Hamada indices* (n, m) [170]. The translation vector is perpendicular to the chiral vector and is defined by the first intersected site that is crystallographically equivalent to the origin site, A. The unit cell of a carbon nanotube is obtained by rolling up the graphene sheet AA'B'B and joining the edge-AA' to edge-BB' to form a cylinder. Three possible classes of nanotube exist, examples of which are illustrated in Fig. 11. The *zig-zag*, $\mathbf{C_h} = n\mathbf{a}$, and *armchair*, $\mathbf{C_h} = n(\mathbf{a} + \mathbf{b})$, nanotubes are defined by the chiral vector's coincidence with the requisite edges of the graphene sheet. In the case of *chiral* nanotubes where $n \neq m \neq 0$, the hexagons of the graphene sheet are

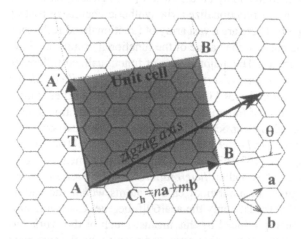

FIG. 10 The definition of a carbon nanotube unit cell: chiral vector, $\mathbf{C_h}$, translation vector, $\mathbf{T}$, the zig-zag axis, and the Hamada indices (n, m) in the vector space defined by the basis vectors, $\mathbf{a}$ and $\mathbf{b}$. (Adapted from Ref. 170.)

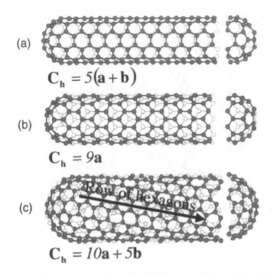

(a)

$$C_h = 5(a+b)$$

(b)

$$C_h = 9a$$

(c)

$$C_h = 10a + 5b$$

FIG. 11 Examples of the three classes of carbon nanotube: (a) zig-zag nanotube; (b) armchair nanotube; and (c) chiral nanotube. (Adapted from Ref. 170.)

oriented along a screw axis as illustrated in Fig. 11c where the arrow highlights the spiraling of the rows of hexagons along the length of the tube. The diameter and length of the nanotube unit cell are defined solely in terms of the Hamada indices as $d_t = C/\pi$ and $l = \sqrt{3}C/d_R$ respectively, where $C = a\sqrt{m^2 + mn + n^2}$ is the circumference of the nanotube (i.e., length of the chiral vector), d_R the greatest common divisor of and $2m + n$ and $2n + m$, and $a = 0.143$ nm the length of the C—C bonds [170]. As illustrated in Fig. 12, the Hamada indices also dictate the conductivity of the nanotube [170]. Computer code for generating the coordinates of the carbon atoms of a nanotube unit cell given the Hamada indices can be found in Saito et al. [171]. Carbon nanotubes of arbitrary length are simply constructed by joining together a number of the unit cells and, as illustrated in Fig. 11, they can also be capped with hemifullerenes [170].

(b) Multiwall Nanotubes. As illustrated in the electron micrographs of Fig. 13, *multiwall nanotubes (MWNTs)* are constructed from a number of concentric carbon nanotubes. Although smaller monochiral MWNTs have been observed [173], the chiral angle of the individual nanotubes within a MWNT are usually distributed [174,175]. A brief survey of the literature indicates that the number of constituent nanotubes of MWNTs, their external and internal diameters, and their lengths are also distributed to a greater or lesser extent. Generally, the number of constituent nanotubes varies from as little as two [172,176] up to ~80 [177] as the external diameter increases from ~2 to ~55 nm, although far larger

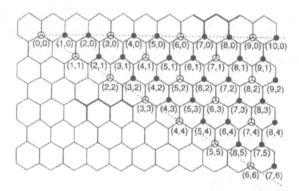

FIG. 12 The Hamada indices of metallic, ○, and semiconducting, ●, nanotubes. (Adapted from Ref. 170.)

MWNTs have been recently produced by Lee et al. [178]. The inner diameter of the MWNTs of Lee et al. are correspondingly large at ~50 nm, while inner diameters as small as 2 nm have been observed [179,180]. MWNTs between 10 and 100 μm in length are now regularly produced [176,177], and Pan et al. [181] have recently produced nanotubes as long as 2 mm. Kiang and coworkers [182] have shown that the intertube separation is inversely related to the diameter of the nanotube, with the separation varying from ~0.41 nm for the smallest known nanotubes to a limiting value of ~0.344 nm for the larger nanotubes (>10.0 nm),

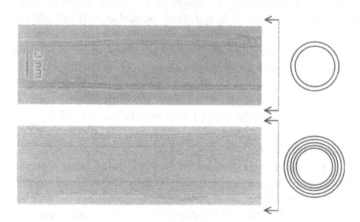

FIG. 13 The first published electron micrographs of multiwall nanotubes. The top and bottom carbon nanotubes consist of two and five concentric nanotubes, respectively. (From Ref. 172.)

a value typical of *turbostratic carbons* (see below for details of such carbons). MWNTs are invariably marred by defects and are almost never perfect cylinders. For example, MWNTs are rarely straight due to the presence of five- and seven-membered rings [174]. The presence of these rings also results in the capping of individual nanotubes to yield *bamboo-like structures* [178] and *nanocones* [183], as illustrated in Fig. 14. These and other defects are discussed extensively in the literature [184–189]. Depending on the methods of production, MWNTs occur as tangled aggregates mixed with nanoparticles and other debris [180,190] or as bundles of largely aligned but separate nanotubes [177,178,191].

(a)

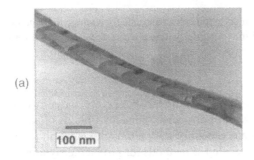

(b)

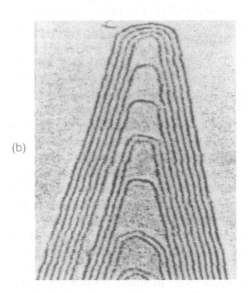

FIG. 14 Capping of individual nanotubes can yield: (a) bamboo-like structures (from Ref. 178) and (b) nanocones (from Ref 183).

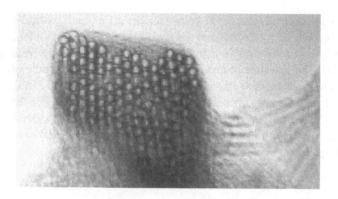

FIG. 15 The first published electron micrographs of a nanotube rope where the spacing between the tubes is about 0.3 nm and the triangular lattice pitch is about 1.7 nm. (From Ref. 192.)

(c) Single-Wall Nanotubes. *Single-wall nanotubes (SWNTs)* are normally produced as extensive tangled mats [192,193] or course filaments/ribbons of *nanotube ropes* [194]. Isolated SWNTs are generally rare, although Dai et al. [195] and Kong et al. [196] have recently developed methods for producing significant numbers of such nanotubes. A nanotube rope is shown in Fig. 15 where the cross-section is observed to reveal individual SWNTs of similar diameter assembled on a regular triangular lattice by van der Waals forces [192]. The uniform nanotube diameter and regular lattice structure means the nanotube ropes are, in effect, crystallites. A brief survey of the literature indicates that these nanotube crystallites are normally 5–35 nm in diameter and can be longer than 100 μm [192,193,197,198]. Recent scanning tunneling microscopy (STM) studies [199] suggest that larger nanotube ropes (20–100 nm) may consist of interwoven crystallites, although no further evidence exists to confirm this finding. Other STM studies have shown that individual SWNTs within a crystallite may twist as they attempt to conform to the complex potential field arising from neighboring nanotubes [200]. SWNT diameters normally range from 1 to 2 nm, with the averages being ~1.4 nm for conventional production methods, and somewhat larger at ~1.7 nm for SWNTs produced from catalytic decomposition of hydrocarbons [194]. The smallest SWNT observed to date is 0.7 nm [201], while SWNTs as large as 4–5 nm have been reported [194,195,202]. Following a review of the literature, Lambin et al. [203] conclude that the diameter and chirality of SWNTs within a crystallite are distributed to some extent. Single-wall nanotubes are believed to possess fewer defects compared to their multiwall counterparts [192], although defects have been observed both directly and indirectly [204,205].

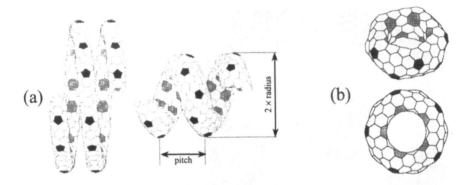

FIG. 16 Hypothesized exotic carbon nanotubes: (a) toroidal nanotube (from Ref. 207); (b) helical nanotube (from Ref. 209).

(d) Exotic Nanotubes. Following the discovery of nanotubes in 1991, some workers hypothesized the existence of toroidal [206–208] and helical [209] nanotubes (Fig. 16) by allowing five- and seven-membered rings to occur in the graphene sheet of the nanotube surface. Up to 30-μm-long helical MWNTs of 20–50 nm diameter were observed by Zhang, Nagy, and coworkers [210–212] soon after these hypotheses were made; an example is shown in the micrograph of Fig. 17. More recently, Biró et al. [213] claim to have imaged a spiral SWNT using STM. The radius of the MWNT helices observed to date varies from 8 to a few hundred nanometers, while their pitch can vary from 20 nm to a few hundred nanometers [211]. The SWNTs helices reported by Biró et al. are much smaller,

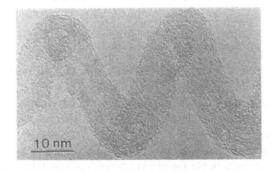

FIG. 17 Micrograph of helical MWNT with radius and pitch about 18 nm and about 30 nm, respectively. (From Ref. 211.)

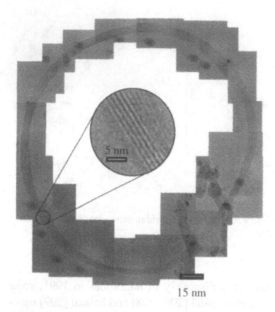

FIG. 18   Micrograph of toroidal MWNT. (From Ref. 214.)

with the tubes being 0.7 nm in diameter, and the helix radius and pitch ~2 nm
and 1–1.2 nm, respectively. Toroidal structures of individual SWNTs and ropes
of SWNTs have been more recently observed [214–217] and were christened
nanotube crop circles [218]. At diameters that vary from 200 to 500 nm, however,
these observed toroidal structures are substantially larger than those hypothesized
by Itoh, Ihara, and coworkers [207]. The constituent SWNTs and ropes of the
observed toroidal structures are ~1.5 nm in diameter and 5–30 nm thick, respectively. The micrograph of Fig. 18 shows one of the toroidal structures observed
by Liu et al. [214].

5.   Carbon Onions

Carbon onions [219], which are also termed *bucky onions, onion skin fullerenes,
multilayered fullerenes, Russian doll fullerenes, hyperfullerenes,* and *nested fullerenes,* are essentially an assembly of concentric fullerenes of increasing size.
An example of a carbon onion is illustrated in Fig. 19, which consists of concentric C_{60}, C_{240}, and C_{540} molecules and is denoted by $C_{60}@C_{240}@C_{540}$. Amorphous
films of carbon onions have recently been produced [221–222]—the structure
of these films suggests they may be suitable for use in adsorption-based devices.

Modeling suggests that carbon onions may be extremely large [223] and that
they are more stable than fullerene structures above a critical number of carbon
atoms, although the exact value of this number is not well known [84,88,224–

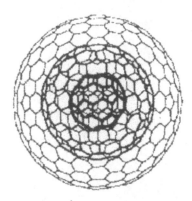

FIG. 19 Model carbon onion C_{60} @ C_{240} @ C_{540}. (From Ref. 220.)

226]. Zwanger et al. [227], on the other hand, use HRTEM observations to argue that they are metastable, with their decay being promoted by the presence of amorphous carbon and oxidative gases. Carbon onions varying from 3 to 1000 nm in diameter have been observed experimentally, with the size depending on the nature of the production process (e.g., Cabioc'h and coworkers have synthesized carbon onions of widely varying sizes using the same method but different conditions: 10–20 nm [221]; 5–8 nm and 5–20 [222]; 50–1000 nm [228]). The size of the innermost shell can also vary. It is commonly believed that C_{60}, the smallest stable *isolated* fullerene, constitutes the smallest innermost shell of carbon onions [99]. However, Oku et al. [229] have recently reported a C_{28} innermost shell, while Selvan et al. [230] have observed internal cavities as large as 2 nm in diameter. Diamond cores varying in size from 2 to 4.5 nm have also been reported recently [231]. It is not clear that a full understanding exists on the spacing between two successive shells. Using HRTEM, Ugarte [219] reports a spacing of 0.334 nm similar to that of bulk graphite, while other groups report values substantially above [228,229] and below that of graphite [230] using the same and additional experimental methods, such as SAD. It is probably more likely, as recently observed by Banhart and Ajayan [231], that the intershell spacing increases from a value well below that of graphite when at the center of the carbon onion to a value approaching that of graphite at the outermost pair, much in the same way as has been observed for MWNTs [182].

By comparing HRTEM micrographs with simulated micrographs of Goldberg polyhedra, Zwanger and Banhart [232] argue that carbon onions are highly likely to be spherical despite the fact that isolated fullerenes are in general nonspherical. They also argue that the individual shells need not be aligned in any way and that they are unlikely to be rotating relative to each other. These arguments are largely supported [233] by simulations that show that the van der Waals interac-

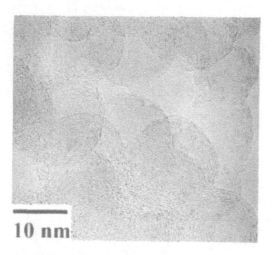

10 nm

FIG. 20 Portion of an amorphous film of carbon onions as produced by Cabioc'h et al. Note the high degree of uniformity of carbon onion size. (From Ref. 221.)

tion energy between the shells of a carbon onion more than offsets the strain energy induced by the curvature of the shells [88,234], and that the electron irradiation used in the manufacture of carbon onions encourages the movement of nonhexagonal rings to minimize the strain energy of the shells [226,235].

Carbon onions were first observed by Ugarte [219] as part of the raw material (more familiarly known as *soot*) from which fullerenes and nanotubes are isolated. These carbon onions could not be isolated and they varied widely in size. Cabioc'h and coworkers [221,222,228] have more recently produced irregularly packed carbon onion films in which the onions are more controlled in size and are the dominant entity; the micrograph in Fig. 20 shows a portion of one such film.

6. Activated Carbon

In its most common form, *activated carbon* (also termed *active carbon* and *activated charcoal*) is a highly microporous material formed by the activation of a char obtained from the carbonization of an organic precursor [2,236]. Organic precursors that have been used include natural materials such as coal and coconut shell, reconstituted natural polymers such as rayon, and synthetic polymers such as organic gels. Carbonization is normally undertaken via pyrolysis in an inert atmosphere at temperatures between 873 and 1273 K. The resultant char is high in carbon but contains little internal surface area; this surface area is subsequently developed in the form of micro- and mesopores (the activation process) by reaction at similar temperatures in oxygen, carbon dioxide, steam, or other activating agents.

The structure and chemistry of activated carbons (ACs) have been the subject of intense experimental analysis since the early 1900s. Much of our current understanding of the molecular structure of activated carbons rests on data obtained from X-ray and electron diffraction, Raman, FTIR and other forms of spectroscopy, and HRTEM. As a result of extensive and careful X-ray studies [237,238], it has long been accepted that ACs consist primarily of sp^2-hybridized carbon atoms. It has also long been known that, depending on the chemistry of the precursor, ACs contain smaller amounts of oxygen, sulfur, and nitrogen. Chemical and spectroscopic analysis indicates that these heteroatoms occur within the ring structures (e.g., pyrone) and as surface functional groups (e.g., lactone) [239]. The exact nature of molecular structure of ACs has been the subject of much more debate in recent times, and it was not until HRTEM was brought to bear in the 1960s and beyond that consensus was, to a large measure (see below), reached. A number of workers [240–245] have contributed to our understanding of the molecular structure of microporous carbons using HRTEM. However, perhaps the most active group in this area has been that of Oberlin, whose vast volume of work has been summarized in two key reviews [32,246]. It is from this work that the current view of activated carbon structure is taken.

One widely used interpretation of the vast amount of experimental data is the *hierarchical structure* shown in Fig. 21. The so-called *basic structural unit (BSU)* is the fundamental building block of activated carbons. This term was coined by A. Oberlin and coworkers [249], but the existence of such a fundamental building block was recognized, although not universally accepted, as long ago as 1928 [250] or, arguably, even before [251]. The BSU consists of a few roughly aligned small polyaromatic-like molecules, which we will refer to here as *layers*. There is an absence of graphite-like order between the layers within a BSU and, due to the presence of the functional groups, the interlayer spacing is generally greater than that of graphite. This disorder in the c direction (see BSU of Fig. 21) is termed *turbostratic* [237]. As illustrated in Fig. 21, in Oberlin's interpretation the BSUs are assembled to form regions of *local molecular orientation* (LMO) [252]—the second level of the hierarchy—which are in turn assembled in space to yield the complex structures illustrated at the right-hand end of this hierarchy.

The extent and height of the regions of LMO, L_a and L_c, respectively, the degree of order within the regions of LMO, and the interlayer spacing, d_{002}, vary widely depending on the chemistry of the precursor and the nature of the heat treatment [252]. In general, the regions of LMO are similar in size to the individual BSUs for carbons prepared at temperatures between 873 K and 1273 K [252]. These carbons are considered *nongraphitic* [37], a term first coined by Franklin in 1951 [238]. Depending on the nature of the precursor and the processing condi-

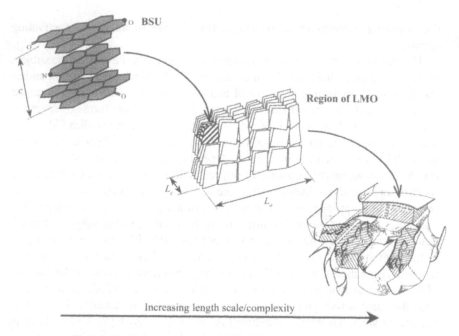

FIG. 21 A hierarchical view of the molecular structure of an activated carbon (AC). The fundamental building block of AC is the basic structural unit (BSU). These are assembled into regions of local molecular orientation, which are in turn assembled into the complex structures normally associated with microporous carbons (right-hand end of hierarchy). (From Ref. 248.)

tions, the BSUs of the unactivated char will be cross-linked by oxygen and sulfur to a greater or lesser extent [248,252]. For those carbons in which the BSUs are highly cross-linked, additional heat treatment sees only slight growth in LMO extent and order [252]. These carbons are said to be *nongraphitizing* [37,238]. Nongraphitizing carbons contain significant microporosity and it is therefore these materials that are used in the preparation of activated carbons. A lack of cross-linking in a nongraphitic carbon allows the BSU elements to rearrange and line up when heated to temperatures beyond 1273 K [252]. This rearrangement leads to a growth in L_a and L_c through coalescence of BSU elements, and increased order within the individual regions of LMO. At sufficiently high temperatures, the regions of LMO become macroscopic in size and essentially graphitic in nature; indeed, it is by this route that artificial graphite is manufactured [41]. It is for this reason that nongraphitic carbons lacking any significant cross-linking are termed *graphitizing carbons* [238]. Examples of the variation of L_a and L_c with heat treatment temperature for typical nongraphitizing and graphitizing carbons are shown in Fig. 22a–b, while the corresponding variations of d_{002} and intra-LMO order for a typical graphitizing carbon are shown in Fig. 22c–d.

Recently, Harris and coworkers [254,255] argued on the basis of HRTEM images that microporous carbons can be built from curved sheets of graphene, where the curvature is made possible by randomly occurring nonhexagonal rings in the graphene sheets (i.e., as in fullerenes and, see below, schwarzites). Although this interpretation of the HRTEM micrographs has not been confirmed by any other group to date, it does not appear to be at odds with that of Oberlin and coworkers, which places no restriction on the nature of the rings within the BSUs or regions of LMO, nor on the minimal size of the LMO.

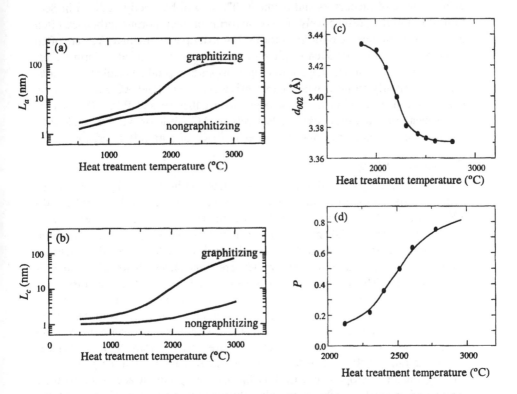

FIG. 22 Typical variations of the average size and measures of internal order of regions of LMO vs. heat treatment temperature: (a) extent of region of LMO for graphitizing and nongraphitizing carbons (from Ref. 253); (b) height of region of LMO for graphitizing and nongraphitizing (from Ref. 253); (c) average interlayer spacing within a region of LMO for a typical graphitizing carbon (from Ref. 252); (d) intra-LMO order for a typical graphitizing carbon (P is the probability of finding two successive layers within an LMO in the AB configuration of graphite; i.e., $P = 1$ for graphite). (From Ref. 252.)

Activated carbons occur in a range of physical forms. *Particulate activated carbon* [2,236] is the oldest and most common form. Particulate carbons less than ~200 µm in size are termed *powdered activated carbon (PAC)*, while those between 200 and 5000 µm in diameter are termed *granular activated carbon (GAC)*. The 1960s saw the development of the *activated carbon fiber (ACF)* [4], which is the second main form of microporous carbon. Individual ACFs are typically 5–10 µm in diameter. The *microporous carbon membrane* [256,257], which was first developed in the early 1980s, is the third main form of microporous carbon. These membranes can be either *unsupported membranes* or, more commonly, *supported flat membranes* or *supported hollow-fiber membranes*. Particulate carbons, ACFs, and microporous carbon membranes may all be prepared using a range of precursors and methods. These will be briefly outlined in Sections II.B and II.C, respectively. The basic forms of microporous carbon are often used in the manufacture of derived forms. For example, particulate ACs are often combined with a binder such as pitch or molasses and extruded to form *pellets* 1–5 mm in size. Larger, shaped, microporous carbon products called *monoliths* are also similarly formed, while inexpensive AC *fabric* and AC *paper* may be manufactured by coating fabric or paper with adhesive and PAC. The size of ACFs mean they are often used collectively as AC fabric or AC *felt*, while they are also used in monoliths and AC paper in place of particulate carbons.

7. Carbon Blacks

Carbon blacks are produced from the incomplete combustion of liquid or gaseous hydrocarbons, and are generally held to consist of discrete carbon entities 10–500 nm in size [258]. Prior to any heat treatment, the carbon particles may be near-spherical (Fig. 23a), melded near-spherical (Fig. 23b), or what has been termed by Oberlin [246] as *statistically spherical* (Fig. 23c), which demonstrates no obviously spherical entities but, rather, graphene sheets distorted in three dimensions [246]. The near-spherical particle seen here in Fig. 23a is not dissimilar to the carbon onions seen in Fig. 20, while the statistically spherical structure in Fig. 23c is reminiscent of activated carbon structures. The fineness of the carbon black entities means they occur as aggregates with external surface areas of up to 150 m^2/g [258,259]. These surface areas are derived from the high surface-to-volume ratio of the carbon black entities rather than through the presence of any significant microporosity. Carbon blacks are classified according to their method of production, which strongly influences particle size. *Furnace blacks* are the most common of these materials with typical diameters of only 10–50 nm [258]. The less common *channel blacks* are also of similar size [258]. Channel blacks, produced in an oxidizing atmosphere, can be microporous with surface areas in excess of 1000 m^2/g. *Thermal black* and now uncommon *lampblack* particles are larger at 100–500 nm, giving correspondingly smaller surface areas

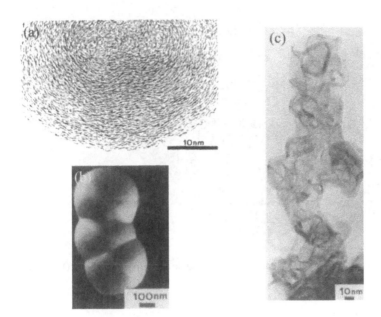

FIG. 23 Carbon black entities [246]: (a) near-spherical carbon black particle; (b) entity of melded near-spherical carbon black particles; and (c) a so-called *statistically spherical* entity that is constructed from continuous distorted graphene sheets which, presumably, contain nonhexagonal rings.

of ~10 m^2/g [258]. Acetylene blacks, which are derived from the gas of the same name, are typically 40–55 nm in diameter [259].

It has long been recognized that the molecular structure of a carbon black entity is not unlike that of activated carbons [246]. As illustrated in Fig. 24, a spherical carbon black particle can be built from concentric layers of BSUs. In the case of a non-heat-treated carbon black, the extent and height of the regions of LMO are of the order of 1.0 nm yielding well-rounded structures as seen in the micrograph (Fig. 23a). As with nongraphitic carbons, the regions of LMO may grow upon heat treatment yielding more polyhedral structures [240,246]. They are, however, nongraphitizable materials [261]; typical values for the LMO extent and layer spacing for heat-treated commercial carbon blacks are [240,259,261]: L_a ~ 1.2–3.0 nm, L_c ~ 1.0–2.0 nm, and d_{002} ~ 0.35–0.36 nm.

8. Carbon Xerogels, Aerogels, and Cryogels

Organic gel is a porous polymeric material formed through sol-gel polymerization of highly cross-linking organic monomers followed by drying [262,263].

FIG. 24   A sketch of a spherical carbon black particle showing the BSUs. (From Ref. 260.)

Evaporative, supercritical, and freeze-drying yield *xerogels*, *aerogels*, and *cryogels*, respectively. The first of these drying methods causes collapse of the mesopores of the wet gel due to surface tension at the gas-liquid interfaces. The other methods avoid such collapse to a greater or lesser extent. Pyrolysis of organic gels yields carbon gels that can in turn be treated to form activated carbon gels.

As illustrated in Fig. 25, the mesoscopic structure of the organic gels can be broadly described as a network of polydisperse polymeric beads with the main differences being the size of the beads and the distance between the network junctions. The mesoporosity is located between the beads, while the microporosity is predominately found within the beads. The size of the beads, their internal structure, and the degree of mesoporosity are in general all influenced by the sol-gel chemistry [264–268]. For example, in the case of the resorcinol formaldehyde system, the bead size and mesoporosity increase as the catalyst level decreases relative to that of resorcinol, while the degree of polymer cross-linking within the beads diminishes [264]. A brief survey of the literature [264–266,269–273] indicates that bead sizes can be varied from 3 to 30 nm by changing the sol-gel chemistry, but they are not greatly affected by either the monomers used or the drying method. The terms *polymeric aerogel* and *colloidal aerogel* in the sense of Schaefer [274] are used by some to differentiate between the structures that result from different sol-gel chemistries. Colloidal gels are built from larger, more spherical beads, as illustrated in the upper part of Fig. 25. Polymeric aerogels, on the other hand, consist of finer, more irregular beads and even fibrous polymeric

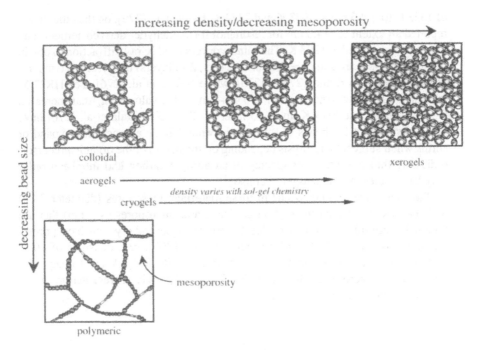

FIG. 25 The different types of gel structures. Colloidal gels and polymeric gels are differentiated on the basis of the bead size (vertical axis). The density of the gel materials increases as the degree of mesoporosity decreases. Aerogels and xerogels are the least and most dense of the gel materials, respectively. Cryogels tend to be somewhat denser than an aerogel produced using the same sol-gel chemistry. The density of aerogel/xerogel materials can be varied by an order of magnitude by changing the sol-gel chemistry.

structures which could be viewed as beads that have failed to grow or which are very small and ill-defined [264] (bottom Fig. 25).

To a large extent, carbon aerogels, xerogels, and cryogels retain the *mesoscopic* structure of their organic counterpart [269,275]. The polymers used in making organic gels yield nongraphitizing carbons when pyrolyzed (see previous section on activated carbons). The beads of the organic materials normally contain some microporosity, and it is this which is substantially developed during carbonization [268] and subsequent activation [273,276]. The structure within the carbon beads is similar to that of activated carbon. However, knowledge of d_{002}, L_a, and L_c, is scant and unclear. For example, using Raman spectroscopy, Pekala and coworkers [269,276,277] estimate that L_a for a resorcinol formaldehyde–derived carbon aerogel increases from ~2.5 nm at a pyrolysis temperature of 1323 K to ~4.5 nm at 2073 K and 2373 K. Even et al. [275], on the other hand, reported an extent of ~4.0 nm for resorcinol formaldehyde–derived aerogels and xerogels carbonized at 1203 K. The same workers used X-ray diffraction to obtain $d_{002} \sim 0.38$–0.39 nm and $L_c \sim 0.6$–0.9 nm (i.e., 2–3 layers) for pyrolysis temperatures between 773 K and 1273 K [275], while Pekala et al. [276] used HRTEM to estimate a larger separation of ~0.404 nm and a value of L_c that increased from 4–5 layers at 2073 K up to 10 layers at 2373 K. Pekala et al. [270] have also reported that the beads of resorcinol formaldehyde aerogel may expand or shrink when subject to pyrolysis depending on the nature of the organic precursor, indicating that the interlayer spacing of an aerogel carbon and intrabead order may be affected by the sol-gel chemistry.

Two pore size regimes occur in these materials: mesopores (diameter 2–50 nm) that occur between the carbon particles, and micropores (<2 nm) that are within the individual carbon particles. Carbon aerogels and cryogels have porosities in the range 40–98%, surface areas of 400–1200 m^2/g, densities of 100–1500 kg/m^3, and pore volumes of 1–3 ml/g [269,276–284]. A TEM image of a typical carbon aerogel is shown in Fig. 26. A schematic representation of the aerogel structure is given in Fig. 27.

FIG. 26 Transmission electron micrograph of carbon aerogel CA-0.4.

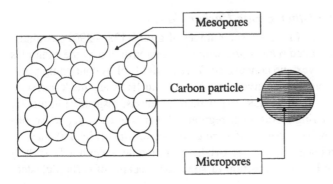

FIG. 27 Carbon aerogel schematic description.

The most important property of carbon aerogels for practical applications is their electrical conductivity. Their resistivity is low, in the range 10^{-2} to 10^{-3} Ωm [278], with the value depending on the density. This property, together with the high surface area, makes them promising materials for applications as electrodes in supercapacitors [279,280] or in fuel cells. Capacities up to 100 F/cm^3 have been reported for carbon aerogel supercapacitors [553], and the stored energy is rapidly available due to the high electrical conductivity. Another potential application is the use of carbon aerogel electrodes for removal of various ions from wastewater [553]. Carbon aerogels have low thermal conductivities (about 0.015 W/mK in air), making them potentially useful as thermal insulators [281]. The aerogel skeleton leads to low sound velocities, in the range 20–300 m/s, and a wide range of acoustic impedances [282]; this latter property makes possible the application of these materials to match the high impedance of ultrasonic transducers with the low impedance of air.

Carbon aerogels have been prepared in microsphere, particulate, film, and monolith forms ([283] and references therein); their sol-gel basis means they are particularly suited to the manufacture of thin films and complex monolithic shapes. As the solvents used in the preparation of aerogels expand upon freezing, only monolithic forms of carbon cryogels and their powders have been reported to date. Although the organic cryogels are in general more dense than their aerogel counterparts, the carbon cryogels generally take on physical properties that are comparable to if not better than those of the aerogels [284]. For example, Tamon et al. [284] found that carbon cryogels possessed micropore surface areas much greater than aerogels produced under the same conditions. It is unclear if carbon xerogels have been produced in anything other than monolith form. Despite having densities greater than the other organic gels, organic xerogels still yield carbon products with comparable surface areas and pore volumes [273,284].

9. Schwarzites and Other Exotic Carbon Solids

Following the discovery of fullerenes, a number of groups [285–290] postulated the existence of the so-called *schwarzites* which, as demonstrated by the examples of Fig. 28, combine convex and concave surfaces to yield solids containing pores of molecular dimensions. The convex and concave surfaces are made possible by the presence of rings smaller and larger than hexagons, respectively, among the normal hexagonal rings. Although regular schwarzites have yet to be observed, several groups have argued that *random schwarzites* (Fig. 29) may be found in the soot produced by the arc discharge fullerene generator [291,292] and in films produced by the deposition of supersonic beams of carbon clusters [121] (Fig. 30).

A vast range of open, potentially porous, molecular carbon solids based on combinations of sp^1, sp^2, and sp^3 hybridized carbon atoms have also been proposed since the 1960s [158,159], some examples of which are shown in Fig. 31. Although none of these structures yet exist in reality, there are active research programs aimed at their production via synthetic chemistry routes [159].

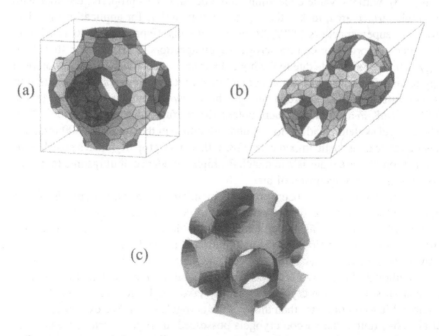

FIG. 28 Examples of hypothesized schwarzites: (a) and (b) are unit cells of some of the first schwarzites to be suggested [285,286,289] (from Ref. 289); (c) a section of a schwarzite built from heptagons and hexagons totaling 1080 carbon atoms. (From Ref. 290.)

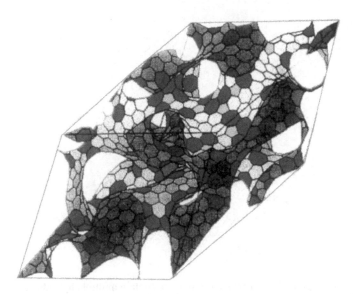

FIG. 29 The random schwarzite model of Townsend et al. (From Ref. 289.)

B. Precursors

Almost any carbon-containing substance can be used as a precursor for the production of carbon materials. As Table 1 illustrates, the range of precursors that have been reported in the literature is vast and spans all the possible states of matter. Given the often strong relationship between the characteristics of a carbon

FIG. 30 TEM micrograph of a thin film produced by supersonic cluster beam deposition [293]. The film consists of, among other things, curved graphene sheets embedded in an amorphous matrix of carbon; note the voids in the film. The makers of this film have suggested that it is reminiscent of a random schwarzite. (From Ref. 293.)

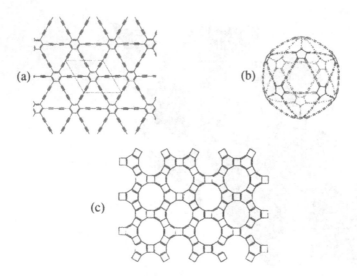

FIG. 31 Examples of hypothetical carbon allotropes that contain multiple hybridization: (a) *graphdiyne* which consists of a 0.25 nm diameter pore space defined by sp^2 hybridized carbons in benzoid rings linked by sp^1 hybridized carbons in buta-1,3-diyne units (from Ref. 294); and (b) *fullerenediyne*, C_{180}, the 0D analogue of graphdiyne (from Ref. 295). Other interesting structures which have been hypothesized but not yet synthesized or observed in nature include the zeolite-like structure of Balaban et al. [296] illustrated in (c) which would quite clearly contain porosity when assembled as a molecular solid. (From Ref. 296.)

material and the molecular structure of its precursor [344–348], it is useful to classify the precursors on the basis of their molecular structure. Such an approach would also benefit any attempt to develop a systematic approach to the mimetic modeling of carbon materials. Accordingly, the various precursors of Table 1 have been divided into a number of broad classes, where the polymer group is recognized as including only *synthetic* polymers. As with any such classification scheme, most of the classes require further subdivision. For example, it is well known that lignite and bituminous coal lead to very different carbon structures when subjected to heat treatment. Therefore, it is anticipated that the precursor classification scheme outlined here will evolve.

In order to adopt a mimetic approach, it is necessary to have an understanding of the molecular structure of the precursor, and an ability to model the structure and its evolution under the relevant processing conditions. The structure of the simpler precursors such as graphite and the lower molecular weight liquid and vapor compounds is generally well understood. The crystal *structures* of cellulose are reasonably well understood, and there is emerging understanding of their in

TABLE 1 A Partial List of Precursor Materials for the Carbons of Section II.A with Some Representative References[a]

Precursor	Class	Phase	Carbon product	Usage	Ref.
Graphite	G	S	Diamond	F	44
			Carbyne	R	297
			Fullerenes	F	298
			Nanotubes	F	299
			Carbon onions	F	230
Anthracite	G/C	S	Activated carbon	O	300, 301
			Fullerenes	R	302
Semianthracite	C		Activated carbon	O	301
			Fullerenes	R	303
Bituminous coal	C	S	Activated carbon	F	304, 305
			Fullerenes	R	302, 303
			Nanotubes	R	306
Subbituminous coal	C	S	Activated carbon	O	307, 308
			Fullerenes	R	303
Coal (unknown)	C		Nanotubes	R	309
			Carbon onions	R	309
Lignite	C/H	S	Activated carbon	F	300, 310
			Fullerenes	R	303
Peat	H	S	Activated carbon	F	305
			Fullerenes	R	311
Shale	Ar	S	Activated carbon	R	312
			Fullerenes	R	313
Petroleum pitch	Ar	Liq/S	Graphite	F	41
			Activated carbon	F	314, 315
			Furnace black	O	258
Coal tar pitch	Ar	Liq/S	Graphite	F	41
			Activated carbon	F	315, 316
			Furnace black	O	258
			Fullerenes	R	317
Low MW aromatics (e.g., naphthalene)	Ar	Liq/S	Lampblack	O	258
			Furnace black	F	258
			Channel black	O	258
			Fullerenes	R	318–320
Low MW alkanes (e.g., natural gas)	Al	V	Carbyne	R	321
			Channel black	F	258
			Thermal black	F	258
			Furnace black	F	258
			Fullerenes	O	320

TABLE 1   Continued

Precursor	Class	Phase	Carbon product	Usage	Ref.
Acetelyne	Al	V	Carbyne	R	322, 323
			Acetylene black	F	258
			Fullerenes	R	319, 320
			Nanotubes	F	324
Wood	L/Ce	S	Activated carbon	F	325, 326
Coconut shell	L/Ce	S	Activated carbon	F	328, 329
Nut shells[b]	L/Ce	S	Activated carbon	O	305, 327
Fruit stones[c]	L/Ce	S	Activated carbon	O	328, 329
Cellulose (Rayon)	Ce	S	Activated carbon	F	315
Lignin	L	S	Activated carbon	R	336
Polyfurfuryl alcohol (PFA)	P	S	Activated carbon	O	337
Polyvinylidene chloride (PVDC)	P	S	Activated carbon	O	338
Polyacrylonitrile (PAN)	P	S	Activated carbon	F	315, 316
PVDC-PVC (Saran)	P	S	Activated carbon	O	339
Polyimide (e.g., Kapton)	P	S	Activated carbon	F	340
			Fullerenes	R	341
Phenolics	P	S	Activated carbon	R	342
Polypyrrolone	P	S	Activated carbon	R	343
Waste tires	?	S	Activated carbon	R	330, 331
Biological sludge	?	Liq	Activated carbon	R	332, 333
Industrial sludges[d]	?	Liq	Activated carbon	R	334, 335

[a] Given the immense number of references for some of the precursors, the citations here are necessarily selective. Note that the list is ordered so that there is a relatively smooth transition between the different classes (e.g., graphitic → anthracite, lignite → peat). Key for *class*: G, graphitic; C, coal; H, humic; L, lignitic; Ce, cellulosic; Ar, aromatic; Al, aliphatics; P, polymeric; ?, unknown mixture of waste materials. Key for *state*: S, solid; Liq, liquid; V, vapor. Key for *usage* (determined by the number of reports in the literature relative to other materials of their class): F, frequently reported; O, occasionally reported; R, rarely reported.
[b] Includes peanut, almond, walnut, hazelnut, brazil nut, pistachio nut shells.
[c] Includes apricot, olive, cherry, plum and peach stones, and grape seeds.
[d] Includes oil, solid carbon, solvents, paints, etc.

situ mesoscopic structures [349–351]. There is, however, considerable debate over the molecular structure of the noncrystalline solids including coal [352–354], lignin [355], wood [356,357], and synthetic polymers [358], and the structures of the constituent molecules of coal and petroleum pitch [359].

Despite the lack of full understanding of the molecular structure of many of the precursors, a combination of new analytical techniques with chemical information theory and molecular modeling has, in recent years, lead to considerable advances in our understanding of these structures. Of particular note are the SIGNATURE approach of Faulon [360–364] and the method of Nomura and coworkers [365,366], which take experimental data on a material's structure and its fragments and generate a library of model molecular structures that are "statistically equivalent," in some sense, to the real structure. This approach has been used to suggest libraries of *possible* structures for a number of coals [365–370] and kerogen [360], the major organic constituent of shale. Faulon and coworkers have also considered the structure of lignin [371] and the lignocellulosic structure of a softwood [356]. The molecular structure of asphaltenes, an important constituent of crude oil and their pitch, has also been considered using the SIGNATURE method [372], while other methods have been used to propose molecular structures of asphaltenes [373] and the heavier end of hydrocarbon mixtures in general [359]. The molecular structure of components of the liquid extracts of coal, which are relevant to coal tar pitch, have similarly been elucidated by Zhang et al. [374] and, most recently, Nabeel et al. [375].

C. Manufacture of Porous Carbons: A Molecular Perspective

Any mimetic approach clearly requires an understanding of the production process, particularly from a molecular perspective. We, therefore, briefly review the methods of manufacture of the carbons of Section II.A, with emphasis on current understanding of the precursor-to-carbon process at the molecular level.

1. Activated Carbons and Carbon Molecular Sieves of All Forms

The processes used for the manufacture of microporous carbon particles, fibers, and membranes are multistage with specifics that vary with the nature of the precursor and the final product. Given the emphasis of this review, the specifics of the manufacturing processes are not considered here; the reader is instead referred to the published literature [376–380].

The key processes for the production of microporous carbons are carbonization of the precursor to yield an intermediate material high in carbon and activation of this intermediate to produce the final carbon product. There are, broadly speaking, two means of undertaking these processes. The first is termed *physical activation*, where the precursor is first carbonized in an inert atmosphere at 873–1273 K, and then activated separately in an oxidizing atmosphere such as air, carbon dioxide, or steam at similar or slightly higher temperatures. The alternative approach is *chemical activation*, where a mixture of the chemical activating agent and the precursor is heat treated at temperatures between 673 and 1073 K in an inert atmosphere to effect carbonization and activation simultaneously. There are, of course, variations on these sequences of events. For example, groups have reported use of *one-step physical activation* where carbonization/activation occur simultaneously in an oxidizing atmosphere (e.g., [328]), *two-stage chemical activation* where the activating chemical is added to the char (e.g., [381]), and *physical-chemical activation* where an oxidizing atmosphere is used with chemical activation [382]. The evolution of the molecular structure of the carbons during carbonization and activation will be reviewed below.

The manufacture of microporous carbons involves many other processes. For example, precursors, carbonized intermediates, and final products may all be subjected to mechanical milling and sizing; blending with other solids or liquid binders; high pressures during briquetting, tableting, or extrusion; impregnation with chemicals; and washing in acid or alkali. Some of these processes are known to affect the molecular structure of the precursor, carbon intermediate or final product—those that are known to have an effect will also be briefly reviewed in the following.

(a) The Process of Physical Activation.   Carbonization is the process of producing a carbon-rich solid char from a liquid or solid carbon-containing precursor. In the case of microporous carbon manufacture, the aim is to also produce a char that will resist the tendency to approach the structure of graphite at high temperatures and that will permit growth of a significant level of microporosity when exposed to oxidizing gases at moderate temperatures. The transformation of a precursor to such a char involves profound changes in its chemistry and molecular structure. These thermotransformations for the vast majority of the precursors have been studied by a number of workers [243–245,383,384] with the most comprehensive work being that of Oberlin and her group summarized in [32,246,252]. The following outline of the carbonization process follows that recently outlined by Oberlin [32]; this process does, however, match elements of that reported independently by others, most notably Marsh and Crawford [244], and Gómez-Serrano and coworkers [383,384].

From a chemistry perspective, the carbonization stage of the physical activation process sees the loss of the aliphatic and many of the noncarbon species of the precursor leaving a highly aromatic char. This char will typically contain high levels of carbon with small fractions of oxygen, sulfur, nitrogen, and hydrogen depending on their relative abundance and mode of occurrence in the precursor. The aliphatic and noncarbon species are driven off in a particular order as the temperature rises to the maximum of the process. In the case of a natural solid precursor such as coal and wood, it is well known that the physically bound moisture is the first casualty of heat treatment. This *drying stage* is followed by a gradual expulsion of aliphatic hydrocarbons that are physically and chemically associated with the aromatic components of the precursor and, if they exist, any O, N, and S containing functional groups that are *weakly bonded* to the aromatic framework (i.e., depolymerization). This so-called *primary carbonization* stage ends once the aliphatic hydrocarbons have been completely expelled from the precursor. This is then followed by the *secondary carbonization* stage where the aromatic-bound CH groups are released as low molecular weight gases, and the aromatic molecules increasingly link and cross-link via C—C, C—O, and other bonds.

The changes in the microstructure during the carbonization stage of the physical activation process are equally dramatic. The freeing of the bulky and restricting aliphatic molecules from and between the aromatic molecules during primary carbonization allows the latter to move in the "suspensive medium" formed by the aliphatic molecules. As the aromatic molecules move, they first associate to form largely disoriented BSUs, each a few molecules high. Depending on the mobility of the BSUs, these may "demix" into nanoscale regions of LMO. These regions may grow, become more ordered, and even coalesce with neighboring regions depending on how long the BSUs remain mobile. The degree and period of mobility depends on the amount of aliphatic phase present, the rate at which it is removed, and the prevailing temperature. It is also strongly affected by the propensity for cross-linking to occur between the aromatic molecules. Oxygen and certain sulfur moieties act to promote cross-linking, while hydrogen terminates dangling bonds at the edges of aromatic molecules, thus inhibiting cross-linking. As made clear in Section II.A.6, because microporous carbons require small regions of LMO, suitable precursors often possess a high $(O + S_s)/H$ ratio, where S_s is the sulfur that is stable above 1973 K. The secondary carbonization phase, which is not normally conducted when seeking to produce activated carbons, involves solid-state growth of the extents of LMO and increasing order. Carbons that are highly cross-linked at the end of the primary carbonization stage normally show little further ordering, while the reverse is true for carbons derived from oxygen-deficient precursors.

The carbon obtained from the thermocarbonization of a precursor contains only low levels of microporosity. Physical activation involves substantial development of the microporosity by removing carbon via reaction with an oxidizing gas, typically steam, carbon dioxide, or oxygen. The mechanism of such gasification reactions have been studied for many years [385,386]. In the simplest terms, gasification proceeds by the progressive removal of "active sites" of carbon by the gas molecules. Debate has long centered on what exactly are these active sites. While our understanding is still not complete, substantial progress has been made in identifying the active sites and the associated reaction mechanisms in the presence and absence of catalytic elements [387–389].

(b) The Process of Chemical Activation. The chemical activation process has been used extensively with lignocellulosic materials [326,390–394] and bituminous coal [304,393,395,396]; other coals have also been considered, but to a lesser extent [310,393,397]. Although a wide and diverse range of activating

agents have been studied since the early 1900s [376b, pp. 182–183], the most common chemicals used today are H_3PO_4, KOH, and $ZnCl_2$.

The change in the molecular structure of a precursor during chemical activation has received far less attention than that of physical activation. The most extensive study appears to be that of Derbyshire, Jagtoyen, and coworkers who have focused in great detail on a H_3PO_4-hardwood system [326,391,392]. These workers found that the acid greatly accelerates the carbonization process in comparison with thermocarbonization [326]. At temperatures below 423 K, the acid promotes bond breakage leading to dehydration and extensive depolymerization of the biopolymers with consequent evolution of H_2O, CO, CO_2, and CH_4. Beyond this temperature, the acid increasingly promotes cross-linking in favor of bond breaking and even acts as a cross-linker. The enhanced cross-linking leads to much reduced levels of tar evolution, while the presence of the acid physically prevents the collapse of the microporosity during carbonization. Removal of the acid after carbonization yields a microporosity commensurate in size to the level of acid used; it is the occurrence of this porosity that eliminates the need for subsequent activation in an oxidizing gas.

The effect on coal also appears to be similar to that observed for lignocellulosic materials [304,395], although the surface area is smaller and tends to decrease with increasing rank. This is in line with the idea that the resistance of coal molecules to chemical attack increases with rank [391,393]. The other popular chemicals—$ZnCl_2$ [390,396] and KOH [396]—also appear to lead to similar types of behavior during carbonization/activation. They inhibit the production of tar evolution and they maintain the microporosity during carbonization. There are, however, some differences, including the nature of the noncondensable gases evolved [396]. It is, therefore, unclear if the mechanisms for the different activating agents are the same. Sodium hydroxide was found to be unsuitable for an activation agent [397]. Some workers have also shown recently that the chemical activation agent strongly affects the surface chemistry of the carbon [394].

(c) Milling of Precursors, Carbon Intermediates, and Microporous Carbon Products. As final carbon products must invariably be of a specific form and size, their precursors and carbon intermediates are often subjected to mechanical milling and subsequent reconstitution. Processes such as crushing, chopping, and classifying do not affect in any real way the molecular structure of the final carbon product. Grinding, on the other hand, is known to affect the microstructure of both precursors and carbon.

Begak et al. [398] found that the interlayer spacing and LMO size of coals ground in air under high loading increased relative to those ground under low loading. They also found that the aromatic character of the coal, surface oxygen species, and number of free radicals increased while the amount of aliphatic hydrogen decreased. As will be seen below, these changes are not dissimilar to those seen in the initial stages of carbonization. Therefore, we suggest that grinding of a precursor is likely to lead to a carbon that is more highly cross-linked and which will, therefore, be more microporous. This hypothesis is supported by further experiments of Begak et al. [398] who found that carbons derived from heavily ground coal are consistently more microporous than those obtained from lightly ground coal, irrespective of the coal rank. Taraba [399] found that milling of coal promoted oxidation when undertaken in air, and loss of carbon and solid-bound oxygen when executed in an inert atmosphere.

In the case of carbon solids, a number of groups [400–402] have found that the interlayer spacing and extent of local molecular order decrease dramatically when they are subject to heavy grinding. Salver-Disma et al. [402] also found that the degree of disorder and its irreversibility under heat treatment at 3073 K increases with the energy of the grinding process; on the basis of their experimental evidence, they hypothesized a mechanism for the microstructural evolution of the carbon.

(d)   Blending of Precursors and Carbons.   By their very nature, many precursors that lead to highly microporous carbons do not soften when heated. In order to form briquettes and extruded shapes from these carbons, it is normal to blend them or their carbonized intermediates with a liquid or softening precursor to act as a binder. The most common softening precursor used is bituminous coal, while a range of liquid binders are in common use including coal and petroleum pitches, molasses, and corn syrup. Irrespective of the binder used, the blend in the form of a briquette or extrudate is subjected to carbonization and subsequent activation to yield the final product. The graphitizing nature of binders means it is expected that the adsorptive capacity of any final product will be reduced somewhat by pore blocking [403]. This is a purely physical effect and one that occurs at a length scale beyond that of interest to us here. Sakurovs [404,405] recently elucidated a more relevant effect in a series of experiments showing that the volatiles of a binder material, solid or liquid, may very well enhance the softening of the (other) coal precursors involved. It was hypothesized by Sakurovs that vapor from the binder might transfer to the other precursors and act as a solvating agent. More recently, Sakurovs [406] confirmed that the softening of a coal may in fact be reduced by addition of a charcoal. The work of Sakurovs indicates that the molecular structure of the final carbon may be very different depending on whether the binder is added to a raw precursor or its carbonized intermediate.

(e)   Demineralization of Precursor via Acid/Alkali Washing.   Some natural precursors such as coal are demineralized using acid and/or alkali to reduce the ash content of the eventual carbon product [407]. It is not altogether clear to what extent such treatment affects the molecular structure of the coal. Observing changes in the elastic properties of an HCl-demineralized coal, Krzesinska [408] concluded that its molecular structure was changed, although no statement was made on the nature of the change. A number of groups have used solid-state

NMR [409–411], extraction experiments [412], FTIR, and UV fluorescence [413] to argue that acid treatment does not affect the aromatic structure of coal to any significant extent. However, they did find that the aliphatic component is chemically affected and diminished. Given the important role played by the aliphatic component during carbonization, acid cleaning may affect the carbonization pathway to some extent.

It is generally accepted that minerals can act as catalysts for the breakdown of coal and the cracking of tars within coals [414], and for the consumption of coal when reacted with oxygen, carbon dioxide, and other oxidizing gases [415,416]. Their removal from coal and perhaps other precursor types may, therefore, affect the eventual carbon by modifying the carbonization and activation pathways. Once again, there is mixed evidence that such modifications occur. A number of groups [417,418] found, on a dry-ash-free basis, that acid treatment of a variety of coals had little effect on the structure of the activated carbon product. Samaras and coworkers [419], on the other hand, found that the microporosity and surface area increased with the degree of demineralization of the coal precursor.

(f) Preoxidation of Precursor. Caking coal precursors are often subject to preoxidation to reduce the degree of softening, which is known to lead to the production of graphitizing carbons. The oxidation is undertaken either through introduction of air under mild heating [420,421], or via chemical treatment with boric acid [376d,376g] or other chemicals [376d]. The first of these treatments clearly affects the precursor by increasing the level of oxygen in favor of hydrogen, thus enhancing the propensity for cross-linking during carbonization. The means by which boric acid affects the softening behavior has not been studied greatly but, in the case of coal precursors, it is thought that it causes esterification of hydroxyl groups (i.e., it effectively increases oxygen in favor of hydrogen) [376d].

(g) Modification of Carbon Pore Structure and Surface Chemistry. The pore size distribution and surface chemistry of microporous carbons can be modified by adding chemicals to the precursors or changing the preparation process [257]. However, the nontrivial nature of the precursor-carbon pathway means that it is difficult to unambiguously relate such additions or modifications to the pore size distribution and chemistry of the final structure. It is, therefore, common to adjust the pore size and surface chemistry of a carbon after it has been manufactured. Pore size and surface chemistry may also be modified during the use or regeneration of a porous carbon.

The main method for controlling the pore size distribution of existing porous carbons is the deposition of organic compounds from the vapor phase [422,423]. The mechanism for this deposition process has been investigated in some detail by Kawabuchi and coworkers [423–425]. Hydrocarbon deposition may also be used to adjust the surface chemistry of the carbons [424]. However, the most popular method for adjusting the surface chemistry is gas or liquid phase chemical treatment of the carbon. Many examples of such treatments and their uses, ranging from chlorine gas to metal impregnation, are discussed by Bansal et al. [426]. Oxidation of the carbon surface is widely practiced for a variety of technological reasons and has therefore been subject of considerable study. Common oxidizing agents include hot oxygen–carrying gases [427] and aqueous solutions of $(NH_4)_2S_2O_8$, H_2O_2, HNO_3, and H_2SO_4 [427,428]. It has been shown that such treatments can affect the available accessible porosity and the pore size distribution [427,428]. Use of oxygen plasmas has been recently suggested as a means of modifying the surface chemistry without affecting the pore structure or texture to any significant degree [429].

2. Fullerenes

The *main* methods of fullerene manufacture can be broadly divided into two groups. The first involves vaporization of carbon fragments from a solid carbon precursor into an inert, quenching, atmosphere where they combine and subsequently deposit to form fullerenes among other carbon entities, including carbon nanotubes; this deposit is normally termed *soot* because of the dominant thermal decomposition reactions. The second group of methods produces fullerene-containing soot deposits through a chemical vapor deposition (CVD) route with the chemical vapor being formed by reacting aromatic or aliphatic compounds in an oxidizing or inert atmosphere. These two groups of methods are briefly reviewed here with particular emphasis on the postulated fullerene formation mechanisms. More details on the technology and procedures of the carbon vaporization methods may be found in several reviews [74,298,430–432], while similar information for the most intensively studied CVD route, hydrocarbon combustion, may be found in the reviews of Richter et al. [74] and others [319,320,433]. For perhaps the first time, fullerenes were very recently produced via a chemical synthesis route [82], in this case C_{20} from dodecahedrane ($C_{20}H_{20}$). The reader is referred to this publication for details regarding the procedure and mechanism.

(a) Fullerenes by Carbon Vaporization. Fullerenes were first *deliberately* produced by the *laser ablation method*, which, in its original form, uses a high-energy laser to vaporize the solid carbon precursor at room temperature [63,434]. This, the original variant of the method produced very low levels of fullerenes that could not, as a consequence, be isolated in quantities sufficient for study in the solid state. Yields were substantially boosted by carrying out the laser ablation in a furnace at temperatures in excess of 1273 K [435]. Although this so-called *laser-furnace method* is not the most commonly used method due to its relative complexity and expense, it does possess some advantages over the other methods. For example, because the carbon precursor need not be conducting, a far wider range of carbon materials may be used compared with the other conventional

techniques. Precursors that have been used with success include coal, coke, carbon black, kerogen, peat, shale, coal tar pitch, polyaromatic hydrocarbons (PAHs), and synthetic polymers (see [317] and references therein). The combination of precursor flexibility, laser, and furnace also provides greater control over the nature of the fullerene product that may be produced by the method [99].

The most common technique of fullerene production is currently the *arc discharge method*, which was first developed by Haufler et al. [436] as a derivative of the *resistive heating method* of Krätschmer et al. [437] (which is commonly termed the *Krätschmer-Huffman method*). The basic arc discharge method produces fullerenes by consuming a carbon electrode in a plasma located between it and a second carbon electrode some 1–10 mm away. The plasma is initiated and sustained by applying an AC or DC voltage drop across the gap to cause arcing; this is where the method inherits its name. In the Krätschmer-Huffman method, no gap exists between the electrodes, and carbon vaporization is caused by the high temperatures that result from resistive heating. While the arc discharge method is currently the highest yielding method, its basic form does not offer the same level of control over the product profile as laser ablation [99]. Variants on the arc discharge method have been proposed in recent years to improve product control [438,439], including an approach that feeds carbon particles into the plasma as the precursor rather than using the carbon electrodes [440].

A number of other carbon vaporization methods have been proposed from time to time. *Inductive heating* has been used with some success to effect carbon vaporization [441], but has seen little take-up due to its energy inefficiencies. A number of groups [442–444], including Bunshah et al. [445], developed the use of ion beams to produce fullerene soot, while the latter group have also demonstrated the use of electron beam vaporization with some success. Although the ion sputtering and electron beam methods were shown to yield soot rich in C_{70} and higher fullerenes [445], they have not been widely adopted. None of the carbon vaporization methods mentioned so far are suitable for mass production of fullerenes. Solar radiation–induced carbon vaporization methods are currently under investigation as a viable method of mass production [446–448].

It is now generally accepted that the plasmas of the various carbon vaporization methods contain carbon clusters of various sizes and shapes. The smaller clusters (~ 10 or fewer carbons) are linear, while the larger clusters are generally monocyclic and polycyclic. As will become clear below, it is also now widely accepted that these clusters and fullerenes are assembled from the smallest of carbon clusters (up to C_3). However, the exact path between these smallest of clusters and the larger cyclic clusters, and the role of the larger nonfullerene clusters in fullerene production, is still hotly debated. A variety of mechanisms for fullerene assembly via the carbon vaporization route have been proposed since the mid-1980s. The *pentagon road* [83,435], *fullerene road* [449], *ring coalescence* [450] mechanisms and their developments since [81,451] are currently the most popular hypotheses. These and older mechanisms have been reviewed a number of times over the past decade [83,452,453], and new mechanisms are still being proposed from time to time (e.g., [454]) (see also Chap. 1).

There appears to be substantial evidence that supports the ring coalescence/fullerene road mechanisms over the pentagon road mechanism [452,455]. Despite this fact, there is still no agreement on the exact mechanism responsible for fullerene production via the carbon vaporization route. For example, Goroff [453] argues that neither the pentagon road nor the fullerene road mechanism need be correct. More recently, Xie et al. [456] discovered bowl-like clusters alongside fullerenes in soot derived from laser ablation of $C_{12}H_{18}$, in contradiction with earlier studies [457] which are widely used as a basis for discrediting the pentagon road mechanism [452]. Shvartsburg et al. [458], on the other hand, recently used high-resolution ion mobility measurements to identify, for the first time, precursor clusters that are key to the fullerene road mechanism.

While the fullerene assembly mechanism for the carbon vaporization methods has received considerable attention during the past decade or longer, the same is not true about the physical location of the fullerene assembly process within the apparatus. One of the few studies that appears to address this issue explicitly is that of Belz et al. [459] who argue, at least for the arc discharge method, that fullerenes assemble outside the plasma from small clusters ejected from the plasma. On the other hand, Gao and Gao [454] sketch a mechanism where the fullerenes are assembled at the cathode and within the plasma region, while others [460,461] find that their simulation results agree best with experiment when formation is assumed to occur within the laser or ion beam induced plasma. As the fullerene assembly process is strongly affected by the *local* cluster density and temperature (inert gas type/pressure are also important) [447,462,463], it is clear that the current lack of understanding of where fullerene formation physically occurs will hinder any attempt to develop mimetic molecular models.

(b) Fullerenes by Chemical Vapor Deposition. Fullerenes were first produced by CVD in 1987 via low-pressure oxygen-lean hydrocarbon *combustion* [464]. Four different hydrocarbon precursors were used with varying success: benzene, acetylene, 1,3-butadiene, and naphthalene. Fullerenes were isolated from the soot of a benzene flame a few years later by Howard *et al.* [465], and much work has followed since [320,433]. A variety of hydrocarbon precursors have been investigated ranging from benzene, acetylene, and naphthalene, on the one hand, to natural gas, petroleum, and camphor, on the other [320]. Fullerenes from C_{60} to C_{600} have been detected [433] and, unlike carbon vaporization methods, the amount of C_{70} can easily exceed that of C_{60} with C_{70}/C_{60} ratios ranging from 8.8 to 0.26 [320]. Isomers of C_{60} and C_{70} that are less stable than that normally observed have also been isolated using the combustion method [320]. The production of fullerene derivatives and low yields (~0.5% [320]) means combustion is not yet a preferred method [320]. However, combustion does offer one of the

more feasible routes for cheap mass production and, hence, work aimed at improving the yield through use of metal catalysts is continuing with some success [466].

Presumably in order to avoid fullerene destruction by oxygen, Taylor et al. [467] investigated fullerene production by *pyrolysis* of naphthalene in argon at 1500 K. However, yields were no better than with the combustion method, reflecting the observations of Homann [319] that O_2 does not cause excessive fullerene loss. The method has been extended to other aromatic precursors [318,468, 469] and acetylene [470] without any significant improvements in yield. Pyrolysis methods using lasers instead of direct heat have also been reported [471, 472] with yields similar to the latter. It was shown that addition of N_2O [473] and CCl_4 [474] increased fullerene yields by acting as a scavenger of liberated hydrogen atoms.

CVD-based fullerene production by feeding hydrocarbons into plasmas [475,476] has also been studied, with the most recent efforts giving yields of ~2% using C_2Cl_4 [477]. It is believed that the high yield may arise from the relatively weak C—Cl bond (compared to that of C—H). It is argued by these investigators that plasma-based CVD methods may well offer a viable means of mass production.

The mechanism of fullerene production in CVD-based methods is expected to be different from that of the carbon vaporization methods due to the vastly different nature of the precursors and processes. Two main mechanisms have been proposed for the formation of fullerenes during combustion: the *zipper mechanism* [478] and that of Pope et al. [479]. Although the former mechanism has recently been enhanced by Homann [319], it is still not clear which of these mechanisms, if any, is valid [320]. Little work has been undertaken to elucidate the mechanisms associated with the pyrolysis approach, which, given the absence of oxygen, could be somewhat different from that of the combustion method. The only mechanism proposed specifically for the pyrolysis method is that of Taylor et al. [467]. However, some believe [318] that the mechanism cannot explain the formation of fullerenes other than C_{60}, a serious flaw given the ease with which C_{70} can be formed by this method.

(c) Fullerene Extraction and Purification. No existing method of fullerene production yields a pure product (i.e., a single-fullerene isotope). In fact, while clearly dependent on the specific method and conditions used, the soot obtained from standard methods is typically only 5–20% fullerenes with the remainder being nanotubes, multishell polyhedra, amorphous carbon, and graphite. The isolation of a "pure" (see below) fullerene product normally involves two steps. The first, commonly termed *extraction*, yields a fullerene-enriched solid residue from the soot. This residue typically consists of multiple fullerene isotopes. The second step, *purification*, involves the isolation of a specific fullerene isomer or isomers from the extract of the first step.

Extraction is normally undertaken by sublimation or solvent extraction. The former, which is conducted at moderate temperatures (~ 673 K) under low pressure or vacuum [480,481], has not seen significant use to date. However, this may change, as it is environmentally friendly in comparison with the more commonly used solvent extraction technique [481]. A variety of solvent extraction techniques are used. Solvent extraction commonly occurs in two stages: the first substantially concentrates the fullerenes, while the second removes the lighter hydrocarbons carried across in the first stage [298]. The solvent used in the first stage may be varied to enhance extraction of a particular fullerene over others. More details of this and other aspects of solvent extraction may be found in the review of Théobald et al. [482].

Chromatography is the standard means of purification and is capable of yielding high-purity batches of C_{60} and C_{70}, or mixtures of higher fullerenes. A number of different chromatography methods have been used, including liquid chromatography and gel permeation chromatography. Once again, the reader is referred to the reviews [482,483] for details on chromatography purification. Other purification methods have been used instead of chromatography, including crystallization [484–486], complexation [487], and an interesting distillation process based on multistage sublimation that combines extraction and purification in a single unit [488].

It should be noted that, irrespective of the extraction and purification methods used, the final product is never 100% pure [489]. Typical impurities include solvent molecules, oxygen adducts, and undesirable fullerene isomers (e.g., C_{70} in C_{60}). Unfortunately, even at very low levels, these impurities can profoundly affect the structure and properties of fullerenes in the solid phase (see [123] and references therein). For example, Heiney and coworkers [122,490] found that low levels of undesirable isotopes and solvent can dramatically affect, often in unforeseen ways, phase transition behavior, including the phase transition temperatures.

(d) Exohedral Fullerenes, Endohedral Fullerenes, and Heterofullerenes. Endohedral fullerenes can be produced using a solid carbon-heteroatom precursor with any of the vaporization methods used to produce empty fullerenes. The arc discharge and laser-furnace methods are both capable of producing macroscopic quantities of metallofullerenes [139], with the former being the most popular [139]. Inductive heating has also been demonstrated [491,492], but has seen little take-up. The extraction and purification of endohedral fullerenes from soot is extremely challenging [141]. As with empty fullerenes, the main approach involves use of either solvent extraction or sublimation followed by liquid chromatography, particularly high-performance liquid chromatography. Greater details may be found in the reviews [139,141].

Endohedral fullerenes have also been produced in submacroscopic quantities by a number of other methods, including laser vaporization of fullerite-metal oxide mixtures [493], extremely high-pressure heating of gas/fullerene mixtures [137], low-temperature K-C_{60} plasma [494], ion implantation of fullerite films [495,496], and neutron irradiation of fullerite in a gaseous atmosphere [497] or of an intercalated fullerite [498]. Chemical synthesis routes are also being actively pursued [499].

As expected, the mechanism for the formation of endohedral fullerenes is less than well understood; to our knowledge, no review of this subject exists. In the case of solid vaporization methods, it is taken for granted that the heteroatoms are incorporated within the fullerene shell during the course of the fullerene assembly process. Assuming that the presence of a heteroatom does not greatly disturb the fullerene assembly process except through the imposition of a minimal size of fullerene [500], the pentagon road, fullerene road, and ring coalescence mechanisms previously outlined should all still be relevant. However, retention of the heteroatom or heteroatoms during assembly is clearly an issue that must be considered. Saito and Sawada [501] propose a *parasitic mechanism*. Jarrold and co-workers [502,503] used data obtained from ion mobility experiments to argue that the ring coalescence route is also applicible for the formation of endohedral fullerenes. These workers also proposed a basis for predicting if the metal atom will occur inside or outside or even in part of the shell.

Production by methods other than the arc discharge or laser-furnace almost certainly involves the heteroatom passing through the shell of already existing fullerenes. A number of mechanisms have been suggested for this process, all of which appear feasible and valid under conditions applicable to one or more of the less common production methods. The first involves the atom squeezing through a ring in the shell, the *ring penetration mechanism*, while the second involves the temporary rupture of a C—C bond, which has been termed the *windowing mechanism* [504,505]. Which of these occurs for a particular system is likely to be dictated by factors such as system temperature, energy and nature of the atom-fullerene collision, and size of the atom relative to the rings of the fullerene shell. It is clear, however, that either the system temperature and/or the energy of impact must be reasonably high, suggesting that these mechanisms are likely to occur during, for example, ion implantation or laser vaporization of fullerite-metal oxide mixtures. A third suggested means for atoms entering an already formed fullerene is via an autocatalytic route in which two fullerenes bond temporarily to allow formation of a window through which heteroatoms may enter the fullerenes [506]. Such bonding is known to occur at higher pressures, making this a possible mechanism for the high-pressure/temperature method. The creation of a temporary window in a similar way by nonautocatalytic routes has been proposed as a means of producing endohedral fullerenes in bulk; this proposal has been recently reviewed by Rubin [507].

Heterofullerenes have been prepared using the laser-furnace method [137, 493,508] and the arc-discharge method [509,510]; this work has been briefly reviewed a number of times [511]. Little comment has been made on the possible mechanism for heterofullerene formation via the solid vaporization route. The only exception is Jarrold and coworkers, who argue that the ring coalescence mechanism is valid for all types of metal-containing fullerenes and that the specific type of metal-containing fullerene depends largely on the electronic properties of the metal, with the precursor cluster size also playing a role [512]. For example, lanthanum *preferentially* (i.e., not exclusively) forms an endohedral fullerene due to its having only three relatively weakly bound valence electrons, while niobium preferentially forms heterofullerenes due to its five valence electrons of higher ionization energy [512].

It is felt by some [165] that synthetic chemistry may offer the best hope for the efficient mass production of heterofullerenes. Macroscopic quantities of the dimer and aducts of $C_{59}N$, an aza[60]fullerene, was first produced by Hummelen et al. [162] in 1995, and by others [163] soon after. This breakthrough motivated much work in the area, which has been recently reviewed [164,165]. At present, only the dimer and aducts of aza[60]fullerenes and aza[70]fullerenes can be produced via the synthetic chemistry route. However, the existance of an underlying methodology [513] means it is likely that additional heterofullerenes will be produced in the coming years.

One means of producing macroscopic quantities of exohedral fullerenes or clusters of fullerenes is via the covaporization of fullerite and heteroatom source; the heteroatom source may initially be mixed with [514] or separate from [144] the fullerite source. A large variety of metallic [144,514] and nonmetallic [147,515] exohedral fullerenes have been manufactured in this way. A vast array of metallic and nonmetallic endohedral fullerenes have been produced via the chemical synthesis route [149]; this work has been reviewed extensively in recent years [149,513,516,517], and the reader is referred to these reviews for greater details.

(e) Fullerites. As already indicated, fullerite was first produced using the now famous Krätschmer-Huffman technique [437]. Crude fullerite produced by this and similar techniques is powdered, contains a mixture of different fullerenes and residual solvents, and consists of crystals that are small (of the order 10 nm). A variety of methods have therefore been developed to produce fullerites from the crude material that are of high purity and of the desired form, including large single crystals and thin films. An overview of the former is contained in the previous section. A brief overview of the methods aimed at producing fullerites of the desired form—which may be broadly divided into vapor phase and solution phase techniques—is given here.

The main solution phase techniques for the production of fullerite thin films are *solution casting* (also termed *drop coating*), *electrodeposition*, *Langmuir-Blodgett* assembly, and *self-assembly*; brief reviews of these techniques may be found in [518,519]. The solution casting method, which is by far the most popular of the solution phase approaches, simply involves coating of a substrate by a fullerene-solvent solution followed by evaporation of the solvent [520]. The electrodeposition technique entails electro-oxidative deposition of fullerene anions onto an electrode from a fullerene salt solution [521,522]. Early attempts to form Langmuir-Blodgett fullerite films were frustrated by the hydrophobic nature of fullerenes [518,519]. However, the burgeoning field of fullerene chemistry means that it is now possible to form thin fullerite films via both the Langmuir-Blodgett and self-assembled routes that see functionalized fullerene molecules physisorbed and chemisorbed, respectively, on the substrate.

The vapor phase methods, which have been reviewed a number of times [298,519], all essentially rely on the sublimation of the fullerite from a source material with subsequent deposition on a substrate. These methods are used to produce both thin films [523–526] and large single crystals [527–529]. Evaporation is achieved by vacuum only as in the *vacuum deposition* method [526], heat only as in the *vapor transport* class of methods [524,528,529] or a combination of both as in *thermal gradient* [523,527] and *hot wall epitaxy* [525] methods. A variety of substrate materials have been used, including mica, silver, iron, and gold, and the substrate may be at room temperature or heated, with the latter approach generally yielding larger crystals [530]. Molecular beam methods have also been widely used to produce fullerite films [531,532].

(f) Carbon Nanotubes. As with fullerenes, MWNTs are produced via carbon vaporization and chemical vapor deposition. SWNTs are also produced via carbon vaporization, but require the presence of metal catalysts. More recently, SWNTs have been produced via a chemical vapor deposition route [194,533–535]. As all of these methods have been briefly described above, and the reader is referred to the many reviews that now exist on this topic for specific details [299,324,536,537]. In many of the production methods, the nanotubes are produced alongside fullerenes, carbon onions, amorphous carbon, and/or metal catalyst nanoparticles. The fraction of nanotubes produced via the various methods can be optimized to a certain extent (e.g., [538–540]), but associated separation and purification are invariably required [540]. As the nanotubes are not soluble and do not vaporize, separation and purification has proved far more challenging than for fullerenes. However, methods have been developed, and the reader is referred to the review of Duesberg et al. [541] for additional details. Chemical vapor deposition methods provide a route to the production of relatively large quantities of highly pure MWNTs [537] that do not require separation or purification.

The mechanism of nanotube growth has been the subject of much study. On reviewing the proposed mechanisms, it is clear that the mechanism is profoundly affected by the presence or absence of a catalyst, the physical form of the catalyst when present (i.e., fixed or free clusters, or dispersed atoms), and the method of production (i.e., vaporization or CVD). It is clear that MWNTs dominate in the absence of any catalyst, and it is generally accepted that growth occurs by the mechanism first proposed by Iijima et al. [183], which has since been elaborated by a number of workers [187,542–544]. In the case of CVD methods in which catalysts are used, it is generally thought that growth occurs via the *base growth* and/or *tip growth* mechanisms, depending on the mobility of the catalyst particles [189,195,545–547]. Matters are less clear at this stage in the case of carbon vaporization methods where Smalley and coworkers [192,548] proposed the so-called *scooter mechanism*, which has since received support from independent simulation studies [549], while Kiang and Goddard [550,551] argue that their *polyyne ring nucleus mechanism* is more consistent with the experimental evidence and their own simulation results [551].

(g) Carbon Onions. Carbon onions were first produced by Ugarte [219,220] via irradiation of soot obtained from an arc discharge with an electron beam 10–20 times that normally used in electron microscopes for tens of minutes. Ugarte proposed a relatively straightforward mechanism for the formation of carbon onions [220]. However, more recent work of Banhart and coworkers [227,552] suggests that carbon onion formation is far more complex, and that carbon onions may in fact be continually evolving and transforming into other forms of carbon.

The Ugarte method cannot be used to produce macroscopic quantities of carbon onions. Cabioc'h and coworkers [221,222] have more recently produced carbon onion via a carbon ion implantation method. This method first involves the implantation of carbon under low-pressure conditions into a thin copper or silver substrate held at 773–1273 K by a 120-keV carbon ion beam. The carbon onion film is then recovered by evaporation of the metal at 1073–1173 K under near-vacuum conditions. A mechanism for the formation of the carbon onions via carbon ion implantation has been proposed by Cabioc'h et al. [221].

3. Carbon Aerogels

Inorganic aerogels are usually formed through sol-gel polymerization involving the hydrolysis and condensation of metal alkoxides, such as tetramethoxysilane [553,554]. The best known of these materials is silica aerogel, obtained from the polymerization of silicon alkoxide, followed by drying under supercritical conditions; such drying avoids the collapse of pore space that occurs during subcritical drying, due to surface tension at the gas-liquid interfaces. Organic reactions that proceed via a sol-gel transition also exist. Organic aerogels have been synthesized from resorcinol formaldehyde, phenol resorcinol formaldehyde, phenol furfural, melamine formaldehyde, polyurethanes and polyureas. The most widely studied of these is the aqueous polymerization of resorcinol (1,3-dihydroxybenzene) with formaldehyde, with sodium carbonate as a catalyst. This results in polymer clusters, with formaldehyde forming covalent bridges between the resorcinol rings; these clusters cross-link to form a gel. The gels thus produced are supercritically dried using carbon dioxide to produce resorcinol formaldehyde aerogels. These resorcinol formaldehyde aerogels are mesoporous and composed of interconnected aromatic polymer beads having diameters that are typically 7–10 nm. The resorcinol formaldehyde aerogel can be pyrolyzed in an inert atmosphere to form carbon aerogel. The final structure of these carbon aerogels is a three-dimensional network of interconnected carbon particles, of diameter 5–10 nm.

Such carbon aerogels were first synthesized by Pekala and coworkers [555,556]. During the last decade there have been several studies of the influence of reaction variables on the structure and properties of carbon aerogels [267,283,557–563]. The structure of both the resorcinol formaldehyde and carbon aerogels was found to depend strongly on the resorcinol/formaldehyde (R/F), resorcinol/catalyst (R/C), and resorcinol/water (R/W) ratios [267,553,559, 561,563]. The R/F ratio largely determines the density of the aerogel. The R/C ratio has been found to strongly affect the carbon particle size and surface area, large R/C ratios (small catalyst concentrations) leading to larger particles, larger mesopore sizes, smaller surface areas, and a more disperse pore structure [267,553,559]. Increasing the R/W ratio for a given R/C leads to smaller mesopores and mesopore volumes [267,559].

III. EXPERIMENTAL INFORMATION

Although activated carbon has a complex structure, it has much merit as an adsorbent in practical applications. Hence, we must elucidate the relationship between its structure and properties in order to design better activated carbons. However, no simple approach to activated carbons can reveal the complex structures of pores, pore walls, and three-dimensional networks. An integrated approach using methods such as gas adsorption, X-ray diffraction, small-angle X-ray scattering (SAXS), electron microscopy, X-ray photoelectron spectroscopy, X-ray absorption spectroscopy, Raman spectroscopy, chemical and electrochemical analyses, and electromagnetic property analysis is needed. Each method has an inherent role in the characterization of activated carbon (Table 2). As this review is mainly concerned with the modeling of activated carbon structures, we emphasize the fundamental information that is useful for the modeling that can be determined by each of the methods. Thus, gas adsorption, X-ray diffraction, HRTEM, and analysis of surface chemistry provide the adsorption isotherm, electron radial distribution function of the pore walls, a microstructural image of pores and pore walls, and pore wall chemistry, respectively. This structural information is indispensable to the modeling.

TABLE 2 Most Often Used Methods for Characterization of Activated Carbon Surface

Method	Detection	Ref.
Small angle X-ray scattering	Total surface area (open and closed pores), mean pore size or crystallite size	
X-ray diffraction	Mean crystallite size, carbon-carbon pair distribution function	237–580
Transmission electron microscopy	Two-dimensional images of the pore structure, matrix correlation function	582–604
Gas adsorption	Porosity, pore size[a] distribution,[a] surface area[a]	605–666
Boehm titration	Type of oxygen functional groups (four types)/their number	668, 675–680
Potentiometric titration	pK_a of functional groups/their number	680, 683, 684, 690–692, 695, 696, 711
Calorimetric titration	pK_a of functional groups/their number	720
Temperature programmed desorption (TPD)	Type of oxygen functional groups (strong or weak acids)/their number	697–701
Fourier transform infrared spectroscopy	Type of functional groups/(number?)	680, 684, 700, 704, 714–718
X-ray photoelectron spectroscopy	Type of functional groups/ amount of heteroatoms (number of groups?)	702, 704–713
Immersion calorimetry	Number of primary adsorption centers (oxygenated groups)	721–723
Flow adsorption	Average polarity	724, 725
Inverse gas chromatography	Average acidity	727–729

[a] The introduction of an approximate model or theory is needed to extract this data from the experiment.

A. Diffraction Techniques

X-rays have a high transmittance for carbon materials. The transmittance of electromagnetic waves depends on the mass absorption coefficient of the material, its density, and the optical pathlength. The transmittance of MoKα radiation for carbon materials of 1 mm thickness and of 1 g cm^{-3} in density is 95%. This transmittance is fairly large compared with electrons, indicating that X-ray techniques can provide information on the bulk structure of carbon materials. However, carbon materials absorb even X-rays mainly due to the photoelectron effect, and it is important to correct for the incoherent scattering for an accurate analysis. On the other hand, the absorption of neutrons by carbon is negligible and the incoherent scattering is minor. Although neutron diffraction is effective for structural studies, neutron facilities are very limited. Here the structural analysis of activated carbons with X-ray diffraction is described. A review on neutron scattering of carbon materials was published recently in a preceding volume [564].

1. Turbostratic Structure and Size of Nanoscale Graphitic Crystallites

X-ray diffraction has been used to elucidate the structure of carbon materials such as carbon black. The X-ray diffraction patterns for carbon black are composed of (001) three-dimensional and (hk) two-dimensional reflections. This diffraction profile indicates the presence of the unit structure of graphitic layers, but the graphitic unit has a turbostratic structure (see Section II.A.1 and Fig. 1) whose graphitic layers are parallel but rotated around the c axis. The interlayer spacing d_{002} is determined using the Bragg equation:

$$d = \frac{\lambda}{2 \sin \theta} \tag{1}$$

where λ is the X-ray wavelength and θ is the scattering angle for the peak position. The positions of the (001) reflections indicate that the c-spacing d_{002} of carbon black is larger than that of graphite.

The crystallite size along the c-axis, L_c, and the size of the layer planes, L_a, are determined from the half-width of the diffraction peak using the Scherrer equation [237]:

$$L = \frac{K\lambda}{B \cos \theta} \tag{2}$$

where L is L_c or L_a, B is the half-width of the peak in radians, and K is the shape factor. The quantities L_c and L_a are named stack height and stack width, respectively. The (002) and (10) peaks are used to calculated L_c and L_a, respectively. The shape factor K depends on the lattice dimension and the values $K = 0.9$ and $K = 1.84$ are used for calculation of L_c and L_a, respectively. The L_c and

L_a values so determined are in the range of 1–3 nm. Accordingly, the higher order structure of nanographitic crystallites of carbon black, which was proposed by Franklin [238], must be the frame structure of carbon black.

As X-ray diffraction arises from the periodic arrangement of atoms, the information obtained for activated carbons is not concerned with the pore structure but the nanographitic frame. The basic structure of nanographitic units in activated carbon obtained from X-ray diffraction is similar to that of carbon black. Figure 32 shows the X-ray diffraction patterns of the pitch-based activated carbon fiber (ACF) A20 using CuKα radiation. This X-ray diffraction pattern of A20 has features that are common to other activated carbons. The two broad peaks observed near $2\theta = 24°$ and 42° are assigned to the (002) and (10) reflections. Generally speaking, the diffraction pattern of activated carbon has only (001) and (hk) peaks of the graphite structure; it has no (hkl) peaks of the graphite.

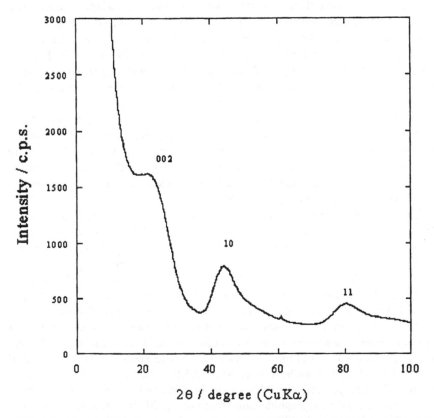

FIG. 32 X-ray diffraction pattern of pitch-based ACF (A20) measured with CuKα.

Therefore, activated carbon has a turbostratic structure similar to that of carbon black. When the shape and intensity ratio of (10) and (11) peaks can be determined by careful experiments, turbostratic distortions can be discussed [565]. The (002) peak provides the d_{002} with the aid of Eq. (1). The d_{002} value, which is listed in Table 3, is in the range of 0.34–0.36 nm. The d_{002} value of activated carbon is greater than that of graphite (0.335 nm) except for activated mesocarbon microbeads, MCMB-40. In the case of the pitch-based ACF, A10, the change of d_{002} with high-temperature treatment in Ar for 1 h is shown in Table 3. This table lists not only the structural parameters from the X-ray diffraction, but also the surface area determined by N_2 adsorption at 77 K with the aid of the subtracting pore effect method [572]. The d_{002} of A10 does not seriously vary through the high-temperature treatment, although the micropores disappear completely after heat treatment at 2500 K. The (10) peak in Fig. 32, shows the characteristic shape of the two-dimensional diffraction peak having a tail on the wide-angle side. This result indicates the turbostratic structure of nanographitic units in activated carbon as well as carbon black. The stack height L_c and stack width L_a determined by use of Eq. (2) are also shown in Table 3. The L_c of activated carbon is of the order 1 nm, corresponding to bi- or trilayers of graphene sheets. On the other hand, L_a is ~1–3 nm, being larger than L_c. In the case of pitch-based ACF, the L_c and L_a values become slightly smaller as the surface area increases. The activation procedure of mesocarbon microbead (MCMBs) decreases the L_c, but it does not cause L_a to vary. The high temperature treatment of A10 below 1673 K does not explicitly increase the L_c and L_a; the L_c and L_a clearly increase by the heat treatment at 2500 K and the microporosity is lost. Therefore, activated carbon has graphitic unit structures of 1 to ~3 nm in size from the X-ray diffraction. In activated carbon these nanographitic units must be connected mainly by sp^3 bonding to form a three-dimensional structure, which is similar to the Franklin model for nongraphitizable carbon black [238].

The following corrections are necessary for the above-mentioned analyses. The observed X-ray scattering intensity at θ, $I_{obs}(\theta)$, of the carbon sample depends on the polarization factor, which is associated with the monochrometer, the geometrical factor for normalization of X-ray irradiation volume, the parasitic scattering mainly due to the scattering by the X-ray cell windows, and the X-ray absorption factor attributed to absorption of X-rays by carbon and by the gas phase X-ray path. These corrections depend on the experimental system.

As activated carbon has pores and a nonuniform electron density structure, it shows an intense scattering in the low-angle region. This is called small-angle X-ray scattering (SAXS). Figure 33 shows the SAXS spectra of two pitch-based ACFs, A5 and A20. Both have a marked SAXS, which affects the wide-angle X-ray diffraction pattern. The SAXS must be corrected in order to get accurate information on the nanographitic unit structures [573]. We assume that the observed X-ray intensity, $I_{obs}(\theta)$, can be expressed by the additive form:

TABLE 3 Nanographitic Structures and Surface Area of Selected Carbons

Sample	d_{002} (nm)	L_c (nm)	L_a (nm)	Surface area ($m^2 g^{-1}$)	Ref.
Coconut shell activated carbon	0.38	1.1	2.6	940	570
Coal-based activated carbon	0.37	1.0	3.0	1100	570
Cellulose-based ACF (KF1500)	0.36	0.7	1.6	1540	566, 567
Polyacrylonitrile-based ACF	0.36	1.1	2.2	870	567
Phenol resin–based ACF	0.41	0.9	2.6	1960	569
Pitch-based ACF (A5)	0.36	1.0	1.8	900	568, 569
Pitch-based ACF (A15)	0.35	0.9	1.8	1550	568, 569
Pitch-based ACF (A20)	0.34	0.8	1.3	1770	568, 569
Pitch-based ACF (A10)	0.35	0.8	1.8	1190	568
A10 treated at 1473 K in Ar	0.34	0.8	1.7	780	568
A10 treated at 1673 K in Ar	0.35	0.9	2.0	500	568
A10 treated at 2500 K in Ar	0.35	1.3	2.3	nil	568
MCMB[a]	0.34	3.8	0.9	nil	571
Activated MCMB-4	0.37	1.2	0.9	570	571
Activated MCMB-30	0.37	0.9	0.9	1790	571
Activated MCMB-40	0.32	0.8	0.9	3090	571

[a] Mesocarbon microbead.

Note: Because the peaks are very broad, the absolute values of the X-ray diffraction parameters may not have definite physical meaning, but their relative values are thought to be significant.

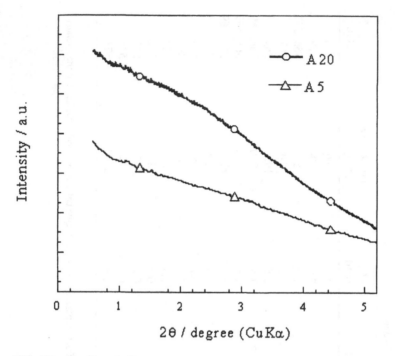

FIG. 33 Small angle X-ray scattering of pitch-based ACFs (A5 and A20) with CuKα.

$$I_{obs}(\theta) = I'_{obs}(\theta) + I_{sax}(\theta) \tag{3}$$

Here $I_{sax}(\theta)$ is the SAXS intensity and I'_{obs} is the corrected intensity. Activated carbon has no long-range periodic structure. Accordingly, the corrected total intensity, I'_{obs}, does not have any peak below 15°. The X-ray scattering intensity at $2\theta < 15°$ can be attributed to the SAXS. A log-log plot of X-ray diffraction intensity in this range versus scattering angle for most parts of activated carbon gives a straight line. Therefore, the SAXS is expressed as

$$\ln I_{sax}(\theta) = a \cdot \ln \theta + b \tag{4}$$

The constants a and b can be determined by the least-squares method using the observed data. Thus, the corrected intensity, $I_{obs}(\theta)$, is derived by the subtraction of $I_{sax}(\theta)$ from $I_{obs}(\theta)$.

Activated carbon adsorbs a large quantity of water molecules in the micropores under high humidity conditions. Bulk liquid water has a broad peak in X-ray diffraction patterns around $2\theta = 29°$ (CuKα) [573]. Hence, an activated carbon sample must be dried carefully prior to the X-ray diffraction experiments. It is necessary to dry and keep the activated carbon sample in a vacuum X-ray diffraction cell [574]. It was shown that water adsorbed in the micropores of activated carbon has a structure different from that of bulk liquid water [575], shifting the peak to a smaller angle ($2\theta = 27°$). The X-ray diffraction pattern of the bulk liquid water cannot be used for the correction.

In addition to the above factors, the following must be taken into account. The defective structure of nanographitic layers can induce broadening of the diffraction peaks [576], but the separation of this strain factor from the crystallite size effect is not easy in the case of activated carbon. The setting of slits also causes broadening of the diffraction peaks. This effect is removed by subtraction of the peak width of highly crystalline silicon from that of the mixture of the reference silicon crystal powder and the carbon samples. If the activated carbon sample contains mineral impurities over 1% in weight, the X-ray diffraction pattern is affected and the diffraction peaks due to the impurities must be corrected accordingly [577].

2. Radial Distribution Function Analysis

The electron radial distribution function or two-particle distribution function of activated carbon can be derived from the X-ray diffraction pattern. This function provides information on the interatomic structure of activated carbon. The electron radial distribution function shows the average coordination number or atomic density at a distance r from a central carbon atom. This method was developed by Zernicke, Debye, and Warren in the 1930s [578]. The basic transformation of X-ray diffraction patterns to the electron radial distribution functions is as follows:

The scattering angle, 2θ, is transformed to the scattering parameter, s.

$$s = \frac{4\pi \sin \theta}{\lambda} \tag{5}$$

The scattering parameter s can be used as the unit of the reciprocal lattice space that is independent of the X-ray source. A short-wavelength X-ray enables the spectrum to be measured over a wide s range. The upper limits of the measurable s range are 80 and 170 nm^{-1} for CuKα ($\lambda = 0.1542$ nm) and MoKα ($\lambda = 0.07107$ nm), respectively. Figure 34 shows the X-ray diffraction pattern of a pitch-based ACF using MoKα radiation. Here the abscissa is expressed by s. The diffraction peaks of two-dimensional graphite crystallites having high indices, such as (21) and (30), appear in the wide s region, which are not observed in the X-ray diffraction measured by CuKα. These peaks include structural information such as atomic distance and arrangement of graphene sheets.

The observed X-ray scattering intensity, $I_{obs}(s)$, of the carbon samples at s after the above-mentioned corrections consists of coherent and incoherent X-ray scatterings, $I_{coh}(s)$ and $I_{inc}(s)$, as given by

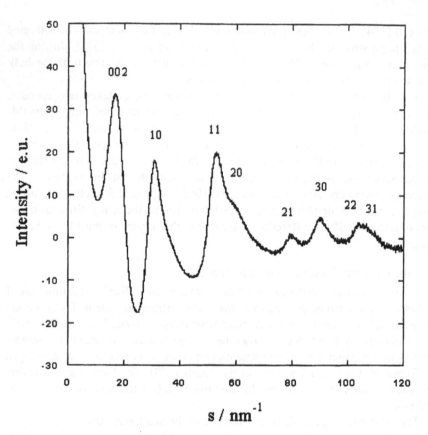

FIG. 34 X-ray diffraction pattern of pitch-based ACF (A20) measured with MoKα.

$$I_{tot}(s) = I_e\{I_{coh}(s) + \Phi(s) \cdot I_{inc}(s)\} \tag{6}$$

where the coherent scattering is due to interference of waves scattered from two different atoms in the material, and the incoherent scattering is from single carbon atoms and is due to the Compton effect. The incoherent scattering can be esti-mated from the scattering factor of a carbon atom and the atom density of carbon samples [579]. The coefficient $\Phi(s)$ is the ratio of the detectable intensity by the detector to incoherent scattering, which depends on the experimental procedure. It can be evaluated using the established relations. The coherent scattering, $I_{coh}(s)$, is fitted to the self-scattering factor [579] in the large s region to determine the normalization factor, I_e, using the following equation:

$$I_e \approx \frac{\sum_i f_i(s_{max})^2}{I_{coh}(s_{max})} \tag{7}$$

where s_{max} is the maximum s value in the scanning diffraction angle. Thus, $I_{coh}(s)$ leads to the structure function, $s \cdot i(s)$.

$$s \cdot i(s) = s\left\{ I_{coh}(s) - \sum_i f_i(s)^2 \right\} \qquad (8)$$

There are various expressions for the radial distribution function that are obtained by the Fourier transformation of the structure function. In the case of activated carbon, the electron radial distribution function (ERDF) $4\pi r^2[\rho(r) - \rho_0]$ defined by Eq. (9), which is suitable for liquids and noncrystalline solids [579], is preferable.

$$4\pi r^2[\rho(r) - \rho_0] = \frac{2r}{\pi \Sigma z_j^2} \sum_s s \cdot i(s) \cdot \exp(-Bs^2) \cdot \sin(sr) \cdot \Delta s \qquad (9)$$

Here $\rho(r)$ and ρ_0 are the electron density at a distance r from a central atom and the average density, respectively, and Z_j is the electron number of a component carbon atom j. The exponential term $\exp(-Bs^2)$ is called the convergent factor and is used for diminishing the Fourier noise.

As the ERDF expresses the distribution of the electron clouds, the information of the nanographitic structure other than the c spacing and the crystallite sizes can be obtained by ERDF analysis. Figure 35 shows the ERDFs of Madagaskar graphite, coconut-shell activated carbon, cellulose-based ACF (KF-1500), and pitch-based ACF (A20). The fundamental feature of the ERDF is associated with the two-dimensional sheet structure. Figure 36 shows the model structure of the two-dimensional carbon network of the graphite sheet. The coordination number at an interatomic distance around a central carbon atom of this two-dimensional sheet is determined geometrically as follows: 3 at 0.142 nm, 6 at 0.246 nm, 3 at 0.284 nm, 6 at 0.375 nm, 6 at 0.425 nm, 6 at 0.492 nm, and 6 at 0.511 nm. The ERDF of the graphite has peaks at 0.151, 0.254, 0.437, and 0.500 nm and a shoulder at 0.37 nm. Although the peaks at $r > 0.3$ nm are affected by the stacking and turbostratic structures, the position of the main peaks up to 0.5 nm approximates the interatomic distances of the two-dimensional sheet. Thus, the ERDF leads to information on the two-dimensional sheet structure of carbon. A detailed analysis of the peak position and peak amplitude of the ERDF provides information on the two-dimensional network sheet size, the turbostratic structure, and the ratio of carbon atoms forming the graphene structure, provided we can determine an accurate ρ_0 value and correct for the SAXS effect exactly for activated carbon. Since the above ERDF data have these experimental limitations, the comparison of the ERDF of activated carbon with that of graphite is helpful in understanding the structure of activated carbon.

The ERDF of graphite has a strong amplitude structure even above 1.5 nm due to the highly crystalline structure. On the other hand, the amplitude of the

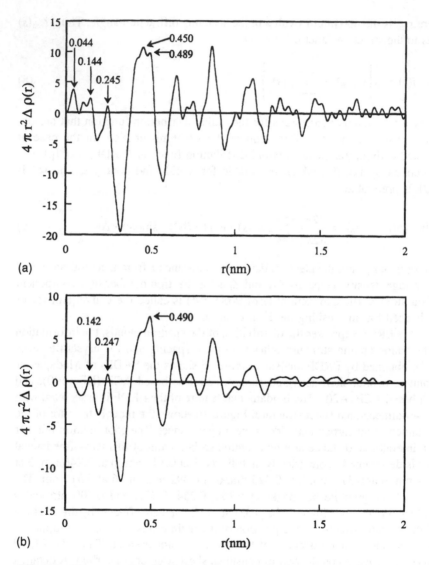

FIG. 35 Electron radial distribution functions of activated carbons and graphite. (a) Pitch-based ACF (A20), (b) cellulose-based ACF, (c) coconut-shell based activated carbon, and (d) Madagaskar graphite.

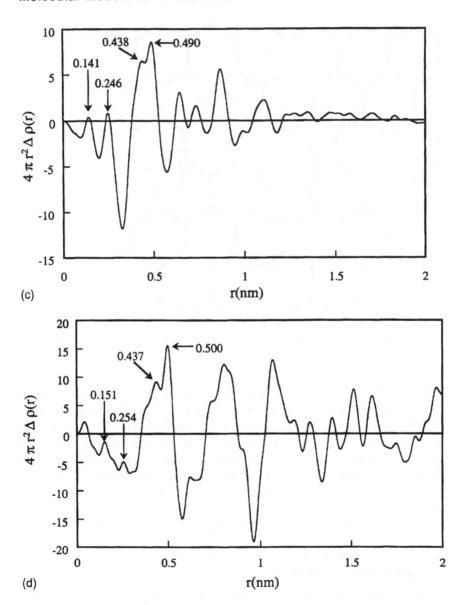

(c)

(d)

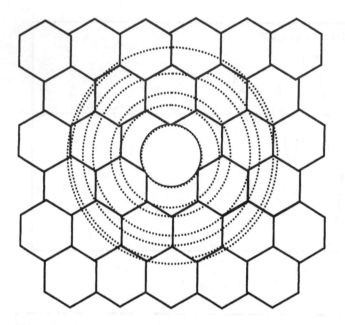

FIG. 36 The coordination structure of a graphene sheet.

ERDF of activated carbon samples in the distance range of $r > 0.5$ nm is smaller than that of graphite and attenuates between 1.0 to 1.5 nm. The smaller amplitude of activated carbon stems from imperfect graphene structures. The ERDF extension limit, which is close to the stack width, is associated with the size of the graphene structure. Therefore, the imperfect graphene sheets stack to form the turbostratic structure of 1–2 nm in width according to the ERDF analysis. As to the higher order structure of mutually stacked graphene sheets, accurate density determination and the development of X-ray diffraction experiments of higher accuracy for activated carbon are necessary.

B. High-Resolution Transmission Electron Microscopy

Analytical techniques such as X-ray diffraction and X-ray photoelectron spectroscopy (XPS) are powerful tools for elucidation of atomic and chemical structures of activated carbon. As activated carbon is essentially a noncrystalline solid, the characterization provided by X-ray diffraction is still limited, as described in this chapter. XPS cannot provide information on the bulk structure of activated carbon, although XPS determines the surface chemical structure of activated carbon. In order to understand the structure of activated carbon at the atomic level, we must combine information obtained from several different techniques. One of the indispensable experimental tools for characterization of activated carbon is transmission electron microscopy (TEM).

Scanning electron microscopy (SEM) is also useful in carbon science, and the images are rapidly and easily obtained. However, the images are limited to the surface structures. On the other hand, the TEM is usually used to study the internal microstructure and crystal structure of samples, which are thin enough to transmit electrons. Moreover, recent instrumental developments have made it possible to perform high-resolution measurements on the atomic scale. Clearly, the application of TEM techniques to the study of pore structures will allow a much deeper understanding and correlation between structures and properties.

In this section, an introduction to TEM will be given with emphasis on high-resolution transmission electron microscopy (HRTEM) and its application to the study of porous structure. The TEM technique will be treated from both theoretical and experimental points of view, and selected examples from the recent literature will be introduced.

1. Mechanism of Image Formation

(a) Image Formation in TEM. The transmission electron microscope is similar to the conventional light optical microscope in terms of optical principles. In all of optics, the Abbé theory is applicable to the mechanisms of image formation and resolution. Figure 37 [580] shows the electron beam path diagrams in the formation of the TEM image (primary image) formed by the objective lens, which

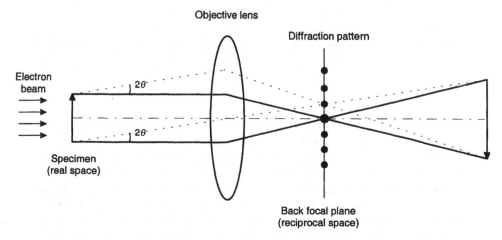

FIG. 37 Optical ray diagram with an optical objective lens showing the principle of the imaging process in a transmission electron microscope. (From Ref. 580.)

determines the resolution obtainable. When the primary beam of wavelength λ impinges at an angle θ on the crystal lattice planes with interplanar space d, constructive interference occurs, i.e., Bragg diffraction. Only when the angle θ between lattice planes and the incident beam direction obeys the Bragg equation does constructive interference occur. The beams deflected by the Bragg diffraction are focused on a spot in the back focal plane by the objective lens, indicated by spots in Fig. 37. The direct beam (zero maximum with reference to Bragg diffraction) passing through the aperture creates an enlarged image of the sample in the image plane of the objective lens.

In order to obtain the diffraction pattern on the final image screen, it is merely necessary to completely remove the objective aperture and to focus on the back focal plane by reducing the excitation of the intermediate lens. In this way, the focused plane is not the plane of the enlarged image but rather the plane of the diffraction pattern. Using a selected area aperture at the level of the first image, a diffraction image from areas down to about 300 nm can be obtained. If the transmission electron microscope has the facility to form a fine illumination probe, smaller areas down to a few nanometers may be selected for diffraction analysis. Each diffraction spot represents the reciprocal lattice point of the associated family of lattice planes. Thus, the back focal plane and image plane are called the reciprocal space and the real space, respectively. The Fourier transformation presents a real space as a reciprocal space mathematically.

(b) Image Contrast. As described above, TEM is often compared to optical microscopy because there are certain analogies to the conditions in light optics, such as applicability of the laws of geometrical optics and the Abbé theory. However, the transmission electron microscope is different from the optical microscope in terms of the formation of the image contrast. The brightness of a particular image point in light optics depends on the absorption of the light in the different sample areas. In the transmission electron microscope, the sample is so thin as to be more or less transparent to electrons. Thus, for TEM, absorption plays only a minor role, and scattering and diffraction are mainly responsible for image contrast. Therefore, the processes during passage of the electrons through the sample must be considered in more detail.

During passage of an electron wave through a specimen, scattering of the electrons occurs at the atoms in the specimen. For example, suppose that the specimen consists of very light atoms A together with atoms B with a higher atomic number (Fig. 38). The electrons can penetrate the area A of the specimen almost without scattering, while the electrons incident in the area B are deflected comparatively much more strongly by scattering because B consists of heavier atoms with greater charge. The objective aperture is so narrow that most of the electrons scattered by the heavy atom B are intercepted by the objective aperture. The few electrons that pass through the aperture produce an image of the area that includes B atoms. The intensity distribution of the electrons reaching the

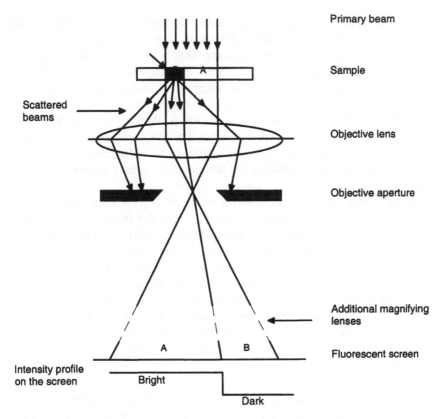

FIG. 38 Image formation in amorphous samples consisting of very light atoms that contain atoms with a higher atomic number. (From Ref. 581.)

fluorescent screen is determined by the number of electrons either passing or being stopped by the objective aperture, depending on the amount of local scattering. The intensity distribution results in relative darkness in the image area from B because many electrons have been lost from sample area B due to the strong scattering. Although it may appear that absorption phenomena occur in the specimen, in fact true absorption of electrons in the sample practically does not occur. Therefore, it is called scattering absorption, and the resulting contrast is called absorption contrast or scattering contrast.

For amorphous specimens, the scattering of electrons at the individual atoms could explain the image formation, while for crystalline materials the wave nature of the electrons must be considered. Due to the periodic atomic arrangement in the crystal lattice, periodic path differences which give rise to phase differences occur between neighboring scattering centers, resulting in the maximal possible constructive interference, i.e., Bragg diffraction. Thus, the image contrast reflects the diffraction condition in the specimen. This is a special case of scattering contrast and is called the *diffraction contrast.* The image formed by the diffraction contrast is essentially different from that formed by an optical lens. In the image formation of the conventional optical lens, an optical transformation corresponding to the Fourier transformation occurs twice, while it occurs only once in the image formation by the diffraction contrast. The information obtained from the image with the diffraction contrast is related to the diffraction of a group of atoms. There is no direct relationship between the contrast and the location of individual atoms.

When the diffracted beam is intercepted by the objective aperture, the direct beam passing through the aperture creates an enlarged image of the sample in the image plane of the objective lens (Fig. 39a). The diffracted beam is screened by the objective aperture so that these parts become dark. Thus, in such conditions, the image produced from the sample on the screen is the brightfield image. On the other hand, the darkfield image is produced when the contours of an object are imaged not by the direct beam but by any diffracted beam (Fig. 39b). This is achieved by moving the objective aperture so that the desired diffracted beam passes through the aperture. This causes the change in the image contrast, i.e., light and dark reverse their contrast, and only the regions that are oriented correctly for diffraction are bright.

(c) HRTEM Image. When both the direct and the diffracted beams pass through the objective aperture and contribute to imaging, an HRTEM image is formed on the image plane (Fig. 39c). This is called the crystal lattice image or

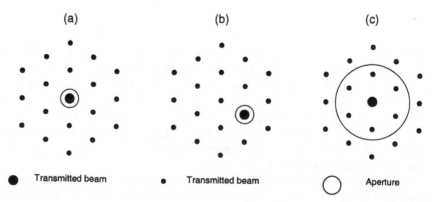

FIG. 39 Three observation modes in electron microscopy using an objective aperture. The center of the objective aperture is assumed to be set to the optical axis. (a) Bright field image; (b) darkfield image; (c) high-resolution electron microscopy. (From Ref. 580.)

the lattice image. The lattice images formed by diffracted beams with a direct beam give information about microstructures at the atomic scale. Therefore, the arrangement of atoms in a material can be seen directly on the screen as a fine variation in the crystal potential [582,583].

The improvement in the resolving power of TEM results in the possibility of using diffracted waves with large scattering angles. The image formed by many diffracted beams with the primary beam gives information about the structural properties in a crystal more precisely. When some imaging conditions are satisfied on observing a specimen, the contrast of a lattice image corresponds to the atomic arrangement (the crystal structure image or structure image). Therefore, the many-beam lattice image has become useful as the most important method of HRTEM.

For a weak-phase object [584], the image intensity is

$$I(r_i) \approx 1 - 2\sigma \, \Delta Z FT[V_u \sin 2\pi\chi(u)] \tag{10}$$

where σ, ΔZ, FT, V_u, and $\sin 2\pi\chi(u)$ are the interaction coefficient, the distances between phase gratings, the Fourier transform, the Fourier coefficient of the potential, and the phase change while passing through the electron lens, respectively. Therefore, the image contrast depends on $\sin 2\pi\chi(u)$, which is called the phase contrast transfer function. The total phase difference $2\pi\chi(u)$ is given by

$$2\pi\chi(u) = \pi\varepsilon\lambda u^2 - \frac{1}{2}\pi C_s \lambda u^4 \tag{11}$$

where u, ε, and C_s are the inverse of a distance in the crystal lattice ($1/d$), the defocus value, and the spherical aberration constant. The function $\chi(u)$ indicates the plane with identical phase of electrons passing through a lens and is called an aberration function. With a constant C_s for a particular electron microscope, the $\sin 2\pi\chi(u)$ curve changes its state only depending on the defocus value. Figure 40 shows the transfer function for $C_s = 1.0$ mm, $\lambda = 0.00251$ nm (for the accelerating voltage of 200 kV) at different values of defocusing. For the most suitable defocus value, the transfer function assumes almost constant value near either $+1$ or -1 over as wide a range as possible. Activated carbon consists of nanographitic stacking layers with an interlayer spacing of about 0.34 nm, as shown in Section II.A.6. Thus, the optimal defocus point is 30 nm.

If $\sin 2\pi\chi(u)$ is equal to 1, the image intensity will be

$$I(r_i) \approx 1 - 2\sigma V_p(r_0) \tag{12}$$

Therefore, the projected potential of the crystal, $V_p(r_0)$, is reflected in the image. The project potential depends on the electron density of the substance; the crystal site with heavy atoms having a high potential will be imaged as a dark area. Thus, information about the crystal structure can be obtained from the HRTEM image.

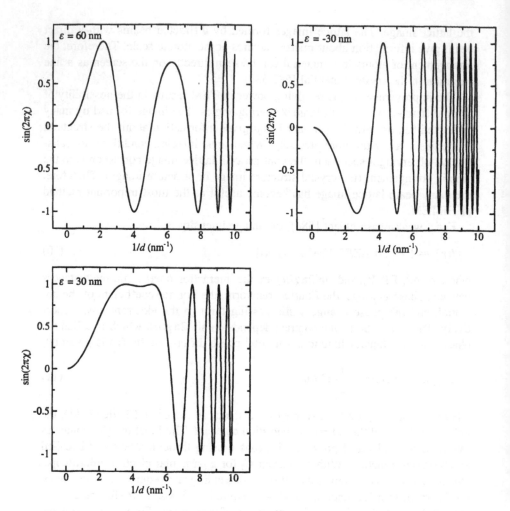

FIG. 40 Phase transfer function $\sin(2\pi\chi)$ against u. $E = 200$ kV and $C_s = 1.2$ mm.

2. Characterization of Activated Carbon by HRTEM

(a) TEM Image of Activated Carbon. Physical and chemical properties of activated carbon depend on the size, arrangement, and surface chemistry of the nanographitic units. The X-ray diffraction method is not a suitable tool for elucidation of the microstructure of nanographitic units because X-rays cannot be focused by a lens. Accordingly, TEM has been devoted to characterization of activated carbon. The simplest usage of an electron microscope is to obtain the enlarged image [585]. Initially in the study of porous carbons by electron microscopy,

attention was only paid to the meso- and micropores because of the limitation of the resolution [586]. The electron microscope acts as does an optical microscope with high resolving power, but the information obtained from the image is restricted to morphological determination. With improvement of instrumentation, it is possible to study the microstructures of carbon materials by HRTEM [587–589]. To be used for the study of the microstructure, an image-forming system in TEM must be considered in detail theoretically. The theoretical considerations of the resolution in crystal lattice planes was already discussed extensively by Millward and Jefferson in Vol. 14 of this series [590]. In Vol. 22 of this series, TEM studies on microstructures in various carbons was reviewed by Oberlin [30]. Oberlin also obtained the well-defined information on d_{002} spacing between the aromatic sheets of carbonaceous material by using a darkfield technique [591,592].

Since the HRTEM images provide sufficient contrast between a pore and the surrounding carbon matrix, the images can give not only qualitative information, such as pore shape and its orientation, but also quantitative information such as pore size and its distribution [593–596]. Fryer investigated the micropore structure of some disordered carbons by HRTEM operated at different acceleration voltages [593]. There was good agreement between the pore structures observed by HRTEM and gas adsorption measurements on the same carbon. Marsh et al. studied the pore structures of some carbons with high surface area [594]. They deduced the size and shape of meso- and microporosity of carbons from micrographs. These results suggested that the meso- and micropores consist of cage-like voids 1–5 nm in diameter, with the cages separated by walls of 1–3 carbonaceous layers in thickness. Innes et al. studied the correlation between micropore distribution obtained from molecular probe methods and from HRTEM [595]; they indicated slit-shaped micropores. Gas adsorption and dye adsorption from solution indicate the presence of slit-shaped micropores in activated carbon [596]. Therefore, more discussion of pore characterization with HRTEM and molecular adsorption is needed.

In particular, more quantitative information about the pores, in addition to the structural information on individual pores, is desirable. Therefore, image analysis techniques have been applied to the TEM observation of activated carbons.

(b) Image Analysis. By using an optical diffractometer, an image corresponding to the diffraction contrast image of a TEM is obtained. This is called the optical filtering method, which is useful to select a definite phase from a complex matrix. For the weak-phase object, the Fourier transform of the HRTEM image intensity, i.e., the amplitude of optical diffraction, is

$$FT[I(u)] \approx \delta(u) - CF_{hkl} \sin[2\pi\chi(u)] \tag{13}$$

where δ, C, and F_{hkl} are the delta function, a constant depending on the crystal, and the structure factor, respectively [584]. Therefore, the amplitude of the optical diffraction spot can be related to the crystal structure. The lattice parameter and the symmetry of the crystal are obtained by diffraction patterns.

Oberlin et al. have applied optical diffraction techniques to the transmission electron micrograph of amorphous carbon [587]. The optical diffraction pattern gives only two arcs. This pattern should give the distribution of the interlayer spacing because the image corresponds to interference between the (002) beam and the incident beam. Thus, Oberlin et al. deduced that a strong parallel orientation of the fringes is present in the micrograph and evaluated the misorientation to 15°. The optical diffraction method is useful not only for analyzing the structural properties but also for the correction of astigmatism of an electron lens. Huttepain and Oberlin applied the optical diffraction patterns of activated carbons to improvement of the lattice fringe techniques by comparing the patterns with the transfer function curves for each sample [597]. They chose the appropriate values in the transfer function of the objective lens so as to analyze the various activated carbons, giving their pore structures.

When the image intensity is digitized by a system with an image recording tube or image scanner, and the Fourier transformation is done numerically by a computer, we can obtain more detailed information about the structural properties of materials. Oshida et al. and Endo et al. have investigated the micropore structure of pitch-based ACF by using HRTEM combined with a sophisticated computer image analysis [598–600]. The HRTEM images are digitized and sent to a computer for analysis. The qualitative parameters of structure are determined using a binary image obtained from the digitized TEM picture. The two-dimensional fast Fourier transform (FFT) method is used to carry out the frequency analysis and to obtain the pore distribution of the ACFs. In order to get a real-space image associated with a specific frequency, the inverse FFT (IFFT) is carried out. Figure 41 shows the integrated spectrum of the power spectrum around the center point from the FFT spectrum images of the original images. The location of a peak along the abscissa corresponds to the most probable pore width in a certain size range, and the height of the peak is related to the pore density. Therefore, the peaks in the power spectrum correspond to the pore distribution as a function of the pore width of the ACF. In Fig. 41, the result of the image analysis is compared with that obtained by the adsorption method on the ACF. The pore size distribution obtained by image processing is similar to that evaluated by physical adsorption data.

Oshida et al. [598,599] recommended application of the frequency extraction method, including FFT, a masking step, and IFFT to distinguish pores of different widths. The binary images obtained by the frequency extraction method suggested that the pore structure has a fractal characteristic. They obtained fractal dimensions of 1.75–1.91 for ACFs, which are close to those determined with the molecular probe method by Sato et al. [601,602], if one is added to their results.

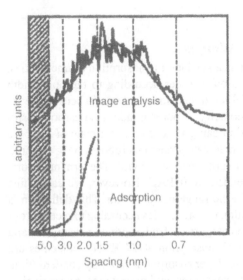

FIG. 41 The pore size distributions of ACF from the TEM image and from the N_2 adsorption method. Here the abscissa is expressed by the spacing determined from the image analysis, but it corresponds to the pore width. The continuous curve is the Gaussian distribution fit to the data. (From Ref. 600.)

Yoshizawa et al. applied the image analysis technique to the HRTEM images of activated carbons for their local regularity by stacking layers [603]. The microtexture of the activated carbon prepared by KOH activation has been investigated by HRTEM observation, together with the measurement of the N_2 adsorption isotherm at 77 K [604]. The change in the microtexture of the activated samples is quite different from that of the nonactivated samples. The orientation of carbon texture is somewhat perturbed at 673 K during the activation. With increase of heat treatment temperature up to 873 K, some twisted and continuous fringes can be recognized. The change in the fringes is remarkable for the sample activated at 1073 K. Furthermore, the structural change in the activation processes is characterized by evaluating the size distribution of the stacks in each digitized image in order to get quantitative information on the pores from the image of the activated carbons. The changes in the number of fringes and the length of fringe with heat treatment temperature were quantitatively examined. The change in the number of fringes is closely associated with the change in the surface area.

Thus, image analysis can provide detailed information on the microstructures and pore structures of activated carbon. Comparison of the characterized results by HRTEM with other techniques is a hopeful way to understand activated carbons.

C. Gas Adsorption

1. Pore Classification and Gas Adsorption

The pore structure of activated carbon is a key factor in adsorption characteristics. IUPAC [605] proposed the classification of pores according to the pore width w (the shortest distance parameter of three-dimensional geometry) because determination of the three-dimensional geometry of pores in activated carbon is still very difficult, and molecular adsorption can give a reliable estimate of the pore width w. The pores are divided into three categories: macropores ($w > 50$ nm), mesopores (2 nm $< w < 50$ nm), and micropores ($w < 2$ nm). The term nanopore, which is often used, is not recommended by IUPAC. However, we can define nanopore as a pore whose width is in the range of 1.5–5 nm. This definition is convenient due to the inherent adsorption characteristics according to recent progress in adsorption science. The three categories of micropores, mesopores, and macropores are introduced based on N_2 adsorption at 77 K. N_2 molecules are adsorbed by different mechanisms (multilayer adsorption, capillary condensation, and micropore filling for macropores, mesopores, and micropores, respectively). Thus, these three categories are suitable for the description of adsorption mechanisms of N_2 at 77 K. The critical widths of 50 nm and 2 nm are chosen for empirical and physical reasons. The pore width of 50 nm corresponds to the relative pressure P/P_0 of 0.96 for the N_2 adsorption isotherm at 77 K from the Kelvin equation; see Eq. (14) below. Here P_0 is the nitrogen vapor pressure. Adsorption experiments above $P/P_0 = 0.96$ are difficult, so that 50 nm is the upper limit of the pore width that can be evaluated from N_2 adsorption. The critical width of 2 nm corresponds to $P/P_0 = 0.39$ from the Kelvin equation, where unstable behavior of the N_2 adsorbed layer (tensile strength effect) is observed [606]. Thus, the capillary condensation theory cannot be applied to pores having width smaller than 2 nm. Accordingly, the pores to which the Kelvin equation can be efficiently applied are named mesopores. The macropores and micropores are greater and smaller than mesopores, respectively. The micropores have two subgroups: ultramicropores ($w < 0.7$ nm) and supermicropores (0.7 nm $< w < 2$ nm). The statistical thickness of an adsorbed N_2 layer on solid surfaces is 0.354 nm. The maximal size of ultramicropores corresponds to thickness of a bilayer of N_2 molecules. Although ultramicropores are very important in the molecular sieving effect, evaluation of their ultramicroporosity is very difficult due to blocking near the pore entrance. Ultramicropores smaller than 0.35 nm, which may be named ultrapores, cannot be accessed by an N_2 molecule (see Table 4).

The pores communicating with the external surface are called open pores, and are accessible to molecules or ions in the surroundings, as described above. The pore structure depends on the structure of the raw materials and on the conditions of the carbonization and activation; insufficient carbonization and activation leave

TABLE 4 Classification of Pores

Type of pore	Factor
	Width
Macropore	$w > 50$ nm
Mesopore	2 nm $< w <$ 50 nm
Micropore	$w < 2$ nm
Supermicropore	0.7 nm $< w <$ 2 nm
Ultramicropore	$w < 0.7$ nm
Ultrapore	$w < 0.35$ nm
	Accessibility to surroundings
Open pore	Communicating with external surface
Closed pore	Not communicating with surroundings
Latent pore	Ultrapore and closed pore

closed pores that do not communicate with the surroundings. Also, severe heat treatment often collapses the open pores to produce closed pores. A closed pore is not associated with adsorption and permeability of molecules, but it influences the mechanical properties of carbon materials. As to classification of pores by the accessibility to the surroundings, open and closed pores are not enough to describe carbon pores. An open pore whose width is smaller than the probe molecular size is regarded as a closed pore (here termed an ultrapore). Ordinarily, an N_2 molecule is used as the probe molecule. Such effectively closed pores and chemically closed pores should be designated latent pores. Ruike et al. [607] proposed a more precise characterization method based on density, N_2 adsorption, water adsorption, and SAXS measurements, which can determine the microporosity and the size and volume of the latent pores.

Another concept of pore structure, porosity, is also important. The porosity is defined as the ratio of the pore volume to the total solid volume; three kinds of densities—true density ρ_t, apparent density ρ_{ap}, and He replacement density ρ_{He}—can be used to determine the open-pore porosity, $\phi_{op}(= 1 - \rho_{ap}/\rho_{He})$ and closed-pore porosity, $\phi_{cp}[= \rho_{ap}(1/\rho_{He} - 1/\rho_t)]$. Nevertheless, determination of these densities is not easy. Although the true density is given by the unit cell structure using X-ray diffraction for well-crystalline materials, the true density of activated carbon cannot be determined by X-ray diffraction. Murata et al. proposed an accurate determination method for the true density of activated carbon using a high-pressure He buoyancy determination method [608].

2. Molecular Probes

Molecular adsorption in pores of activated carbon depends on the balance between the interaction of a molecule with the pore and the intermolecular fluid-

fluid interaction. The selection of the probe molecule and adsorption conditions is important in revealing the pore structures of activated carbon. The probe molecule must have the necessary requirements of sufficiently small size, stability, easy adsorption measurement, and applicability of an established adsorption theory. N_2 adsorption at 77 K with typical physical adsorption has been the representative characterization method of the pore structure of activated carbon. This is because experimental convenience is guaranteed and analytical theories are available. The effective size of the N_2 molecule on the solid surface is 0.354 nm, and hence N_2 adsorption can describe the subnanostructure of solid surfaces without chemical destruction. However, the N_2 molecule is a diatomic molecule and the shape is not a perfect sphere. The N_2 molecule has a quadrupole moment that helps determine the preferred molecular orientation on the pore wall, therefore, N_2 molecules are strongly adsorbed at the pore mouth, blocking further adsorption. Hence, N_2 adsorption at 77 K is not appropriate for characterization of ultramicropores. Recently, it was shown that N_2 adsorption at ambient temperature is very effective for determination of the pore width of the pore mouth of molecular sieving carbon [609].

H_2O, Ar, CO_2, and He are also used as probe molecules for the analysis of activated carbon. The molecular shape of the Ar molecule is spherical, and it has no dipole or quadrupole moment. Liquid Ar is more expensive than liquid N_2 but is the cheapest liquid of inert elements. Also, the adsorption isotherm can be measured at liquid N_2 temperature. Ar is often used for characterization of micropores of activated carbon. Also, some activated carbons have metallic impurities, where even N_2 molecules are strongly chemisorbed. In such cases Ar must be used. The quadrupole moment of CO_2 is much greater than that of N_2, and CO_2 adsorbs strongly in micropores even near room temperature, so that CO_2 has been used for characterization of activated carbon. Rodriguez-Reinoso and Linares-Solano showed the effectiveness of CO_2 adsorption [29]. H_2O is adsorbed only in the micropores of carbon. Hanzawa and Kaneko examined H_2O adsorption on activated carbon aerogel whose micro- and mesoporosities are well controlled [610]. The maximal amount of H_2O adsorbed corresponds to the micropore volume for pores of width 0.7 nm. H_2O vapor is not adsorbed below $P/P_0 = 0.3$, and the amount of adsorption abruptly increases above $P/P_0 = 0.5$. However, the adsorption mechanism of water in carbon micropores has been a subject of debate and much study. Gubbins et al. have applied the GCMC simulation technique to understand the mechanism [611,612], and Kaneko et al. applied in situ X-ray diffraction [613,614], in situ SAXS [616], and heat-of-adsorption measurements [615] to test the cluster-mediated adsorption mechanism that was originally proposed by Dubinin et al. [616]. As an H_2O molecule can be efficiently adsorbed even in small ultramicropores, H_2O is applicable to characterization of molecular sieve carbon [609]. The adsorption of H_2O below $P/P_0 = 0.1$ is sensitive to the concentration of surface functional groups and thus gives infor-

TABLE 5 Small Probe Molecules and Adsorption Conditions

Factors	He	H_2O	N_2	Ar	CO_2
Molecular area (nm^2)[a]	0.117	0.125	0.162	0.138	0.142–0.244
Molecular diameter (nm)[a]	0.26	0.27	0.37	0.34	0.39–0.45
Adsorption temp. (K)	4.2	Room	77	87	Room
Interaction	Weak	Dipole	Quadrupole	Weak	Quadrupole

[a] JO Hirschfelder, CF Curtiss, RB Bird. In: *Molecular Theory of Gases and Liquids*. John Wiley & Sons, New York, 1954, p. 1110. AL McClellan, HF Harnsberger. *J Colloid Interface Sci* 23:577 (1967). IMK Ismail. *Langmuir* 8:360 (1992).

mation about the degree of hydrophobicity. For example, H_2O vapor is not adsorbed on fluorinated ACF even near $P/P_0 = 1$ [617]. Irreversible H_2O adsorption can be correlated with nitrogen amounts controlled by the chemical vapor deposition of pyridine [618]. Thus, H_2O adsorption can determine the concentration and state of hydrophilic sites on activated carbon. Kaneko et al. developed a new porosimetry using He at 4.2 K [619,620]. Helium adsorption at 4.2 K is effective for evaluation of ultramicropores of activated carbon. A comparative study with N_2 adsorption at 77 K provides a description of the micropore structure. However, an inherent analytical theory for He adsorption must be developed. Table 5 summarizes properties of these probe molecules and adsorption conditions.

Organic molecules are also helpful in revealing the pore structure of activated carbon. Surface fractal analysis using different organic vapor molecules gives information on the fine structure of the pore wall of activated carbon [621,622]. Sato et al. showed that the fractal dimension of the micropore walls of ACF is close to 3 and decreases to about 2 after high-temperature treatment in Ar [623]. A comparative study using several organic vapor molecules can determine the pore size distribution, as shown by Carrott et al. [624; see also 625].

3. Determination of Adsorption Isotherm

The adsorption isotherm consists of a series of measurements of the adsorbed amount as a function of the equilibrium gas pressure at a constant temperature. The amount of adsorption can be determined either gravimetrically or volumetrically after degassing of the sample solids above 373 K. The evacuation temperature should be changed according to the pore structure of the sample. An evacuation temperature above 523 K is necessary for carbon samples having ultramicropores. In the case of commercial adsorption equipment, the pre-evacuation at 373 K is not enough due to the small evacuation rate, although a low degassing temperature is preferable so as not to change the pore structure. The equilibration time must be measured in advance. Commercial adsorption equipment based on the volumetric method has several advantages, such as high sensitivity, low cost, and high performance. New volumetric equipment uses highly sensitive pressure gauges, such as Baratron gauges, to realize pressure measurements down to $P/P_0 = 10^{-6}$. However, the dead volume correction maintaining a constant liquid N_2 level and the high vacuum quality should be strictly controlled so as to get reliable data. If not, a cumulative error of more than 20% is often produced. A gravimetric method must have a sensitive balance such as a quartz spring, an electronic balance, or a magnetic suspension balance in addition to highly sensitive pressure gauges. The gravimetric method can independently determine both the adsorbed amount and the equilibrium pressure, although the costs are high. The weight changes that occur due to buoyancy as the density of the surrounding adsorptive changes should be corrected for. This involves subtracting the weight of the sample in vacuo from the weight at the adsorption conditions. Strictly speaking, it is better to measure the weight change under an atmosphere of adsorptive at supercritical conditions, without adsorption. However, the sample does not make contact with the walls of the adsorption equipment, and so the temperature of the sample does not coincide with the bath temperature owing to the low thermal conductivity in the very-low-pressure range. The adsorption temperature must be checked prior to the adsorption measurement.

The amount of adsorption, W_a, depends on the gas equilibrium pressure P, adsorption temperature T, adsorptive (gas), and amount of the adsorbent. Parameters other than P are fixed when measuring the adsorption isotherm and W_a is expressed as a function of P. Physical adsorption is the predominant process for a vapor below the critical temperature, which has saturated vapor pressure P_0. The pressure is expressed by the relative pressure P/P_0. Here W_a may be expressed by the mass of gas (usually mg) or the volume reduced to STP [usually cm^3 (STP)] per unit mass of the adsorbent (usually g).

4. Type of Adsorption Isotherm and Surface Structure of Carbon

IUPAC recommended six types of adsorption isotherm instead of the five BDDT groups, as shown in Fig. 42 [625]. The type I isotherm corresponds to the so-called Langmuir isotherm. In the case of physical adsorption, the type I isotherm represents the presence of micropores where molecules are adsorbed by micropore filling. Ordinary activated carbon has this type of adsorption isotherm for N_2 at 77 K. Figure 43 shows high-resolution N_2 adsorption isotherms for pitch-based ACFs at 77 K as a function of the average pore width, which was evaluated by the SPE method [626,627]. Here the abscissa of Fig. 43a is expressed as P/P_0, whereas that of Fig. 43b is expressed as the logarithm of P/P_0. The smaller the pore width of ACF, the larger the ratio of the amount of adsorption below $P/P_0 = 10^{-3}$ to the total adsorption. Although the W_a vs. P/P_0 expression has

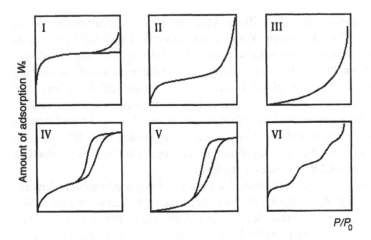

FIG. 42 The IUPAC classification of adsorption isotherms.

no clear difference at the initial part, the adsorption isotherms expressed by the log (P/P_0) shows a clear difference in the adsorption behavior. Sing and coworkers [628] introduced the α_s plot to analyze the adsorption isotherm, proposing a two-stage mechanism of primary and cooperative (or secondary) fillings. However, they did not analyze sufficiently adsorption isotherms below $\alpha_s = 0.5$,

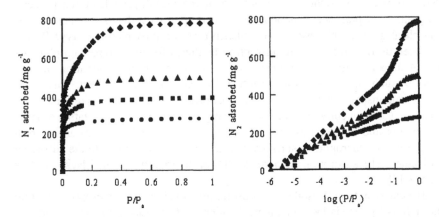

FIG. 43 Adsorption isotherms of N_2 on pitch-based activated carbon fibers having different pore widths at 77 K. Average pore width <± 0.1 nm>/nm: ●, 0.7(A5); ■, 0.9(A10), ▲, 1.0(A15); and ◆, 1.1(A20). The indication in the parentheses is the name of the pitch-based ACF.

which is important in micropore filling. Thus, no clear physical model for the two-stage mechanism was given. Kaneko and coworkers [626,627] introduced the high resolution α_s plot over wide α_s ranges from 0.01 to 2.4, using the adsorption isotherm from $P/P_0 = 10^{-5}-10^{-6}$. They showed the presence of two upward deviations from the linear α_s plot below $\alpha_s = 1.0$. The upward deviations from linearity below and above $\alpha_s = 0.5$ are called filling and cooperative swings, respectively. The filling swing is ascribed to an enhanced monolayer adsorption on the micropore walls, and the cooperative swing comes from filling in the residual space between the monolayer on both micropore walls. This mechanism was supported by GCMC simulation studies [629].

Also, Kaneko and coworkers showed accelerated bilayer adsorption in micropore filling of He at 4.2 K [619,620], pore entrance–enriched micropore filling of n-nonane [630], dipole interaction–mediated micropore filling of SO_2 [631], and cluster-mediated filling of water [632] on the basis of a detailed analysis of their adsorption isotherms. These mechanisms of micropore filling are based on no physical evidence except for adsorption isotherms, although molecular simulation and density functional theory studies have contributed greatly to understanding the adsorption mechanism [633–641]. In particular, Gubbins and coworkers have studied the micropore filling systematically with sophisticated molecular simulations, showing the adsorption mechanism [636–638]. Iiyama et al. applied in situ X-ray diffraction [613,614] and SAXS [573] techniques to elucidate the micropore-filling mechanism in activated carbon.

The type II isotherm is the most familiar; the multilayer adsorption theory by Brunauer, Emmett, and Teller was originally developed for this type of adsorption [642]. Hence, this isotherm is indicative of the multilayer adsorption process, suggesting the presence of nonporous or macroporous surfaces. Nonporous carbon black gives an N_2 adsorption isotherm that is representative of type II, and has been used as the standard adsorption isotherm for the comparison plot. Although the adsorption isotherm near $P/P_0 = 1$ has important information on macropores, such an analysis is not practical for accurate measurement. This is because condensation on the walls of the apparatus begins near the saturated vapor pressure. Consequently, mercury porosimetry is used for characterization of large mesopores and macropores [304,643–646]. The type III isotherm arises from nonporous or macroporous surfaces that interact very weakly with the adsorbate molecules. The adsorption isotherm of SO_2 on nonporous carbon black is of this type [647].

The classical type IV isotherm has useful information on the mesopore structure through its hysteresis loop, i.e., noncoincidence of the adsorption and desorption branches. Vapor is adsorbed on the mesopore wall by multilayer adsorption in the low-pressure range, and then at higher pressures vapor is condensed in the mesopore space below the saturated vapor pressure P_0. This is the so-called capillary condensation. Capillary condensation has been explained by the Kelvin equation:

$$\ln\left(\frac{P}{P_0}\right) = -\frac{2\gamma V_m \cos\theta}{r_m RT} \qquad (14)$$

Here the mean radius of curvature of the meniscus, r_m, of the condensate in a pore is associated with the vapor pressure in the pore, P, of the condensate. γ and V_m are the surface tension and molar volume of the condensate; θ is the contact angle (θ is often taken to be nearly zero). As P is usually smaller than P_0, the bulk vapor pressure, vapors condense in mesopores below P_0. The Kelvin equation determines the condensation (P_c) and evaporation (P_e) pressures, which are governed by r_m. When P_c and P_e are different from each other due to the different r_m values for the meniscus on condensation and evaporation, adsorption and desorption branches do not coincide, and adsorption hysteresis occurs. This classical capillary condensation theory predicts that an adsorption isotherm of N_2 in cylindrical mesopores that are open at both ends has adsorption hysteresis of IUPAC HI [605,625]. Recently, silica having regular cylindrical mesopores was synthesized by two groups using different methods [648,649]. This mesoporous silica has a honeycomb structure of cylindrical straight mesopores whose long-range order leads to an explicit X-ray diffraction in a low-diffraction-angle region. Many physical adsorption studies on these regular mesoporous silicas have been reported [650–657], showing the dependence of the hysteresis on the pore width and on the adsorbate molecule. Llewellyn et al. [651] reported that the N_2 adsorption isotherm at 77 K for mesopores of $w > 4$ nm has an adsorption hysteresis. The dependence of the adsorption hysteresis on the pore width was also studied by density functional theory [653]. As the Kelvin equation only takes account the molecule-surface interaction through the macroscopic concept of the contact angle, a more microscopic account of the interaction of a molecule with the mesopore wall should be included for nanopores, including micropores and small mesopores [657,658]. Also, the molecular state of the condensate in mesopores is not the same as the bulk liquid state according to the in situ time correlation function method using IR spectroscopy [669,660]. Thus, the evaluation of the pore size distribution in the nanopore range should be improved in the future for activated carbons.

Carbon aerogel, made of agglomerates of uniform carbon spheres, has relatively uniform mesopores, providing an N_2 adsorption isotherm of type IV, as shown in Fig. 44. The adsorption hysteresis is of type HI. The adsorption and desorption branches of the adsorption hysteresis are very sharp. If carbon aerogel is activated, the adsorption isotherm has a rectangular rising near $P/P_0 = 0$, as shown in Fig. 44. The HI type adsorption hystersis can be associated with the agglomerate structure of uniform spheres, according to the classical Kelvin theory.

The class V isotherm is similar to the class IV isotherm, but with very weak molecule-solid surface interaction. Hence almost no initial uptake is observed.

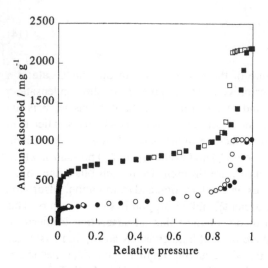

FIG. 44 Adsorption isotherms of N_2 on carbon aerogel samples at 77 K. (○,●) carbon aerogel, (□,■) carbon aerogel activated by CO_2. Open and solid symbols denote adsorption and desorption branches, respectively.

The H_2O adsorption isotherm on activated carbon with few surface functional groups belongs to this type. Figure 45 shows the H_2O adsorption isotherms of pitch-based ACFs at 303 K as a function of the average pore width. All adsorption isotherms are of type V, but they are different from each other. The adsorption uptake in the low P/P_0 region is nil, indicating that surface functional groups are few, so that these ACFs show a hydrophobic nature. The adsorption isotherms have a steep rise at a medium P/P_0 value, and a clear adsorption hysteresis except for A5, which has the smallest pore width ($w = 0.7$ nm). The adsorption hysteresis depends on the pore width; the larger the pore width, the wider the adsorption hysteresis loop. For A20 carbon treated at 1273 K in H2, which has almost no surface functional groups according to the XPS, the rise in adsorption shifts to higher pressures [615] but retains a noticeable hysteresis (this adsorption isotherm is not shown in Fig. 45). Thus, the surface of these ACFs can be regarded as hydrophobic. The adsorption mechanism of water on activated carbon is not sufficiently understood. Questions remain as to why water molecules can be adsorbed in hydrophobic carbon micropores. A water molecule cannot interact strongly with the graphite surface because the induced image dipoles in the graphite surface from the permanent dipole of a water molecule are not stable under the random orientation of physisorbed molecules. Therefore, the electrostatic interaction is small, and the dispersion interaction of a water molecule with the graphite surface is modest, leading to a very small interaction energy. There-

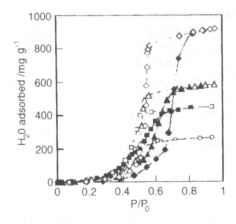

FIG. 45 Adsorption isotherms of H_2O on pitch-based activated carbon fibers having different pore widths at 303 K. Average pore width $<\pm 0.1 \text{ nm}>$/nm: ($\bullet$,$\bigcirc$), 0.7; ($\blacksquare$,$\square$), 0.9; ($\blacktriangle$,$\triangle$), 1.0; and ($\blacklozenge$,$\diamond$), 1.1. Solid and open symbols denote adsorption and desorption branches, respectively.

fore, water molecules are not adsorbed on the graphite surface, even in graphitic micropores, in the low P/P_0 region. It is well known that water molecules form dimers or clusters even in the gas phase. In micropores of activated carbon, highly concentrated water clusters, such as the pentamer $(H_2O)_5$ or hexamer $(H_2O)_6$, whose size is about 0.5–0.6 nm, must be formed due to the compressed conditions [661]. Water molecules are bonded with each other by hydrogen bondings in the cluster, and thereby the electrostatic charges of the water molecules are somewhat compensated. For a single water molecule, the interaction with the pore wall is very weak, but a cluster of H_2O molecules can have a deeper potential well through the dispersion interaction. A cluster that fits the micropore geometry has a deeper potential well. Consequently, water adsorption begins suddenly above the medium P/P_0 where the clusters grow. In the case of a small alcohol molecule, a similar situation applies to adsorption in graphitic micropores [662,663]. The adsorption hysteresis of the water adsorption isotherm is not caused by capillary condensation [610]. Iiyama et al. showed the presence of the different physical processes for adsorption and desorption branches of the adsorption hystersis with in situ SAXS [573]. As H_2O adsorption on activated carbon is essentially important in practical applications, it is important to learn more about the real adsorption.

The type VI isotherm is a stepped adsorption isotherm, which is connected with layering phase transitions of the adsorbed molecular layers, or adsorption on the different faces of crystalline solids [664].

These adsorption isotherms are obtained for vapors. If we use the supercritical gas, the saturated vapor pressure P_0 is not defined and the relative pressure P/P_0 is not used for expression of the adsorption isotherm. Many adsorption isotherms of supercritical gases on activated carbon belong to the Henry or Langmuir type. For example, the N_2 adsorption isotherm on molecular sieve carbon at 303 K is of the Henry type, which can be used for determination of the pore mouth characterization [609]. As the supercritical gas adsorption begins from narrower micropores of activated carbon, the supercritical gas adsorption isotherms are helpful in determining the pore size distribution of micropores [665,666].

D. Surface Chemistry

It is well known that activated carbons or carbon fibers are not built of pure graphene layers. Among atoms other than carbon present in their structure, the most common are hydrogen, oxygen, nitrogen, phosphorus, and sulfur [667–669]. They are incorporated in the aromatic rings as heteroatoms, or arranged in the form of functional groups analogous to those classified in organic chemistry. It is believed that those groups are located at the edges of the graphene layers. Although the groups can have a similar arrangement to that in organic compounds, one cannot assume identical behavior in chemical reactions. Among the many reasons explaining the differences are (1) influence of carbon substrate on edge heteroatoms (steric effects, resonance); (2) interactions between groups in close proximity; (3) side reactions. Examples of possible functionality are presented in Fig. 46. The presence of such groups results in surface chemical heterogeneity, caused mainly by differences in the electronegativity of the heteroatoms compared with the carbon atoms. Such groups as OH, NH_2, OR, and O(C=O)R are classified as electron donors (presence of σ or π electrons), whereas (C=O)OH, (C=O)H, and NO_2 are electron acceptors (presence of empty orbitals). Another classification is based on the strength of Brønsted acids. Acceptor groups, due to their tendency to localize electron cloud density close to them, are weaker Brønsted acids than donors that delocalize their electron pair. However, the final effectiveness of groups as Brønsted acids depends on the charge distribution on heteroatoms. Their effective strength is described using the acid dissociation constant, pK_a. For example, most carboxylic acids have pK_a between 2 and 6 and most phenols between 8 and 10 [668,670,671].

The main cause for carbon basicity is in the absence of oxygen-containing groups at the edges of crystallites [672]. Oxygen-containing groups such as chromene and pyrone also contribute to the basic character of activated carbon surfaces [667–669].

1. Boehm Titration

Due to the huge variety of possible functional groups present on the carbon surface, many specific chemical reactions known from organic and inorganic chem-

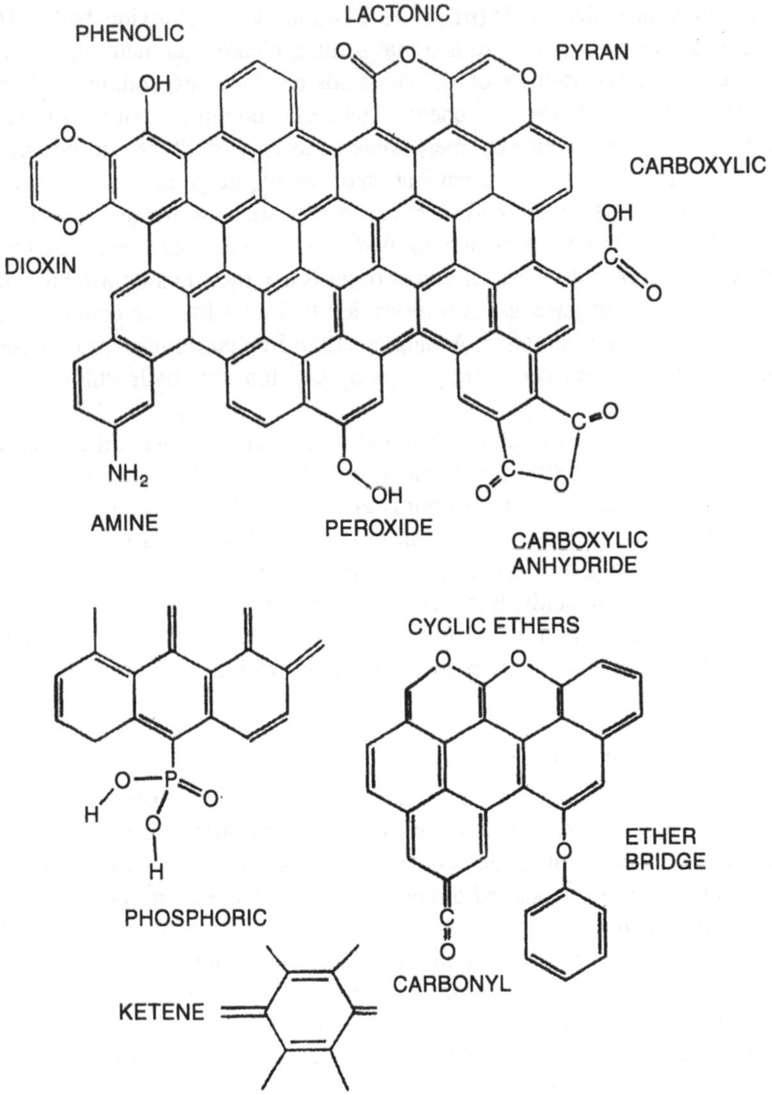

FIG. 46 Examples of functional groups present on the surface of activated carbon.

istry have been proposed to identify them [2,667,669,673,674]. Usually acids or bases of various strength have been used. The method that became the most popular was proposed by Boehm [668]. In this approach a base of certain strength neutralizes only acids having pK_a less than or equal to that of the base. As bases sodium bicarbonate, $NaHCO_3$ ($pK_a = 6.37$), sodium carbonate, Na_2CO_3 ($pK_a =$

10.25), sodium hydroxide, NaOH (pK$_a$ = 15.74), and sodium ethoxide, NaOC$_2$H$_5$(pK$_a$ = 20.58) are used. It is assumed that sodium bicarbonate neutralizes carboxylic acids, sodium carbonate-carboxylic acids and lactones, sodium hydroxide-carboxylic acids, lactones, and phenols, whereas sodium ethoxide will react with all oxygen species, even extremely weak acids (pK$_a$ < 20.58). In practice, Boehm titration is limited to the determination of carboxylic groups, lactones, and phenols [675–680]. Sodium ethoxide is not used very often due to the necessity of performing the experiment in nonaqueous media and oxygen-free conditions. One reason why the sodium salts were chosen as bases is that they do not form precipitates after reaction with gaseous CO$_2$ [668] and their specific interactions with carbon surfaces are minimal [681]. A supplement to Boehm titration using bases is the determination of surface basic groups by titration with hydrochloric acid [682].

Boehm titration is a very reliable method to evaluate the general trends in surface acidity [668,675–680]. It is simple, fast, and usually gives good reproducibility. Its big deficiency is that all groups are classified as oxygen-containing acids. Since the selectivity is based on the value of pK$_a$ of surface species, all other groups containing, for example, nitrogen, phosphorus, or sulfur will be considered as carboxylic acids, lactones, or phenols. Very often in cases where there are not significant amounts of other heteroatoms the rough estimation proposed by Boehm is enough to obtain good correlation with other properties tested [683,684].

2.   Potentiometric Titration

One of the reasons why the titrations of functional groups have been developed was the impossibility of obtaining a reliable estimation of surface chemistry based on potentiometric titration in aqueous media. Until the mid-1990s many attempts were made, but the curves obtained did not reveal any discrete end-points useful for meaningful interpretation [681,684–686]. It was believed that the smooth titration curves represent the continuum of closely interacting functional groups. The evaluations based on the amount of NaOH used to reach a certain pH are similar in nature to Boehm titration [675,687–690]. The derivatives obtained in some studies were difficult for characterization due to the many maxima revealed when no well-defined end-points were present [688].

The first meaningful deconvolution of titration curves with theoretical description of acid-based dissociation on the surface of carbons was described in the mid-1990s [691,692]. In this approach it is assumed that the system under study consists of acidic sites characterized by their acidity constants, K_a. It is also assumed that the population of sites can be described by a continuous pK$_a$ distribution, $f(pK_a)$. The experimental data can be transformed into a proton binding isotherm, Q, representing the total amount of protonated sites, which is related to the pK$_a$ distribution by the following integral equation:

$$Q(\text{pH}) = \int_{-\infty}^{\infty} q(\text{pH}, \text{pK}_a) \, f(\text{pK}_a) dp\text{K}_a \tag{15}$$

First, the solution of this integral equation was obtained using the Rudzinski-Jagiello method (RJ approximation) [693]. Although the distributions obtained showed the presence of peaks associated with various functional groups, the approximation could not fully resolve peaks for very heterogeneous surfaces [692]. A significant improvement was an application of the numerical procedure SAIEUS (Solution of Adsorption Integral Equation Using Splines) [694,695], which uses regularization combined with nonnegativity constraints. The choice of the degree of regularization/smoothing is based on the analysis of a measure of the effective bias introduced by the regularization and a measure of uncertainty of the solution. SAIEUS was tested using simulated data and experimental titration data of organic standards, and it was demonstrated that this method can completely resolve peaks that are less than 1 pK_a unit apart [694,696]. This results in the precise estimation of surface groups having a certain pK_a value [680,695]. Comparison of the two methods mentioned above showed the superiority of the SAIEUS approach in the resolution of peaks [695]. The application of the RJ approximation [691] was further modified using a Gaussian fit to the deconvoluted distribution function in order to quantitatively assess the number of specific functional groups [690].

As a result of the applied procedure, the distribution of surface pK_a is obtained in the experimental window between pK_a 3 and 11. For very heterogeneous surfaces many peaks are present [680,683,684,695]. The number of groups evaluated from potentiometric titration is in agreement with the Boehm method; however, more detailed information is obtained using the deconvolution of titration curves [667,683,690]. Unlike in the Boehm method, in potentiometric titration the specific pK_a of various groups in the categories defined by Boehm can be obtained. For example, the presence of three distinct species having pK_a in the categories of carboxylic acids can be revealed (Fig. 47) [680,683,684]. Due to the fact that many heteroatom configurations can result in similar pK_a [668,670,671], which also depends on the unknown activity of the solution, the exact classification of species is impossible. However, by knowing the elemental composition of the carbon matrix one can deduce information about the nature of the species (nitrogen, phosphorus, or oxygen functional groups). When the changes in a series of carbons are studied this method clearly shows the trend in surface functionality [684].

3. Temperature-Programmed Desorption

Temperature-programmed desorption (TPD) or thermodesorption has been developed to evaluate the active surface area (ASA) of carbons [697]. ASA is a surface area of the reactive edge carbons. Those carbons are expected to react with oxygen groups, resulting in the presence of functional groups. Upon heating, the

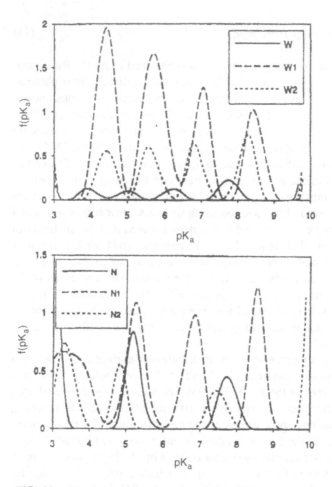

FIG. 47 pK$_a$ distributions for carbons with different degrees of surface oxidation. (From Ref. 680.)

surface groups decompose and CO, CO$_2$, water, and hydrogen are released. The most acidic groups (carboxyls, lactones) are desorbed as CO$_2$ in the temperature range 473–823 K, while less acidic (phenols, carbonyls) and basic (pyrones) groups are desorbed mainly as CO or CO+CO$_2$ in the temperature range 773–1273 K (Fig. 48) [698–700]. The presence of water is a result of the decomposition of carboxylic groups. The TPD data confirm the results of titrations; however, the number of detected groups is usually twice as large [699,701]. A possible reason for this discrepancy lies in the limitation of the titration methods. They are able to detect acids and bases having certain pK$_a$ values, whereas during

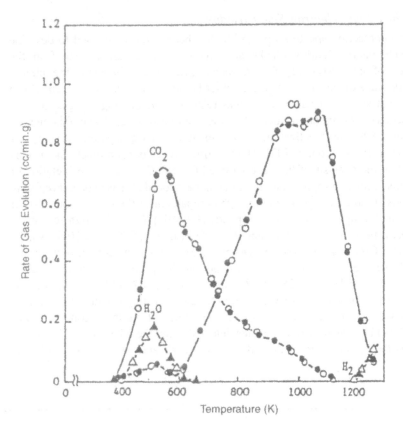

FIG. 48 Gas evolution profiles (CO_2, CO, H_2O, and H_2) for HNO_3-oxidized char. (From Ref. 699.)

thermal desorption all surface groups, regardless of their chemical nature, should decompose.

The TPD method can be used to evaluate the surface chemistry of carbons that were obtained by carbonization at high temperature. In the case of low-temperature carbons, such as those obtained using phosphoric acid activation, the results are meaningful only at temperatures lower than that used in the manufacturing process. At higher temperature carbonization occurs and its byproducts interfere with the analysis [701]. In the TPD method, as in Boehm titration, the species are classified as strong and weak acids. The experiment cannot provide their exact formulas or distribution on the surface. When mass spectroscopy is used as a detection method the decomposition products containing heteroatoms other than oxygen do not affect the interpretation of the results.

4. *X-Ray Photoelectron Spectroscopy*

X-ray photoelectron spectroscopy (XPS) has been extensively used to describe surface chemistry of carbon blacks, activated carbons, or carbon fibers from the beginning of the 1970s [702]. The analysis is based on the changes of the intensities of 1s peaks of carbons, oxygen, or other heteroatoms such as nitrogen. Those peaks are at specific irradiation energies related to the binding energies of core electrons ejected from atoms located on the external surfaces. Using this method the evidence for the localization of oxygen functional groups on the edges of graphene layers was found [707]. The identification of oxygen functional groups is based on the analysis of the deconvoluted O_{1s} and C_{1s} peaks. Interpretation of the results requires a sophisticated mathematical procedure based on curve fitting to various symmetrical peaks along with comparison of the results obtained for the modified sample to the reference spectra (Fig. 49) [704–711]. High-resolution O_{1s} spectra provide information about C=O groups (BE 530.4–530.8 eV), C-OH and/or C-O-C groups (BE 532.4–533.1 eV), and chemisorbed oxygen and water (534.8–535.6 eV). On the other hand, on C_{1s} spectra such groups as alcohols and ethers (BE 286.3–287.0 eV), carbonyls (BE 287.5–288.1 eV), and carboxyl and/or esters (BE 289.3–290.0 eV), can be distinguished. The presence of other species, such as nitrogen, can be determined from the deconvolution of N_{1s} peak [711]. The amount of species detected using the XPS method is usually given as the atomic ratio evaluated from the peak area ratios [704–706]. The

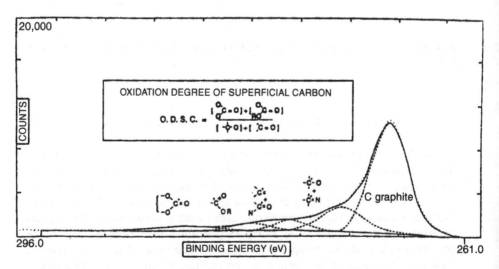

FIG. 49 Definition of oxidation degree of superficial carbon (O.D.S.C.) from ESCA measurement. (From Ref. 707.)

main weakness of the XPS method is the difficulty in identifying the type and distribution of functional groups without ambiguity [668]. This problem has recently been partially solved by metal ion labeling methods or derivatization [712,713].

5. Fourier Transform Infrared Spectroscopy

Fourier transform infrared spectroscopy (FTIR) can provide mainly qualitative information about surface functionality. The analysis is based on the comparison of the wavelength of the peaks obtained to the spectra of known organic compounds [648,700,704,711,714–718]. The interpretation of infrared spectra of activated carbons possessing functional groups was described in detail by Zawadzki [714–716]. According to Zawadzki, surface oxygen groups show their absorption bands at 3600, 1730, 1620, and 1260 cm^{-1}. They represent OH, C-O, C-. . . O, and C-O stretching vibrations. Their relative intensity provides information about the possible arrangement of oxygen into functional groups. For instance, when the band from OH stretching vibrations is broad and is accompanied by intense bands representing C=O (1730 cm^{-1}) and C-O (1260 cm^{-1}), carboxyl structures are present. Bands around 1260 cm^{-1} and 1340 cm^{-1} (OH deformation vibrations) suggest the presence of phenols.

Although qualitative analysis of FTIR spectra is relatively easy [714], quantitative evaluation of surface chemistry using FTIR is an extremely difficult task due to continued background absorption and the possibility of the presence of various functional groups as simple and nonassociated structures [667]. Semiquantitative information about an increase in the number of groups can be obtained using the approach described by Fanning and Vannice [718]. The data analysis is based on the subtraction of the spectra of the initial carbon from the carbon after modification, where an increase in the intensity of certain bands is observed. Using this method, the effect of the background absorption is minimized and the intensity of well-defined peaks can be compared (Fig. 50) [684,700].

6. Calorimetric Titration

In order to see the inventory of species with pK_a beyond the experimental window of titration in aqueous solutions, calorimetric titration (thermometric titration) is applied. In this approach, an acid-base titration is done in a nonpolar solvent [719]. The strength of positive and negative charges on the surface of carbon is measured without the interference of ionized acids. Calorimetric titration uses an enthalpy-sensing probe. Based on the position of the end-points and the amount of titrant consumed, the number and strength of acidic and basic sites is evaluated. As titrants for basic sites, strong organic acids such as trifluoroacetic acid in toluene are used. To evaluate acidic sites n-butylamine in cyclohexane is applied. Using this method one can also obtain information about the water content of

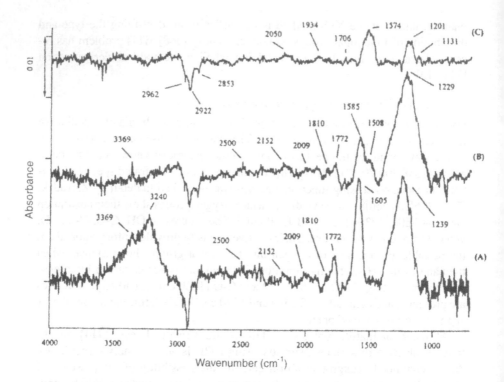

FIG. 50 DRIFT spectra of activated carbons: (A) untreated AC-HNO$_3$; (B) AC-HNO$_3$ after treatment in H$_2$ at 773 K; and (C) untreated AC-ASIS. (From Ref. 700.)

carbons and heats of immersion. However, this method has not been very popular in our pursuit of molecular models of porous carbons.

7. Calorimetry (Immersion Calorimetry and Flow Adsorption Microcalorimetry)

Information about the number of primary adsorption centers of activated carbons, the oxygenated surface groups [720], can be obtained from analysis of the enthalpy of carbon immersion in water. The relationship between these two quantities was proposed by Stoeckli et al. [721]:

$$\Delta H_i(\text{J/g}) = -25.0 \ (\text{J/mmol H}_2\text{O}) \cdot a_o - 0.6 \ (\text{J/mmol H}_2\text{O}) \cdot (a_s - a_o)$$

where ΔH_i, a_o, and a_s are enthalpy of immersion, number of primary adsorption centers, and limiting amount adsorbed, respectively. Using this approach, one can analyze the changes in the degree of carbon oxidation [722,723].

The surface polarity and degree of graphitization can be estimated using flow adsorption microcalorimetry. In this method, heats of adsorption of n-butanol and n-dotriacontane onto carbonaceous materials samples immersed in n-heptane are measured. Based on the relationship between the heat of adsorption and the surface area [724], the resulting surface graphiticity index (SGI) and surface polarity index (SPI) are expressed (in millijoules per square meter of the total surface). The method works well for carbon blacks. For activated carbons with a high degree of microporosity, a considerable part of the surface is not accessible to n-butanol molecules as a result of the molecular sieving effect. The information obtained using this method indicates the affinity of activated carbon surfaces for polar and nonpolar substances [725].

8. Inverse Gas Chromatography

Inverse gas chromatography (IGC) is a rapid and reliable method used to evaluate the acid-base character of activated carbon surfaces. This method is based on the study of physical adsorption of appropriate molecular probes by means of chromatographic (dynamic) experiments [726]. The amounts of injected solutes are very small, and it is assumed that the adsorption is described by Henry's law. From the retention volume the free energy of adsorption, ΔG^0, is calculated. When saturated and unsaturated hydrocarbons are used as the probe molecules, the value of the difference in the ΔG^0 of alkane and alkene is used to study the effect of the π bond interactions with electron acceptor sites on the surface [727–729]. The approach is based on the well-known electronic structure of alkanes and alkenes. The difference in the free energy is called the parameter of specific interactions, ε_π. It represents the average number and strength of the acidic centers present on the carbon surface [728,729].

IV. MOLECULAR MODELS

Realistic molecular models of carbons are needed to interpret experimental data and to aid in the design of carbons for specific applications. At present it is not feasible to develop such models from first principles (e.g., full *ab initio* calculations, or semiclassical molecular simulations based on intermolecular potential models), except for a few crystalline carbons such as graphite or nanotubes. For the great majority of carbons, it is necesary to draw upon experimental measurements of the type discussed in Sections II and III in order to construct a realistic model.

There are two general approaches to the problem of constructing a molecular model of a porous material. The first, which we term *mimetic simulation*, involves the development of a simulation strategy that mimics the synthetic process used to fabricate the material in the laboratory. The second, termed *reconstruction*, seeks to build a molecular model whose structure matches the available experimental structure data.

Mimetic simulation methods have the advantage that they provide insight into the synthesis mechanism and may thus suggest better ways to synthesize the material. Also, they are less subject to bias on the part of the researcher than are reconstruction methods. An example of a mimetic simulation protocol is that used recently to prepare controlled-pore glasses [730,731]. Such glasses are prepared by spinodal decomposition. A mixture of oxides (SiO_2, B_2O_3, Na_2O) is heated to about 1473 K to produce a miscible liquid mixture and then quenched into the spinodal liquid-liquid region at about 973 K. Provided that the initial composition is not too far from the critical mixing value, the mixture starts to separate into two phases, one of which (the B_2O_3-rich phase) becomes a connected phase of roughly cylindrical geometry. If left at this temperature the diameter of the cylindrical region grows. When the desired diameter is reached, the mixture is quenched to room temperature to form a glass, the B_2O_3-rich phase is removed with an acid, and the resulting silica-rich structure is annealed. It turns out that the spinodal decomposition can be easily mimicked using quench molecular dynamics, and the pore morphology and topology is insensitive to details of the intermolecular potentials [730]. Simple Lennard-Jones potentials, with the like pair 1-1 and 2-2 interactions being identical whereas the 1-2 interaction well depth is substantially weaker, are sufficient to produce pore structures that are remarkably similar to those produced experimentally. A binary mixture is equilibrated at a high, supercritical temperature using isothermal molecular dynamics. It is then quenched into the spinodal region and phase separation commences. When the diameter of the cylindrical phase reaches the desired value, the system is quenched to a very low temperature, the cylindrical phase is removed, and the system is annealed. The intermolecular potential parameters of the solid atoms are then adjusted to values suitable for silica. Glasses have been prepared by this method with diameters up to 7 nm and with a range of porosities [730,731]. GCMC simulations of the adsorption isotherms for nitrogen and xenon in these glasses closely match those found experimentally [732].

Such mimetic simulation methods have the advantage that they yield a unique molecular structure; in addition to providing insight into the synthesis, they can suggest improved synthetic routes. The main difficulty with such an approach is that new simulation protocols must be developed for each new material. In the case of most porous carbons, the details of the synthesis process are so poorly understood that mimetic simulation is not practical at the present time. Nevertheless, some parts of the synthesis can be simulated using *ab initio* and semiempirical quantum mechanical methods. Stochastic methods have also been used. These are discussed in Section IV.C.1.

The great majority of methods used to model porous carbons are reconstruction methods. In the case of regular crystalline materials, such as nanotubes, the model can be constructed from X-ray or neutron diffraction data. For disordered carbon materials the construction of a simplified molecular model can be made from visual observation of TEM images of the material. This is the origin of simple geometrical models, such as the slit-pore and wedge-pore models for activated carbons (Section IV.B.2). More realistic reconstruction methods, such as the stochastic reconstruction methods and those based on reverse Monte Carlo techniques (Section IV.B.3), involve construction of models that match experimental structure data, usually scattering or TEM data.

We note that building a molecular model of a disordered material from scattering or TEM data is an ill-posed problem. Thus, the reconstruction methods do not yield a unique structure for such materials. The realism of the structure model obtained will depend on the type of data used, as well as on its accuracy and sensitivity to details of the pore structure.

A. Regular Carbons: Carbon Nanotubes

Due to the regular pore morphology of carbon nanotubes, it has been common to use simple geometrical models when studying the behavior of fluids confined in these materials. In this section, we first review the models that have been developed for single nanotubes. Such models can be used to study the so-called *endohedral* adsorption, or the adsorption that takes place inside the nanotubes. We then consider the modeling of nanotube arrays, which in addition allows the study of fluids adsorbed on the external surface of the tubes (*exohedral* adsorption).

Carbon nanotubes have a cylindrical shape (Section II.A.4). Thus, most models for carbon nanotubes assume that the pore walls are cylinders made up of carbon atoms. In order to completely specify the pore geometry it is necessary to define the length of the nanotube. Nanotubes can be modeled as infinitely long cylinders, open-ended cylinders, or closed-ended cylinders. Most simulation studies of fluids confined in nanotubes are performed in either infinitely long tubes or open-ended tubes of finite length.

Once the geometry of the pore is defined, one has to define the structure of the pore walls in order to have a complete description of the porous medium. If one thinks of a carbon nanotube as a rolled-up graphite sheet, it is easy to realize that there are several possible structures because there are several ways to roll a graphene sheet to form a cylinder. Different ways of rolling graphite lead to carbon nanotubes with different structures. The most common structures are the armchair and the zig-zag or saw-tooth [733–735], illustrated in Fig. 11a and 11b. If the graphite sheet is rolled in a direction that lies between the two axes shown in Fig. 11a and b, different *chiral* structures can be obtained (see Fig. 11c).

When the structure of the pore walls is taken into account explicitly, by specifying the positions of all the carbon atoms that make up the wall, the fluid-wall interaction energy is calculated by the explicit sum of the intermolecular potential over all of the carbon atoms:

$$U = \sum_i u(r_i) \tag{16}$$

where r_i is the distance separating the adsorbed molecule from the carbon atom i. Pair-wise additivity is assumed in this equation.

An alternative is to assume that the pore wall is structureless. In this case, the wall is made up of a uniform distribution of atoms rather than an array of carbon atoms with defined positions. The atomic density is usually assumed to be equal to that of graphite. For structureless pore walls, two different approaches have been developed to calculate the interaction energy of an adsorbed molecule confined in a carbon nanotube.

In the first approach, one assumes a uniform distribution of carbon atoms and integrates the resulting expression over the surface of the tube in cylindrical coordinates. For a single-wall nanotube, if one uses the Lennard-Jones 6-12 potential, the resulting expression is analogous to the Steele 10-4 potential for the interaction between an adsorbate molecule and a graphene layer. Tjatjopoulos and coworkers [736] developed an analytical expression in 1988. It has been used in Monte Carlo simulations (see, for example, [737]) and density functional theory calculations (see, for example, [738]) of fluids confined in carbon nanotubes.

In the second approach, developed by Maddox et al. [739–741], the pore is divided into a grid and the fluid-wall potential is calculated at each point. The procedure is as follows: The coordinates of each carbon atom are specified by transforming the atomic coordinates of a graphite sheet to those of a graphite tube in cylindrical coordinates. The distance from the center of the tube to the tube wall (or the radius of the pore), R, is divided into N distances $r_j (1 < j < N)$. At each point r_j, the interaction energy of an adsorbate molecule at (r, θ, z) with every carbon atom in the tube wall is calculated for several values of the other two cylindrical coordinates (θ, z). The averaged value of the calculated interactions is the interaction potential $u(r_j)$. In the case of open-ended nanotubes, the procedure is similar. However, since the interaction energy is also a function of the distance from the tube opening to the adsorbate molecule, z, the length of the end region of the tube, is divided into M distances z_k, $(1 < k < M)$. The average is performed for several values of θ at each point (r_j, z_k). In this way, one numerically calculates a potential $u(r_j, z_k)$. This method is useful for surfaces for which the atomic structure is known and can be approximated by a smooth surface [740].

The assumption of structureless pore walls is valid when the distance between the atoms that make up the pore wall is small in comparison with the size of the molecule that is being adsorbed, and the temperature is not too low. In such cases, the variation in the fluid wall potential as the adsorbate molecule moves parallel to the nanotube axis is small in comparison with kT. For adsorption studies of very small molecules, such as hydrogen, the corrugation of the pore walls can have an important effect on the adsorption behavior, and individual atoms must be taken into account explicitly when calculating the fluid-wall interaction energy.

To completely specify the structure of the pore wall, it is necessary to set the number of rolled graphene sheets in the wall. When several layers are considered, it is usually assumed that they are coaxial and with an interlayer separation similar to that observed in pure graphite (0.34 nm). The fluid-wall interaction energy in a multilayer carbon nanotube is calculated by adding the interaction energy of each layer with the adsorbed molecule, regardless of the way in which the structure of the pore wall is being modeled. An alternative is to replace this sum by an integral, assuming that the pore wall is a conntinuum in the radial direction. For this case, Peterson et al. [742] developed an expression analogous to the Steele 9-3 potential for the interaction between an adsorbed molecule and a planar continous graphite surface.

As mentioned above, the modeling of fluids in a single nanotube allows the study of the *endohedral* adsorption, or the adsorption that takes place inside the tubes. Maddox et al. [739,740] performed GCMC simulations of argon and nitrogen in SWNT and in double-wall nanotubes. They calculated the adsorption isotherms, density profiles, phase transitions, and heats of adsorption. The simulations were carried out for infinite cylindrical tubes with smooth walls of two internal diameters: a SWNT of diameter 1.02 nm and a double-wall nanotube of 4.78 nm. The fluid-wall interaction energy was calculated using the grid method described above. The smaller tube was too small to exhibit phase transitions, and Maddox et al. observed a type I adsorption isotherm. For the longer tube, they observed layering transitions, capillary condensation, and hysteresis for both adsorbates. In order to study the effects of open tubes, they also performed simulations in an open-ended double-wall nanotube of 4.78 nm diameter and 7.7 nm length (Fig. 51).

The amount of fluid adsorbed at the same pressure is lower for the open-ended nanotube. As the end of the tube is approached, the fluid-wall forces are diminished, and it is possible for a meniscus to appear on desorption of a pore filled with liquid. During the desorption process, the meniscus reduces the amount of fluid adsorbed before capillary evaporation takes place, making the hysteresis loop more rounded. These effects can be very important for tubes of 6–8 nm diameter or less.

Maddox et al. [741] modeled the adsorption of nitrogen, methane, propane, and their binary mixtures at 300 K in an infinitely long double-wall nanotube

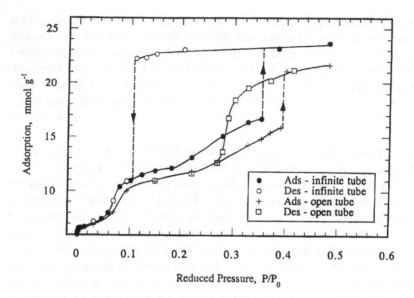

FIG. 51 Detail of the hysteresis loops for argon adsorption at 77 K in a 4.78 nm double carbon nanotube. (From Ref. 740.)

of 4.78 nm diameter. The nanotubes were modeled as in their previous works [739,740]. The component with the stronger fluid-wall interaction is preferentially adsorbed, i.e., methane from the methane/nitrogen mixture and propane from the methane/propane mixture. The separation is greatest at low pressure. They also used nonlocal density functional theory to determine the separation of a trace quantity of propane from bulk methane. They found that the separation is very sensitive to the nanotube diameter and can be improved by several orders of magnitude by finding the optimal size. Finally, they also found that the number of coaxial layers of carbon in the nanotube walls can also be tuned to improve the separation by several orders of magnitude.

Ayappa [735,743] studied the effect of temperature on the adsorption behavior of mixtures of Lennard-Jones fluids in SWNT. In these studies, he modeled carbon nanotubes as infinitely long cylinders with armchair structure and calculated the fluid-wall interaction energy taking into account the interaction potential of the adsorbate with each carbon atom explicitly. The radii of the SWNT in these studies varied from 0.613 nm to 1.904 nm. He observed a transition in selectivity when lowering the temperature of the system. For mixtures of species with different molecular diameters, the energetically favored species is adsorbed in larger tubes. However, at low temperatures or intermediate nanotube diameters, the smaller species preferentially adsorbed. For mixtures of species with similar diameters, the energetically favored species is preferentially adsorbed at all temperatures.

Khan and Ayappa [744] studied the adsorption of two-center Lennard-Jones models of N_2 and Br_2 in SWNT. They contrasted their results with the ones obtained using a simple spherical model for the same fluids. They modeled the carbon nanotubes as infinitely long cylinders with armchair structure, as in the previous studies by Ayappa [735,743]. For the smaller molecule (N_2), the results were very similar for both models. Furthermore, the inclusion of an ideal quadrupole in the molecular model had little effect on the density distribution. However, the difference between both models was more pronounced for Br_2, with significant differences in the density distribution. They also calculated angular density distributions and found that the diatomics adjacent to the wall are usually oriented parallel to the wall. At higher densities, they observed a layer adjacent to the wall with parallel orientation; in the center of the tube the molecules preferred an orientation normal to the wall. This effect is accentuated at lower temperatures.

We now focus on the problem of modeling nanotube arrays, or groups of nanotubes. Modeling nanotube arrays not only allows the study of *endohedral* adsorption, but also *exohedral* adsorption, or the adsorption of the fluid on the external surface of the tubes and in the space between tubes. In the models developed to date, the axes of the tubes in the array are usually assumed to be parallel. Therefore, one only needs to specify the position of the tubes relative to their nearest neighbors in order to determine the topology of the array. In a square array (Fig. 52), the lines that join the tube centers form squares. In the triangular array (Fig. 53), the lines that join the tube centers form triangles. In both cases,

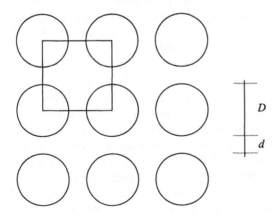

FIG. 52 Square array of carbon nanotubes. D is the nanotube diameter. d is the smallest distance between nanotubes.

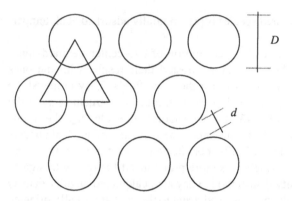

FIG. 53 Triangular array of carbon nanotubes. D is the nanotube diameter. d is the smallest distance between nanotubes.

the only geometrical parameter that can be changed (other than the parameters for each individual nanotube) is the smallest distance between any two nanotube walls. To the best of our knowledge, no attempts have been made to build more heterogeneous models that take into account different kinds of arrays.

Darkrim and Levesque [745] performed Monte Carlo simulations of hydrogen adsorption at 293 K and a bulk pressure of 10 MPa in square arrays of SWNT with saw-tooth structure. They considered nanotubes with diameters ranging from 0.704 nm to 1.957 nm, with the smallest distance between walls of neighboring tubes of 0.334 nm. In this case, the maximal amount of adsorbed hydrogen was found at a diameter equal to 1.174 nm. They next considered the case of nanotubes with a diameter of 1.174 nm with a smallest distance between neighboring tube walls ranging from 0.5 nm to 0.9 nm. They found that for SWNT of 1.174 nm diameter and a smallest distance between neighboring tube walls of 0.7 nm, the presence of the nanotubes increases by 37% the density of hydrogen molecules over that present in the bulk phase. In a later work [746], Darkrim and Levesque also performed simulations of hydrogen adsorption in a triangular lattice. They found that at room temperature the triangular array has a higher uptake than the square lattice because at this temperature only sub-monolayer adsorption occurs on the outside of the nanotube, and the hexagonal array is more compact. Maximal adsorption takes place in the 1.2-nm-diameter nanotube arrays, with a separation distance of 0.7 nm at 293 K. At 77 K, two layers of adsorbed fluid may appear in some cases. Therefore, the maximal uptake at 77 K is obtained for diameters D and separations d that are at least twice as large as those for 298 K. They compared their simulations with experimental results, finding good agreement. The simulations predict a slightly higher adsorption than that shown by the experimental results. The difference was attributed to the homogeneity of the model array.

Wang and Johnson [747] performed GCMC simulations of hydrogen adsorption in SWNT arrays. They included quantum effects through the path integral formalism. They compared their results to classical simulations and found that quantum effects were very important in this case, even at 298 K. In their first study, they performed the simulations in a triangular array of SWNT separated at a distance $d = 0.32$ nm. They used two different nanotube diameters, i.e., 1.22 nm and 2.44 nm. Two different simulations were performed for each case: one inside a single tube and another in an interstice (where three tubes meet). The total hydrogen uptake was then calculated by adding the results. They compared the results with experiments, showing that the simulations do not confirm the large hydrogen uptake observed experimentally. The low uptake is attributed to the excluded surface area in the close-packed array of nanotubes. In a later study, Wang and Johnson [748] studied hydrogen adsorption in different array geometries. They performed the simulations in square and triangular lattices of carbon nanotubes. The separation between tubes ranged from 0.32 nm to 1.2 nm. Three different nanotube diameters were studied: 1.22 nm, 1.63 nm, and 2.44 nm. In this study, the path integral formalism was not used. Instead, they performed classical simulations because they did not expect quantum effects to have a strong influence on the determination of the optimal geometry. They found that the 1.22-nm-tube triangular arrays with a distance between tubes of 0.6 nm and 0.9 nm give the highest volumetric densities of all the tested arrays. Finally, Simonyan et al. [749] modified the model used in previous works [747,748] to include a charge (0.1 e/C) on the nanotubes. They found that the quadrupole moment and induced dipole interactions of hydrogen with the charged tubes lead to an increase in density relative to the uncharged tubes of approximately 10–20% at 298 K and 15–30% at 77 K. In all three works, the conclusion is that the calculated isotherms indicate that nanotube arrays are not suitable adsorbents for achieving the U.S. Department of Energy target of 62 kg of H_2/m^3 and 6.5 wt % H_2 for hydrogen storage and transportation at normal temperatures.

Yin et al. [750] studied nitrogen adsorption in square arrays of open and closed SWNT by using GCMC simulations. The diameter of the tubes ranged from 0.6 nm to 3.0 nm, and the separation between nanotubes varied from 0.4 nm to 3.0 nm. The nanotubes were modeled as structureless cylinders, using an integrated potential to calculate the nitrogen-nanotube interactions. As expected, they found that the amount of nitrogen adsorbed is higher in arrays of open SWNT. The adsorption isotherms for the arrays of closed nanotubes are type I for small nanotube separations, showing that the space available for exohedral adsorption is microporous. With larger tube separations, two-stage isotherms were found. These steps correspond to monolayer formation at lower pressures and condensation of nitrogen at higher pressures.

Williams and Eklund [751] performed GCMC simulations of hydrogen adsorption in finite-diameter SWNT "ropes" in a triangular array. These ropes are arrays of SWNT that exhibit triangular, "honeycomb" coordination. In other words, the arrays are made up of a finite number of SWNTs, unlike previous studies in which the SWNT arrays were infinite in three dimensions. Williams and Eklund modeled the hydrogen-carbon interactions with a Lenard-Jones 6-12 potential. They kept the diameter of the nanotubes and the separation between tubes constant—1.36 nm and 0.32 nm, respectively. However, the diameter of the hexagonal ropes was changed, and they found that the grooves on the external surface of the ropes provide a strong adsorption site, comparable to those available for endohedral adsorption. Based on this observation, they point out that delamination of nanotube ropes should increase the gravimetric hydrogen storage capacity.

Carbon nanotubes may also be modeled as nonrigid structures. When this is the case, the interaction between any two carbon atoms on the adsorbent structure needs to be taken into account explicitly. For example, Mao and Sinnott [752] studied the diffusion and dynamic flow of methane, ethane, and ethylene through arrays of carbon nanotubes. They used a classical reactive empirical bond-order potential developed by Brenner [753].

B. Disordered Carbons

1. Early Models of Carbons

(a) The Fine Scale. Prior to the first powder X-ray studies of 1917, three allotropes of carbon were generally held to exist: diamond, graphite, and amorphous carbon. At this time, amorphous carbon was a catch-all for those carbons that did not possess the characteristics of graphite or diamond. Additional new allotropes (e.g., [754]) and variants of existing allotropes (e.g., [755]) were also (erroneously) proposed from time to time on the basis of what would now be considered very crude experimental evidence (e.g., the "graphitic acid" test). Experimental methods of the time such as calorimetry, gravimetric analysis, and crude chemistry, offered scant basis for hypothesizing molecular structures, even for graphite or diamond (e.g., [754]). The advent of powder X-ray scattering prompted considerable study of the various forms of carbon. The molecular structures of diamond [756] and graphite [757,758] were determined and agreed [759] within a decade of the first powder X-ray studies. However, the molecular structure of "amorphous carbon" was the subject of much greater debate, and several competing molecular models were suggested in the first 60–70 years of the 20th century before computers were introduced. Despite their antiquity and the limited data on which they are founded in some cases, many of these models are still remarkably relevant; they are, therefore, briefly reviewed here.

The first conception of a molecular structure for amorphous carbon was advanced by Debye and Scherrer [757] who, in 1917, studied diamond, graphite, and a number of different chars using powder X-ray diffraction to conclude that amorphous carbons are just finely divided graphite, and not a third allotrope of carbon. They further described this finely divided graphite as benzene-based "carbon molecules" of approximately 30 atoms. Debye and Scherrer made no further statements regarding the structure of these molecules, their configuration in space, or their mode of association, so that we could hardly term it a molecular model at this stage. However, their hypothesis was a significant advance on the state of knowledge as they were essentially arguing that amorphous carbons are not *truly* amorphous because they possess at least some order, albeit at molecular length scales. While this view was to be challenged by some in the future, we now know that this idea is essentially correct for the class of chars studied by them. Asahara [764] used powder X-ray diffraction on more than 20 different types of "amorphous" carbon to confirm the conclusions of Debye and Scherrer, and to show that the size of the constituent graphitic crystallites is a function of the precursor used.

Lowry and Bozorth [250] made the next major advance in the molecular modeling of noncrystalline carbons by suggesting that the lack of three-dimensional character in the diffraction patterns (e.g., absence of 004 and other reflections) was caused by a shifting of the graphite planes within a graphitic crystallite relative to each other. A similar model was advanced by Arnfeld [761] in 1932. This lack of order in the *c* direction of the crystallites was confirmed beyond any reasonable doubt some years later by Warren [237,762] who coined the term *turbostratic* to describe this mismatch. These turbostratic building blocks form the basis of many of the modern molecular models of noncrystalline carbons and are essentially equivalent to the BSUs of Oberlin and coworkers. It was also during this period that greater detail emerged about the specific sizes of the BSUs and how these varied with the type of precursor and preparation conditions used [237,763,764].

Publication of the work of Debye and Scherrer [757] sparked considerable debate over the exact nature of the molecular structure of noncrystalline carbons, and whether or not amorphous carbon was an allotrope in its own right. Roth [765], for example, argued that carbon black could not be composed of molecular sized graphite crystallites as the difference in the heats of combustion of carbon black and graphite could not be explained by the size of the crystallites predicted from the powder X-ray data. Ruff and coworkers [766] and Clarke [767] claimed that although Debye and Scherrer's molecular structures may exist, they only arise as a completely amorphous structure transitions to graphite under appropriate heat treatment and termed this transition structure as *paracrystalline*. They also controversially claimed [768] that only truly amorphous carbons may act as adsorbents—a claim that was quickly and convincingly rebutted by Lowry and Bozorth [250]. Later workers argued that noncrystalline carbons may exist that range from the completely amorphous to structures that demonstrate long-range graphite-like structure [769,770]. As we now know, this view is a valid one. Perhaps the most interesting of the early molecular models of truly amorphous carbons is that of Riley and coworkers [771,772]. Their molecular model, which is illustrated in Fig. 54, is built by tilting alternate rings in a graphite layer about their common axes and attaching the free valences to the similarly tilted rings in the layers directly above and below. This structure would yield a connected pore space reminiscent of that of the zeolites. Greater disorder would be introduced by the presence of heteroatoms. The structure of Riley and coworkers is reminiscent of the schwarzites hypothesized by Townsend et al. [289] following the discovery of fullerenes (see Fig. 28).

(b) Models of Pore System.   Attempts at modeling the structure of noncrystalline carbons beyond the fine structure began in the 1930s. Wesselowski and Wassiliew published an extensive discussion [773] on the possible molecular structures of a wide variety of carbons. Chaotic (i.e., truly amorphous), axial, radial, and spheroidal molecular structures were all proposed; similar classifications of the molecular structure of carbons followed some 60 years later in the well-

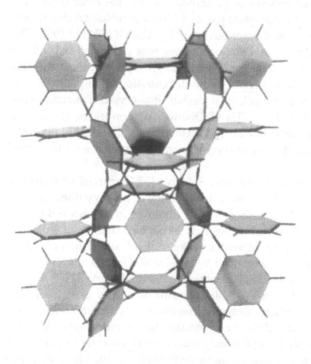

FIG. 54   The amorphous carbon structure of Riley and coworkers [771], which is perhaps one of the first examples of the schwarzites. (From Ref. 771.)

known review of Oberlin [246]. Using SAXS for the first time, Krishnamurti [774] proposed that charcoals possess no order beyond that at the fine scale and that the BSUs are randomly oriented in space. Biscoe and Warren [237] similarly used SAXS to find that the BSUs in carbon blacks occur as clusters of 20–80 nm, as illustrated in Fig. 55; the random BSU orientations within the clusters is assumed with little evidence. The proposed models of Krishnamurti and Warren are both known to be incorrect for the materials they studied; it is clear that the development of molecular models on an ad hoc basis is problematic when using SAXS data only. Franklin [238] used X-ray diffraction to propose her famous models for carbons that demonstrate graphitizing and nongraphitizing behavior. The model of Fig. 56a illustrates the relatively low level of cross-linking between the BSUs that is typical of nongraphitic graphitizing carbons, while the model of Fig. 56b shows the substantial degree of cross-linking in a nongraphitizing carbon and the substantially more open pore space that occurs because of this. These general features of such qualitative molecular models are still largely relevant, and as a consequence these figures are often reproduced in the literature even today.

While the late 1950s saw a general acceptance of the concepts of the turbostratic BSU and the idea that noncrystalline carbons demonstrate varying levels of cross-linking depending on the nature of the precursor and heat treatment sever-

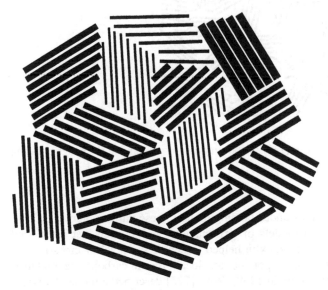

FIG. 55 The model of Warren for carbon black showing the turbostratic crystallites [237]. Note the random orientation of the crystallites. (From Ref. 237.)

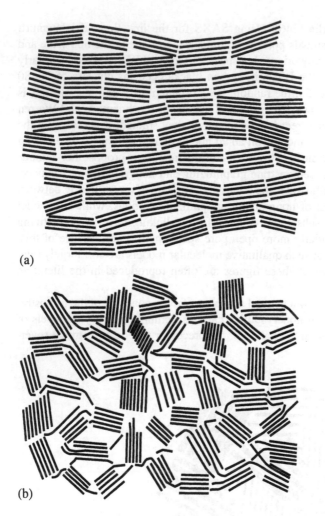

(a)

(b)

FIG. 56 Models of Franklin [238] for (a) graphitizing carbon and (b) nongraphitizing carbon. (From Ref. 238.)

ity, there was still a lack of understanding on how these various elements could be brought together. The models of Krishnamurti and Warren just mentioned were wrong. The model of Franklin helped to decide the level of cross-linking and, therefore, the nature of the porosity and local order in carbons. Beyond that, no models existed at the time. In 1958, Wolff [775] described a molecular structure that consists of collections of regions of stacks of BSUs where the micropores occur within the regions, while the interregion voids define the macropore volume. Using transmission microscopy, Oberlin and her many coworkers were to confirm the validity of this molecular model and expand on its details from the 1970s onwards. This work and the resulting models have been reviewed in Section II.A.6.

2. Single-Pore Models

The most commonly used models for porous carbons are based on simplifications of the complex geometry observed from structural experiments such as TEM. In order to build a simple geometrical model, it is necessary to assume a structure for the pore walls. The pore walls are then arranged to obtain the desired pore shape. All of the models discussed in this section rely on two basic assumptions:

- The pore walls are made up of stacked parallel graphene layers.
- Each graphene layer is made up of a homogeneous arrangement of carbon atoms. This excludes localized defects such as rings of five or seven carbon atoms and chemical heterogeneity.

In the remainder of this section, we first review the different models of pore walls that have been developed under the assumptions mentioned above. Then we show how the pore walls can be arranged in order to build pores of different geometry. Finally, we show how these simple models of pores can be used to characterize real porous carbons by means of a pore size distribution. We also point out the advantages and disadvantages of this approach.

We first consider the structure of a single graphene layer. In graphite, the carbon atoms are disposed in hexagonal rings (see Section II.A.1) with a C-C distance of 0.142 nm and a bond angle of 120°. However, in simple geometrical models, the structure of a graphene layer is strongly related to the way in which the interaction energy between an adsorbed molecule and the surface is calculated. It is possible to define two different types of structure for graphene layers:

1. *Corrugated graphene layer.* The interaction of an adsorbed molecule with each carbon atom is taken into account separately.
2. *Smooth graphene layer.* A uniform distribution of atoms is assumed and the interaction energy depends only on the closest distance between an adsorbed molecule and the graphene layer.

For example, suppose that the interaction energy between a carbon atom and an adsorbed molecule is given by a Lennard-Jones potential:

$$u(r_i) = 4\varepsilon \left[\left(\frac{\sigma}{r_i} \right)^{12} - \left(\frac{\sigma}{r_i} \right)^{6} \right] \qquad (17)$$

where r_i is the distance between the adsorbed molecule and the ith carbon atom, ε is the depth of the energy well, and σ is the value of r_i for which the interaction potential is zero. If pair-wise additivity is assumed, the interaction energy can be calculated by the explicit sum over all of the carbon atoms:

$$U = \sum_i u(r_i) \qquad (18)$$

When this expression is used to obtain the interaction energy, the graphene layer is implicitly modeled as a *corrugated* surface.

If now we replace the set of carbon atoms by a uniform distribution, the summation in Eq. (18) becomes an integral [776]. In this case, the graphene layer is implicitly modeled as a *smooth* surface. The integration limits are set in order to obtain a truncated or an infinite graphene layer. For an infinite graphene layer, the well-known Steele 10-4 potential, Eq. (19), is obtained after integration [776]:

$$U(z) = \frac{2\pi\epsilon\sigma^2}{a_s}\left[\frac{2}{5}\left(\frac{\sigma}{z}\right)^{10} - \left(\frac{\sigma}{z}\right)^{4}\right] \qquad (19)$$

where a_s is the surface area per carbon atom and z is the closest distance between the adsorbed molecule and the graphene layer.

Bojan and Steele have also derived expressions for the interaction energy of a molecule with a truncated (or finite) graphene layer [777,778] and for a molecule with a semi-infinite graphene layer [779,780]. They used the truncated layer model to study the adsorption of krypton on a surface made up of an infinite set of straight square-walled grooves cut into an otherwise flat surface [777,778]. They also studied the adsorption of krypton [779] and xenon [781] on stepped surfaces.

All the structural models for graphene layers that have been mentioned so far are summarized in Table 6. It is possible to combine the different models of graphene layers in order to build pore walls of a desired geometry. Two or more pore walls can then be arranged in space to build a pore with a certain cross-section.

The simplest and most widely used model for a carbon pore is the so-called *slit-pore* model (Fig. 57). The pore is made up of two parallel graphite walls. Each wall consists of a series of parallel and infinite graphene layers. The interaction energy between an adsorbed molecule and one of the walls can be calculated as the sum of the interaction energies between each graphene layer and the adsorbed molecule, given by equation Eq. (19). If the interaction with a single layer is given by the 10-4 potential of Eq. (19), it is possible to sum over the layers. This gives the approximate Steele 10-4-3 potential [776]:

$$U(z) = \frac{2\pi\epsilon\sigma^2\Delta}{a_s}\left[\frac{2}{5}\left(\frac{\sigma}{z}\right)^{10} - \left(\frac{\sigma}{z}\right)^{4} - \frac{\sigma^4}{3\Delta(z + 0.16\Delta)^3}\right] \qquad (20)$$

where Δ is the separation between two graphene layers.

TABLE 6 Structural Models of Graphene Layers

Corrugated graphene layer

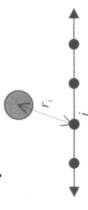

$$U = 4\varepsilon \sum_i \left[\left(\frac{\sigma}{r_i}\right)^{12} - \left(\frac{\sigma}{r_i}\right)^{6} \right]$$

Smooth graphene layer

Infinite
[776]

$$U(z) = \frac{2\pi\varepsilon\sigma^2}{a_s} \left[\frac{2}{5}\left(\frac{\sigma}{z}\right)^{10} - \left(\frac{\sigma}{z}\right)^{4} \right]$$

Truncated
[777,778]

$$U = \frac{4\varepsilon}{a_s} [U_r(z, y_1) - U_r(z, y_2) - U_a(z, y_1) + U_a(z, y_2)]$$

$$U_r(z, y) = \frac{\pi\sigma^{12}}{256} \frac{y}{p} \left[\frac{128}{5z^{10}} + \frac{64}{5z^8 p^2} + \frac{48}{5z^6 p^4} + \frac{8}{5z^4 p^6} + \frac{7}{z^2 p^8} \right]$$

$$U_a(z, y) = \frac{\pi\sigma^6}{8} \frac{y}{p} \left[\frac{2}{z^4} + \frac{1}{z^2 p^2} \right]$$

$$p^2 = z^2 + y^2$$

TABLE 6 Continued

Semi-infinite
[779,780]

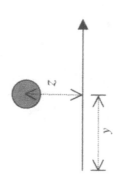

$$U = \frac{4\epsilon}{a_s}[-\sigma^6 I_3(z, y) + \sigma^{12} I_6(z, y)]$$

$$I_3(z, y) = Q_3(z, y) \quad \text{if } y > 0$$

$$I_3(z, y) = \frac{\pi}{2z^4} - Q_3(z, y) \quad \text{if } y < 0$$

$$I_6(z, y) = Q_6(z, y) \quad \text{if } y > 0$$

$$I_6(z, y) = \frac{\pi}{5z^{10}} - Q_6(z, y) \quad \text{if } y < 0$$

$$Q_3(z, y) = \frac{\pi}{8q^4}\left[3 - \left(\frac{z}{q}\right)^2\right]$$

$$Q_6 = \frac{\pi}{1280q^{10}}\left[1008 - 1680\left(\frac{z}{q}\right)^2 + 1080\left(\frac{z}{q}\right)^4 - 315\left(\frac{z}{q}\right)^6 + 35\left(\frac{z}{q}\right)^8\right]$$

$$p^2 = z^2 + y^2$$
$$q^2 = p^2 + yp$$

α_s: Surface area per carbon atom; σ, ϵ, Lennard-Jones parameters. All of the figures are infinitely long in the direction perpendicular to the plane of the page.

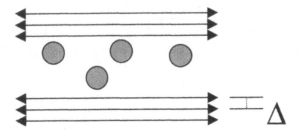

FIG. 57 The slit-pore model.

Although the slit-pore model is the most frequently used simple geometrical model for activated carbons, other pore shapes have also been studied. As mentioned above, the models for graphene layers shown in Table 6 can be easily arranged to obtain the desired shape.

Bojan and Steele [780] performed Monte Carlo simulations to study the sorption of Ar at 90 K in pores with rectangular cross-sections. They modeled each pore wall as a stack of parallel semi-infinite smooth graphene layers (Fig. 58). They fixed one of the dimensions of the rectangle while changing the other one. They showed that the adsorption in these pores could be understood in terms of adsorption on a heterogeneous surface. The corners, which approximate one-dimensional systems, are very strong adsorption sites. On the other hand, the walls are weaker adsorption sites. They pointed out that the stronger adsorption in the corners could complicate the determination of the pore size distribution of a real porous solid by comparing the experimental adsorption isotherm with the ones obtained from simulation in slit-pore models.

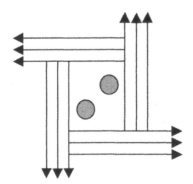

FIG. 58 Pore with rectangular cross-section [780].

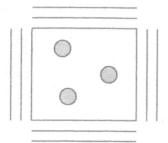

FIG. 59 Pore with rectangular cross-section and truncated walls [782].

Davies and Seaton [782] have also studied the effects of the model pore shape on the characterization of microporous carbons by means of the pore size distribution. Like Bojan and Steele [780], they performed simulations in pores with rectangular cross-sections. However, they modeled the pore walls as stacks of truncated smooth graphene layers (Fig. 59). They performed GCMC simulations to produce methane adsorption isotherms at 308 K in these pore models with different dimensions. Then they used the simulated isotherms to obtain the pore size distribution from the experimental isotherm of a real porous carbon. Finally, they compared this pore size distribution to one obtained using adsorption isotherms generated in a slit-pore model and concluded that the real pore size distribution will be flatter and moved to larger pore sizes than the one determined using the slit-pore model.

Bojan et al. [783] studied the storage of methane in microporous carbons. They performed molecular dynamics simulation of methane in pores with triangular cross-section. They modeled the pore walls as stacks of infinite smooth graphene layers (Fig. 60). The idea behind the model is that partially graphitized

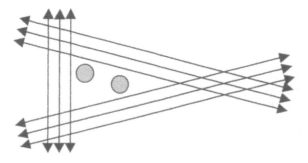

FIG. 60 Pore with triangular cross-section [783].

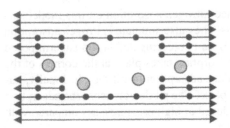

FIG. 61   Slit-pore junction model [784,785].

carbon black often has pore walls that have been deformed by cracking [783]. The resulting pores have a cross-section close to a triangular shape. Therefore, the pores are idealized into isosceles triangles. They performed simulations for a range of apex angles and base lengths. Modeling the pore walls as infinite graphite walls causes an overlapping at the triangle corners and, thus, a somewhat strong interaction.

Sometimes, it is useful to use a combination of corrugated and smooth graphene layers to build pore models for studying specific features of adsorption in microporous carbons. Maddox et al. [784,785] used grand canonical molecular dynamics (GCMD) to study the adsorption of nitrogen at a junction between slit pores of different widths for microporous carbons. In order to build the porous network, they used a hybrid model of corrugated and smooth graphite walls (Fig. 61). The walls consist of an underlying smooth graphite wall, superimposed on which were corrugated graphene layers. In the cavity, only one corrugated graphene layer covered the smooth wall. In the small-pore region, two or three corrugated layers covered the smooth wall. They found severe pore blocking at low pressures for a range of sizes of the small-pore region. The pore blocking was due to the solidification of nitrogen in the small-pore region initiated by the energy barrier at the opening to the larger cavity.

Turner and Quirke [786] also used a hybrid model of corrugated and smooth graphene layers to study the effects of simple pit defects on an otherwise regular graphite surface on nitrogen adsorption. They modeled the surface as a smooth infinite graphite wall with one corrugated graphene layer superimposed on the wall (Fig. 62). In order to create the defects, they removed rings of carbon atoms

FIG. 62   Graphitic surfaces with pit defects [786].

from the corrugated layers. They studied nitrogen adsorption on this model surface for a range of sizes of the pit defects. They also simulated nitrogen adsorption on "deep" and "stepped" defects built from two corrugated layers superimposed on the graphite wall. They found that adsorption takes place at the corners of the pit wall initially, followed by a secondary adsorption to fill the pit. Finally, adsorption on the rest of the surface takes place. The pit defects cause the adsorption to be enhanced at low pressures and reduced at high pressures, as compared with adsorption on a regular graphite surface.

A more realistic way to model the heterogeneity of nongraphitized carbon surfaces was suggested by Bakaev [787]. He developed a model called "rumpled graphite basal plane." The model is obtained by squeezing a graphite basal plane using molecular dynamics simulations. He used the empirical Tersoff potential to describe the interactions between carbon atoms. The degree of squeezing is fixed to reproduce the features of the scattering patterns of nongraphitized carbon black. He performed grand canonical Monte Carlo simulations of nitrogen at 77 K on the resulting surfaces and compared the results with averaged experimental data. The simulated results are in reasonably good agreement with the experimental data, particularly in the BET pressure region.

The studies described above suggest that deviations from the slit-pore geometry and the graphite-like structure of the pore walls significantly change the behavior of confined fluids. However, these conclusions can only be used in a qualitative way when trying to predict the situation in real carbons, since a typical structure of these materials would include a combination of these and other geometries connected in a very complex porous network. In most cases, there is not a straightforward way to combine the type of results described above to accurately predict the adsorption of fluids in real carbons.

3. More Realistic Models

As the structure illustrated in Fig. 21 illustrates, the single-pore models of the previous section omit much detail. Many of the omitted details are known to affect the behavior of fluids, in some cases at such a fundamental level that the single-pore models are simply incapable of reproducing experimentally observed behavior. For example, slit pores impose an artificial maximal limit on the surface area per unit mass and, hence, the adsorbate density predicted by molecular simulation. The only way this upper limit may be removed is to recognize that many microporous carbons are composed of small polyaromatic-like molecules [788]. It has also been shown that the edges of these molecules are far more attractive adsorption sites [789–791], and that diffusion at the micropore level is likely to be most strongly affected by the transition *between* the pores [790]. At longer length scales, it is well known that fluid transport in porous media is profoundly affected by pore system topology (connectivity, deadend pores, pore loops) [792], a factor that clearly is not included in single-pore models. Finally, it has been shown that capillary condensation and pore emptying behavior is fundamentally affected by the presence of deadend pores [793,794] and energetic heterogeneities [795,796]. The inability of single-pore models to capture these and other important phenomena has motivated the development of molecular models of carbons that include, to a greater or lesser extent, the complexity of real carbons.

(a) *The Randomly Etched Graphite Model.* With these considerations, Seaton et al. [797] developed a variation of the slit-pore model to try to account for surface irregularities. The *randomly etched graphite (REG) model*, shown schematically in Fig. 63, is derived by taking a slit-pore representation of a carbon pore and randomly etching away carbon atoms from the innermost surfaces (those planes bordering the pore volume). With periodic boundary conditions in the x and y directions, the model provides an infinite two-dimensional pore topology

(a)

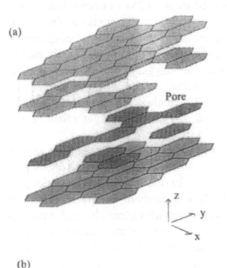

(b)

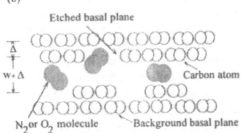

FIG. 63 Schematic diagram of an REG pore: (a) a typical etching pattern; (b) a side view. (From Ref. 797.)

containing channels and pockets near the slit-pore surfaces. The degree of irregularity is controlled by specifying the percent etching (the percentage of surface carbon atoms removed).

As an example of the drastic effects of irregularity, the authors conducted molecular dynamics simulation in REG models with two-site Lennard-Jones parameterized oxygen and nitrogen molecules. The simulations were run for two different nominal pore diameters at varying percentages of etching. Figure 64 shows the observed diffusivities of each molecule type as a function of percent etching. For the 0.209-nm pore the observed diffusivities for both components were negligible at 0% etching. However, at full 100% etching, both diffusivities are essentially equal. At intermediate etchings, the diffusivities of both components decreased; however, the effect on diffusivity for nitrogen was much more pronounced. This occurred because nitrogen is a larger molecule and therefore is more restricted by constrictions than oxygen.

The same type of behavior was observed in the 0.251-nm pore. However, in that case oxygen was much less sensitive to the constrictions. In fact, at 0% etching, the diffusivity of oxygen was greater than at 50% etching. For this particular pore size, oxygen was free to diffuse within the pore region at 0% etching. When etching was present in small amounts, the oxygen molecules were partially obstructed by the channels and pockets in the surface as molecules became "caught" within these crevasses. Beyond 50% etching, the crevasses opened up enough to allow efficient diffusion pathways with much larger effective pore widths.

In both cases, it was shown that the irregularities in the pore surface resulted in enhancement of separation factors due to differences in observed diffusivities between two otherwise similar molecules. In fact, the ratios of diffusivities measured were comparable to experimental values for carbons of similar characteristic pore widths. For this case, the REG model provides some physical interpretation of porous carbon sieving properties and stresses the importance of the nonideal geometries when developing useful models of porous carbon structures.

The REG model accounts for geometrical heterogeneity of the surface and demonstrates the large influence of such heterogeneity on diffusion rates. However, the pores remain slit shaped, and the model takes no account of variations in pore shape, tortuosity, connectivity, or surface chemical heterogeneity. In an attempt to account for these shortcomings, Seaton and coworkers account for topology effects by including their REG model in a network approach (see [798]).

(b) Model of Segarra and Glandt. Segarra and Glandt [799] developed a model to describe the long-range pore topology in a simple fashion. The microstructure was modeled as a collection of cylindrical disks of specific thickness. These disks represented microcrystals of graphite, i.e., a collection of spherical, graphene plates stacked together to form an idealized carbon surface. The thickness of the disks was directly dependent on the number of basal planes in the

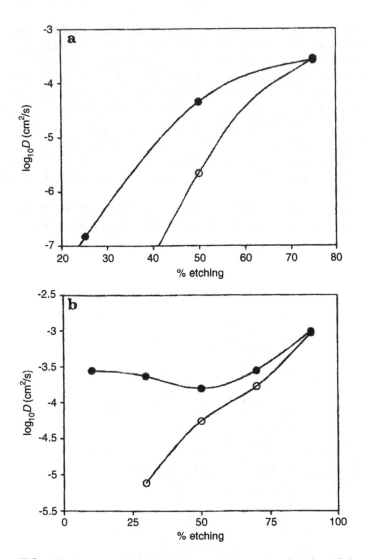

FIG. 64 Single-pore self-diffusion coefficients as a function of the degree of etching for oxygen (solid circles) and nitrogen (open circles). The pore widths are (a) 0.209 nm and (b) 0.251 nm. The curves are to guide the eye. In (a), the simulated diffusion coefficient for nitrogen is barely measurable at 50% etching; the diffusion coefficients of both of the species are 0 below 25% etching. (From Ref. 797.)

stacking; a value of 0.335 nm was used, corresponding to two graphene plates per disk. The disks were also of specific radius, chosen to be characteristic of the porous carbon of interest; values of 0.50, 0.75, and 1.0 nm were used.

The carbon model was generated by randomly orienting a collection of disks into a periodic simulation cell. Canonical Monte Carlo moves were then performed, keeping the disks rigid, until thermodynamic equilibration was achieved [800]. The system was one of hard disks interacting with a pair-wise potential that is infinite for overlapping disks and zero otherwise. The system size was determined by noting when the pair correlation functions of the system properly converged with box length. Figure 65 shows a typical assembly of disks in an equilibrated model.

Once generated, the model can be used to predict adsorption behavior through GCMC simulations. The adsorbate is assumed to interact with the carbon disk

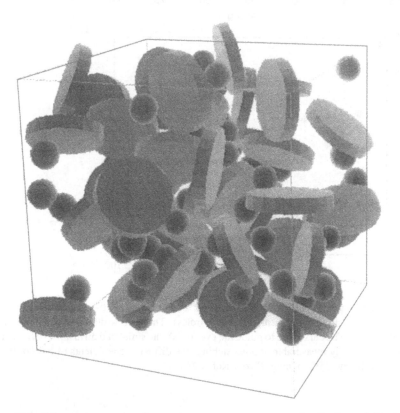

FIG. 65 An assembly of graphitic platelets contains a distribution of dihedral angles between the different carbon surfaces, as well as basal plane edges. (From Ref. 799.)

through Lennard-Jones-type potentials—each carbon atom contributing in a pair-wise additive manner. As an approximation, the carbon disk is replaced by a continuum model in which the carbon atoms are assumed to be distributed within the basal planes of the disk with a uniform, continuous density. The total interaction energy between an adsorbate molecule and each plane then depends on the axial distance the molecule is removed from the surface, and the radial distance it is removed from the center axis of the disk.

Simulated isotherms for methane at 301.4 K gave qualitative agreement with an experimental isotherm obtained from a BPL-activated carbon sample (Fig. 66a). However, the same comparison for ethane resulted in much greater deviation in isotherm shape (Fig. 66b). In both cases, the model isotherms tended to overadsorb at higher pressure, and for ethane this effect was much more pronounced. They calculated the isosteric heat of adsorption at zero coverage and the adsorption second virial coefficient by Monte Carlo integration. For all the platelet radii, these two quantities were lower than the corresponding value for the BPL-activated carbon (obtained experimentally). This indicates that their computer-generated carbons are less heterogeneous than the real material. As a further note, at fixed carbon density, larger platelet sizes resulted in increased uptake for both methane and ethane. We would therefore expect the specific surface area—designated as that portion of the total surface accessible to ad-

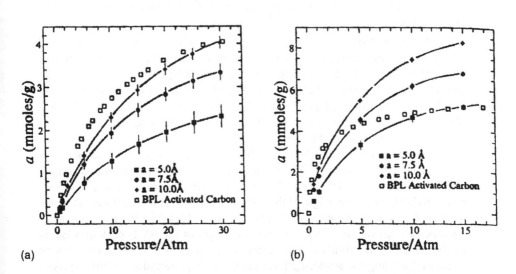

FIG. 66 Comparison of experimental and simulated adsorption isotherms for model carbons of different platelet sizes. (a) Methane at 301.4 K. (b) Ethane at 300.0 K. (From Ref. 799.)

sorbate—to increase with platelet size. The conclusion seems to be that small platelets in the model produce stackings that obscure appreciable portions of the surface.

Segarra and Glandt also proposed a continuum model of surface polarity to account for active, polar sites in carbons. The model assumed that polar sites were distributed uniformly along the edges of the carbon graphene planes. The electric field of a graphene plane was generated by the integral over the plane edge of some fixed linear dipole density—the total electric field being the sum of all graphene planes. The interaction energies for polar fluids were then evaluated by including dipole-dipole interactions directly. Using the simple point charge (SPC) model for water [805], the authors obtained reasonable qualitative agreement with experiment, producing the type V isotherm shape that is characteristic of polar fluids.

The Segarra-Glandt model does introduce a connected pore topology and a range of pore shapes, in contrast to the simple geometry of the single-pore models. However, the geometrical disorder featured in this model is not enough to reproduce the energetic heterogeneity observed in real carbons. It does not include local defects in graphene layers (as rings of 5 or 7 carbon atoms) that are responsible for the curvature of the plates. In addition, all of the platelets have the same size and thickness. This model is a simplified reconstruction based on qualitative observations of experimental structure data. Although it includes some quantitative information, such as the separation between two graphene layers, it does not attempt to be quantitatively consistent with structural information of the material to be modeled.

(c) The Falling Cards Model. As seen by the isotherms from the Segarra and Glandt model, the correlation between adsorption capacity and carbon microstructure is strong. There are many factors involved in this relationship, e.g., carbon density, graphene platelet diameter, platelet alignment, distribution of platelet sizes, curvature of platelets—all of which depend critically on the processing conditions. Dahn et al. [802] proposed a simple model to qualitatively describe some of these relationships.

The *falling cards model* [802] was based on the observation that increased graphene layering during pyrolysis of specific carbon formulations was accompanied by increased pore sizes (Fig. 67). Pore sizes were determined based on SAXS by observing the relative height of the (002) peak (due to constructive interference from scattering between parallel graphene plates). This relative height is calculated as the "X-ray ratio" (R), the absolute height of the (002) peak divided by the baseline height. Similarly, the *radius of gyration* of the pores (R_g) can be extracted by fitting the scattering intensity at small q. Variations in the pyrolysis heating rates and gas flushing rates produced carbon structures exhibiting this relationship for several different precursors.

The authors considered that upon heating, a random arrangement of graphene

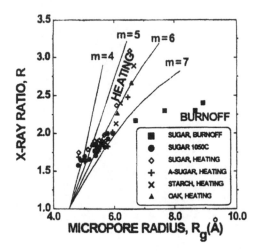

FIG. 67 R vs. R_g for different carbohydrates. The samples are described in the legend. The solid square data points are for samples first heated to 1323 K, followed by burnoff in air at 773 K. The solid lines are the predictions of the falling cards model for pores with walls made of $m = 4$, 5, 6, or 7 graphene sheets. (From Ref. 802.)

plates tends to reorganize by aligning parallel to one another (Fig. 68). Such rearrangements will result in opening of pore regions. The *falling cards* model is a simplification of this process starting with a square lattice of graphene plates (with characteristic length a); see Fig. 69. A certain number of plates are then allowed to "fall" into parallel alignment with adjacent plates. This idealized model is useful in visualizing the pore size/layering relationship and provides simple expressions for calculating pore radius of gyration (R_g) and X-ray ratio (R). Figure 67 shows the predicted relationship for several different three-dimensional pore topologies. The variable m refers to the number of plates around each pore. By varying the number of plates allowed to "fall" and then calculating the appropriate measurements, the authors do a fair job of reproducing the behavior seen experimentally.

(d) Virtual Porous Solids.   The adsorptive properties of a carbon are largely defined by the surface chemistry, geometry, and texture of the porosity. The transport behavior of fluids within microporous carbons, on the other hand, is dictated by characteristics of *both* the individual pores and the pore system topology [792]. Therefore, any serious attempt to study transport within microporous carbons must model volumes large enough to capture the essential character of the pore system topology while still retaining the molecular details at the micropore level. MacElroy et al. [798] have attempted to do this by combining data obtained from a molecular model of a ~4.0 × 1.2 × 1.2 nm volume of microporous carbon with

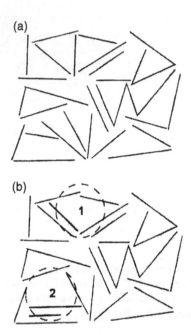

FIG. 68 The falling cards model. (a) A schematic diagram of a microporous carbon. Solid lines represent graphene sheets. (b) Several of the graphene sheets in (a) have "fallen" or rotated to be parallel with neighbors. These sheets are designated by the heavy lines in the dashed circles 1 and 2. (From Ref. 802.)

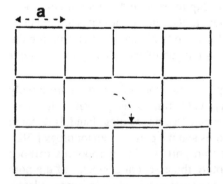

FIG. 69 A two-dimensional square representation of the falling cards model. (From Ref. 802.)

a discrete network model that covers the longer length scales. Once validated, this approach could offer a way of combining pore level and pore system details within a single model suitable for regular engineering design work. An alternative approach that models carbons in molecular detail across all the length scales of interest was developed by Biggs and Agarwal [803–805]; their approach has been used to model volumes of microporous carbon in excess of $100 \times 100 \times 100$ nm^3 in size within 50–250 megabytes of in-core memory. Their approach seeks to capture a microporous carbon in full molecular detail across all length scales of importance, and has been dubbed the *virtual porous carbon (VPC)*. The advantages of the VPC include the fact that the structure can be easily controlled and characterized, and that it is straightforward to perform molecular simulations of fluids within the VPC.

The molecular modeling of large volumes of microporous carbon is made possible by exploiting the idea that these materials can be viewed as hierarchical. Polyaromatic-like molecules combine to form BSUs, which in turn combine to form regions of LMO, which then finally combine to form the mesoporous structure (see Fig. 21). A VPC is built by working with an in-core database of *basic building elements (BBEs)* defined at one of the characteristic levels of this hierarchy. The details stored for each BBE of the database are the coordinates of the constituent atoms relative to a local coordinate frame, and their type. A VPC is built from BBEs of the database by specifying for each BBE of the carbon (1) the index into the BBE database (i.e., the type of the BBE), (2) the translation of the BBE relative to some global coordinate frame, and (3) the matrix that expresses the rotation of the BBE relative to the global coordinate frame. Additional flexibility is allowed in building a VPC by removing atoms from or adding atoms to any of the BBEs in the carbon—this is termed *mutating*.

The length scale at which the BBEs are defined is completely flexible and can vary from a single atom, to a complete BSU, or even an entire region of LMO. However, the definition is strongly dictated (although not strictly constrained) by four considerations:

- The memory required to completely define the BBEs should be significantly greater than that of the $13 \times 4 = 52$ bytes needed to define the type and position of a BBE in the VPC.
- The number of different BBEs within the database must be substantially smaller than the number of BBEs in the VPC.
- The size of the BBE should at least be comparable to the maximum interaction radius of the short-range fluid-solid interactions so as to keep the number of unnecessary fluid-solid interaction tests to a minimum.
- The variety of BBEs should provide sufficient flexibility to allow a VPC to match experimental data such as, for example, the pair distribution function obtained from X-ray scattering and NMR spectra.

A simple comparison shows that designation of polyaromatic-like molecules as BBEs already provides substantial savings in memory while satisfying the other considerations. These molecules, which will here be termed *BSU elements*, typically consist of 10–60 nonhydrogen atoms [806] and are 0.5–1.5 nm in diameter, a size typical of the cutoff radii used for short-range interactions. Comparison of the 200–1000 bytes required to store the position and type of the nonhydrogen solid atoms of any one of these BSU elements with the fixed quantity of 52 bytes under the VPC approach indicates that the memory requirement can be reduced by ~90% when the VPC approach is used. There are also a relatively small number of such BSU elements, but they can be assembled to give a rich combination of BSUs and regions of LMO. The memory requirement may be reduced to only ~3% of the brute force approach if BSUs are used as the BBEs. The trade-off in this case is a loss of some flexibility in defining the molecular structure or, instead, a somewhat larger BBE database.

A number of different BBE databases have been used to build VPCs. The first database used [803–805] consisted of 26 different BBEs based on a five-layer graphitic crystallite with each layer being 4×5 rings in size. The BBEs were differentiated by the removal of those carbon atoms that would have overlapped when adjacent to any of the other 25 different BSUs. More recent implementations of the BBE database have been built on BSU elements. The currently used database is a subset of the polycyclic hydrocarbons enumerated in Clar [807] (the hydrogen atoms are not included and functional groups may be added through mutation). These databases are being used in the assessment of adsorption-based characterization methods on an absolute basis [812]. Work is now underway to generate a more comprehensive BSU element database using graph theoretical methods and genetic algorithms [364,809]. In particular, it is planned to generate databases of all-carbon and other polycyclic structures that (1) are related to elements of the molecular structure of the precursor of the carbon being considered, (2) are constrained to satisfy various spectra of the carbon, and (3) are comparable in size to those reported in HRTEM studies of the carbon. It is anticipated that such databases will provide the basis for the generation of VPCs that are "statistically equivalent" to real carbons.

A number of different methods have also been used to assemble the VPC from the BBEs. The first VPCs [803–805] adopted a simple approach of first generating a completely random solid-void structure of a specified porosity on a cellular tessellation, and then filling the solid cells with one of the 26 BBEs mentioned in the previous paragraph (Fig. 70). The small number of BBEs and the use of this regular cellular structure allowed use of a structure compression algorithm similar to those used in image compression, where patterns in the structure are stored only once; when porosities are significantly less than unity, this reduces in-core memory requirements by an additional order of magnitude compared with the brute force approach. While relatively straightforward to build, the use of an

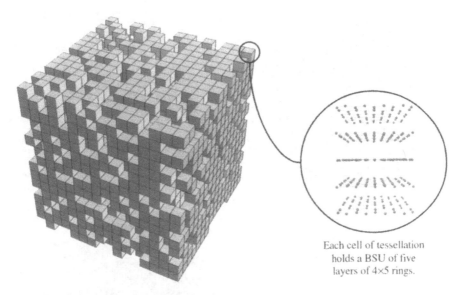

Each cell of tessellation
holds a BSU of five
layers of 4×5 rings.

FIG. 70 The original Virtual Porous Carbon [803–805]. The solid-void structure of a desired porosity is first generated on a regular tesselation, and then the solid cells are filled with BSUs consisting of five layers of 4 × 5 rings. A total of 26 different BSUs were used so as to avoid overlap of carbon atoms in adjacent BSUs.

underlying cellular structure provides significant flexibility with regard to the structures that may be built. For example, as shown in Fig. 71a and 71b, the average pore size at a given porosity and the pore system isotropy can be varied by working with cell clumps, e.g., 2 × 2 × 2 for an isotropic structure with minimal pore size twice that of the simplest structure, and 8 × 8 × 2 for a structure that would tend to have slit-like pores in the *x-y* plane.

One of the main drawbacks of the original VPC model was its reliance on regular BBEs, which impose a regularity on the cross-section of the pores over the length of a BBE. This has been alleviated in more recent work by adopting BSU elements as the BBEs and allowing these to take up arbitrary positions and angles of orientation within the cells as illustrated in Fig. 71c. However, this does not overcome the second main drawback arising from the use of the cellular structure, which is the inability to *precisely* control the molecular structure in general and the pore size in particular. These limitations have been removed in the most recent version of the VPC by removing the reliance on a cellular structure when building the carbon [810] and allowing the BBEs to take up any loca-

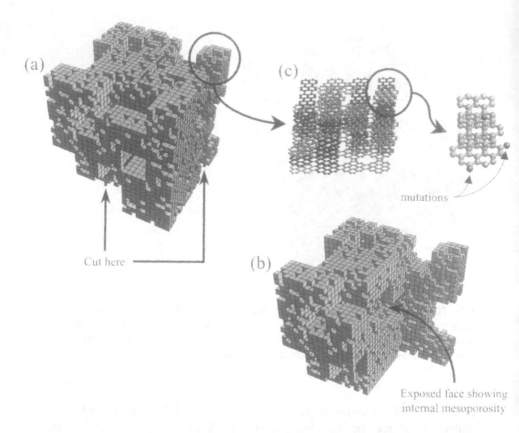

FIG. 71 A more recent version of the Virtual Porous Carbon, which is the basis of current work. (a) The full volume. (b) A view of the internal structure obtained by sectioning the original structure as indicated in (a). The basic approach to the construction of the VPC is similar to that used in the earlier models, but greater flexibility exists in generating the solid-void structure and in the BBEs used. (c) Note the different types of BBEs and the mutations to the BBEs. In this case, the BBEs are drawn from Clar [807], and the solid-void structure is bimodal. The bimodal structure is generated by first randomly designating $N_x \times N_y \times N_z$ blocks of unit cells either void or solid with a given probability (0.3 with $N_x = N_y = N_z = 5$ in this case), and then repeating the procedure for $n_x \times n_y \times n_z$ blocks of unit cells within the larger blocks that were designated solid (probability of 0.3 with $n_x = n_y = n_z = 1$ in this case). This method allows the generation of structures that contain both micro- and mesoporosity for example. Note that these structures are reminiscent of the images presented in Fig. 21.

tion within the simulation volume consistent with reality. As already indicated, work is currently underway to link this flexibility with the capability to build carbons via genetic algorithms that match experimental data at various levels.

As will be demonstrated below, one of the most significant advantages of the VPC approach is the fact that the carbons can be easily characterized in a number of ways. Stereological methods on the full three-dimensional solid [811] or two-dimensional slices [812] can be easily undertaken to yield measures of total porosity, surface area, and pore size. SAXS spectra and associated quantities can also be estimated [813,814], while pore size distributions may be generated using simulated size exclusion experiments. It is also possible to "fill" the pore space defined by the molecular structure of the solid with a Voronoi tessellation [815,816] (Fig. 72) and then use this to determine the accessible and inaccessible pores, and the associated surface areas and volumes. The pore space filled by the tessellation is defined by a critical potential energy value commensurate with the fluid-solid interactions and the simulation temperature. This ensures that the accessible and inaccessible porosities are correctly delimited, and that the volume is representative of that which would be seen by the fluid at that temperature. As the presence of fluid molecules within closed porosity is both physically unreasonable and likely to severely contaminate any results obtained, the identification of the closed porosity at a given temperature is particularly critical.

The first application of the VPC approach was in the study of self-diffusion of atomic and diatomic fluids within microporous carbons as a function of temperature and porosity/pore size [803–805]. The VPCs were all built using the 26-BSU database mentioned above on a regular cellular structure. Velocity autocorrelation functions (VACFs), mean square displacements (MSDs), and diffusion coefficients were generated for the fluids as a function of temperature and porosity. It was found that the VACF was a sensitive probe of the pore surface texture and pore system topology at short and long length scales, respectively. Analysis of the MSD revealed that the time taken to transition from a subdiffusive process, where the path scribed by the fluid molecules is fractal in nature, to Fickian diffusion decreased as the porosity of the structure increased. Even though the cellular solid/void structure was always generated with a porosity above the percolation threshold, the actual percolation threshold of the structure changed with the temperature, with structures at the lower temperatures clearly being effectively below the percolation threshold (i.e., the pore system to all intents and purposes was not percolating, even though the underlying tessellation was). More recently, the ability to generate pair distribution functions for fluid-solid atom combinations either on a domain or local basis has also allowed the study of site/region visitation dynamics and the identification of active adsorption sites.

The VPC model has the merit that it includes pore connectivity and, to some extent, tortuosity. It is based on the BSU/LMO picture of carbon structure due to Oberlin, so that the realism of the final structure will depend on appropriate choices for these building blocks.

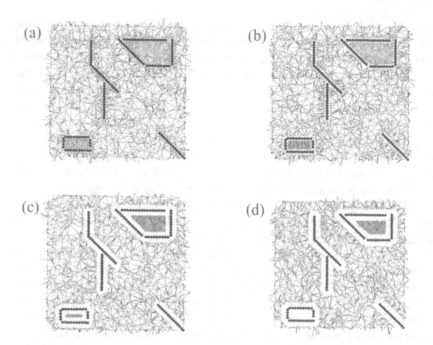

FIG. 72 The pore space at a given temperature can be filled with a Voronoi tessellation. For simplicity, the example shown here is for an imaginary structure in two dimensions where the solid atoms are shown as large black dots. The temperature of the eventual molecular simulation is highest at (a) and lowest at (d). For clarity, the porosity within the cages at the top right and lower left corners of the volume that can be probed by the fluid is shaded. Note how these volumes are accessible at higher temperatures but become disconnected at lower temperatures. The Voronoi tessellation is used to identify the accessible and inaccessible porosity so as to avoid simulation of fluid within the latter in any subsequent MC/MD simulation. The Voronoi tessellation can also be used to gather data on the pore system such as the volumes and surface areas associated with the accessible, inaccessible, and backbone porosity; percolation properties of the pore structure; and fractal characteristics of the pore surfaces. These quantities can be compared with those determined from, for example, adsorption-based characterization methods operating on adsorption isotherms generated from grand-canonical Monte Carlo simulations.

(e) The Chemically Constrained Model. Acharya et al. [817] have presented a model for nanoporous carbons based on chemical constraints. This approach assumes that nanoporous carbons are made up of carbon and hydrogen and that the carbon hybridization is all sp^2, as in graphite. They developed a program in which the user specifies the H/C ratio (number of hydrogen atoms per 100 carbon atoms) and the number of atoms in the structure. The program generates struc-

tures that satisfy these inputs and the constraint of sp^2 hybridization. In order to accomplish this, the program first generates a series of fragments of graphene layers. The size of these fragments is chosen so that the H/C ratio is close to the one specified by the user. A different subroutine (SIGNATURE) identifies the unsaturated carbon atoms and creates bonds between any two atoms in different fragments based on two criteria: (1) connection of nearest unsaturated carbon atoms and (2) connection of two unsaturated carbon atoms irrespective of the distance between the atoms. SIGNATURE [818] is a program that builds molecular structures that meet certain constraints by connecting instances of predefined molecular fragments and interfragment bonds. Those structures generated not possessing a density close to the experimental value are discarded, and the remainder are subject to further refinement to achieve structures with the desired atomic H/C ratio. The further refinement involves removing hydrogens and creating five-, six-, or seven-membered rings by establishing bonds between the resultant unsaturated carbon atoms. This allows for the curvature observed in TEM pictures of many nanoporous carbons. The structures are subject to energy minimization prior to final analysis. Two different strategies are adopted when implementing this structure refinement. The first, termed *one-step refinement*, carries out energy minimization after the final desired structure has been generated. The second *successive refinement* strategy achieves the final energy-minimized structure by carrying out successive hydrogen-reducing/energy-minimizing processes after a small adjustment in the H/C ratio.

Acharya et al. have used the method to build model structures of poly(furfuryl alcohol) at various stages of carbonization. Simple graphene sheets of m rows each of n hexagons were used with a hydrogen atom removed to reveal an interfragment bonding site. Model molecular structures of carbons prepared by carbonizing poly(furfuryl alcohol) at a particular temperature between 673 K and 1073 K were generated by first selecting those graphene fragments which possessed an atomic H/C ratio close to that measured for the carbon. SIGNATURE was then used to generate structures of 500, 1500, or 2500 atoms with H/C ratio and percentage aromaticity equal to those specified as inputs to the program. The resulting structures were refined using the one-step refinement and the successive refinement strategies described above. The densities achieved by the two methods were essentially the same. However, the successive refinement method allowed the building of structures of low H/C ratios, whereas the one-step method did not.

Example structures at four different H/C ratios are shown in Fig. 73. As can be seen, the presence of the five- and seven-membered rings in the otherwise benzoid structures allows the generation of both convex and concave surfaces, which have been observed experimentally for immature carbons [819]. As actually observed for these carbons, the density of the 500- and 1500-atom structures usually increases as the H/C ratio decreases; this trend does not hold for the

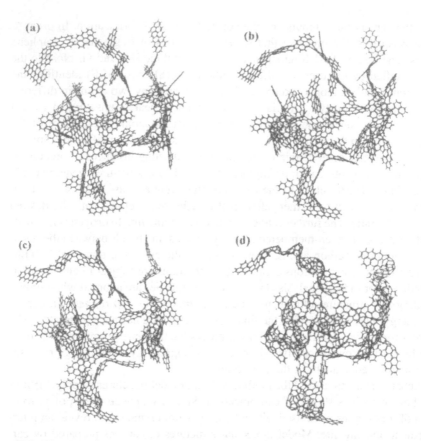

FIG. 73 Structural evolution of the carbon nanostructures as a function of the decreasing H/C ratio (and consequently, increasing temperature). The sequence of images is (a) to (d). (From Ref. 817.)

2500-atom structures, however, which did not perform well, most likely due to computational limitations. Acharya et al. also found that structures with low H/C ratios do not relax even after extensive energy minimization, most likely due to a combination of excessive cross-linking and, once again, computational limitations. The C-C pair distribution functions for model structures at two different "preparation temperatures" are shown in Fig. 74 in addition to that of graphene. This diagram shows that the extent of order decreases as the degree of carbonization increases, which runs counter to that observed experimentally [820].

The way in which the models are generated allows for the formation of local defects that are responsible for the curvature observed in TEM pictures of several

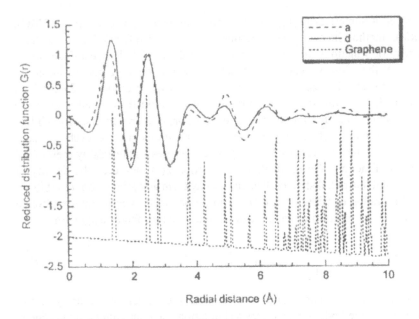

FIG. 74 Radial distribution functions for two of the evolving structures shown in Fig. 73. The simulated distribution for a single graphene sheet is also shown for comparison. (From Ref. 817.)

nanoporous carbons. However, in this approach it is necessary to assume the structure of the fragments that are part of the resulting structures. The authors assumed these fragments to be fractions of graphene layers. For carbons formed at low heat treatment temperature (HTT), the H/C ratio is quite large. This H/C ratio is the input to the program and when it is large it results in a relatively low number of carbon-carbon bonds formed between different fragments. Thus, the resulting structure is made up of small, perfect, graphene-like segments with little cross-linking; almost all segments are made up of six-membered rings. In an experimental study performed by the same group [820], neutron diffraction along with TEM micrographs from a previous study [819] evidence the presence of very distorted plates in nanoporous carbons manufactured at low HTT. This puts in jeopardy the use of small graphene-like segments used in the chemically constrained approach.

(f) The RMC Model. The ultimate goal of a structural model is (1) to reproduce the experimental microstructure of the porous carbon and (2) to capture as much of the long-range *pore topology* as possible. To achieve such goals, the model protocol should incorporate as much of the experimental structural data as possible. One class of methods that have had some success along these lines

is based on a numerical fitting method known as *reverse Monte Carlo* (RMC). The RMC method incorporates experimental structural information in the form of correlation functions into the construction of the model. The result is a structural model that closely matches user-defined experimental characterization.

McGreevy and Putszai [821] introduced the RMC method [822,823] as a means for determining structural coordinates for a fluid from input data in the form of atomic radial distribution functions, $g(r)$, or structure factors, $S(q)$. Unlike crystalline systems such as zeolites, the atomic positions in an amorphous material are not regular. Instead one typically describes these materials by their radial distribution function. We recall that the function $g(r)$ measures the density of atoms located in a radial shell of distance r away from any given atom of arbitrary origin. The radial distribution function can be obtained experimentally through neutron or X-ray scattering techniques via the structure factor $S(q)$ with

$$4\pi r^2 \rho[g(r) - 1] = \frac{2}{\pi} \int_0^\infty qr[S(q) - 1]\sin qr dq \qquad (21)$$

where ρ is the atomic density of the material and q is the elastic scattering vector. With the RMC method, one generates a simulation cell containing atomic centers and then, via Monte Carlo-type moves, systematically rearranges the position of atoms until the simulated $g(r)$ or $S(q)$ matches the experiment. The result is a physical model of the structure in question that can be used in further characterization or as a model matrix for adsorption studies. The RMC method has been useful in the study of amorphous hydrogenated carbons [824,825] and sundry other disordered systems, such as molten germanium, amorphous magnetic metals, disordered crystalline Pb, and doped $AgPO_3$ glasses [826]. It has also been applied to study the structure of liquids, including water [827]. Recently, a molecular dynamics analogue of the RMC method has been reported by Tóth and Baranyai [828].

In RMC the random trial moves can be of single atoms, molecules, or groups of atoms, e.g., graphite microcrystals. The choice is based on experimental information about the structure of the material. In contrast to the usual Monte Carlo procedure, acceptance or rejection of a trial move is not based on the laws of statistical mechanics but on whether the fit to an experimental $g(r)$ of $S(q)$ curve is improved or not. Since many different atomic configurations can lead to the same $g(r)$ or $S(q)$, care is needed to avoid structures that, while matching the experimental data, are unphysical. This nonuniqueness problem can often be overcome by introducing constraints in the fitting, or by fitting to more than one kind of data. In the case of activated carbons, appropriate constraints might be restrictions on the bond angles between carbon atoms, number of nearest neighbors, or arrangement of the atoms in the form of graphite microcrystals.

O'Malley et al. [829] used RMC to analyze the structure of two glassy carbons formed at heat treatment temperatures of 1273 K (sample v10) and 2773 K (sample v25). The trial moves were for individual carbon atoms. They minimized the difference between the experimental and the simulated structure factor $S(q)$ and applied a coordination constraint that assumes a predominance of sp^2 bonding. The quantity to be minimized is then

$$\chi^2 = \sum_{i=1}^{n_{exp}} \frac{[S_{exp}(q_i) - S_{sim}(q_i)]^2}{\sigma_{exp}(q_i)^2} + \frac{(f_{req} - f_{sim})^2}{w^2} \tag{22}$$

where $S_{sim}(q_i)$ is the simulated $S(q)$ and $S_{exp}(q_i)$ is the experimental $S(q)$ evaluated at q_1; s_{exp} is the estimated experimental error, f_{req} is the required fraction of sp^2-coordinated atoms, f_{sim} is the fraction of sp^2-coordinated atoms in the simulated structure, and w is a weighting factor. The simulation starts with an arbitrary random configuration. The error χ^2 is calculated for this configuration. An atom is then randomly selected and moved a random distance, and χ^2 is calculated for the new configuration. The move is accepted with a probabilty:

$$P_{acc} = min\left[1, \exp\left(\frac{-(\chi^2_{new} - \chi^2_{old})}{2}\right)\right] \tag{23}$$

where χ^2_{new} and χ^2_{old} are the errors evaluated at the new and old configurations, respectively.

Their work is concerned with the structure of the graphitic-like ribbons rather than the overall arrangement of the ribbons in space. Thus, the shape and the connectivity of the pores are not included in this model. For this reason, the starting configuration is that of a graphite crystal, which is expected to be similar to the structure of the graphitic-like ribbons of the glassy carbon.

The simulated and target $S(q)$ are shown in Fig. 75. The agreement is quite good with significant deviations only at low q. The authors attribute the deviations to the enhancement of the uncertainty at low q. A structural representation of one of the graphitic-like basal planes is shown in Fig. 76, where the buckled nature of the basal planes is revealed. The bond angle distributions for both samples is shown in Fig. 77. It is centered at approximately 117° for both samples. The peak height is larger in the v25 sample, which was treated at a higher temperature. There is a small peak at approximately 60°, which reveals the presence of three-fold rings in the simulated structure. This is confirmed by the structural representation (Fig. 76a). These authors also exemplified how an extensive structural analysis can be performed on the resulting model to study, for example, the effect of the heat treatment temperature on the topology of the carbons as measured by ring statistics, correlation functions, and so forth.

Application of the RMC method to model the morphology and topology of porous carbons was proposed by Thomson and Gubbins [830] and is another

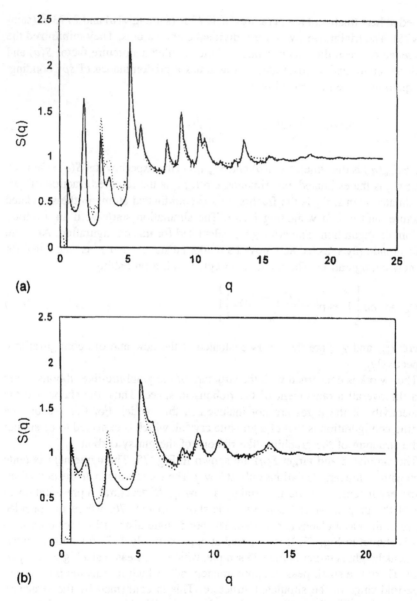

FIG. 75 Structure factor of glassy carbon samples (a) v25 and (b) v10. Experimental data (solid line) and simulations (dashed line). The horizontal axis is in units of $Å^{-1}$. (From Ref. 829.)

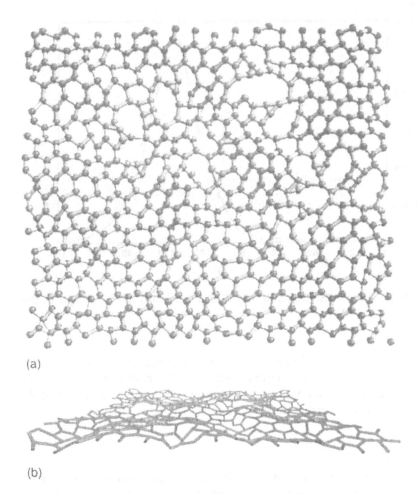

(a)

(b)

FIG. 76 (a) A basal plane of the model corresponding to the v10 sample viewed down the direction of the stacking. (b) A side view on the same basal plane as in (a). (From Ref. 829.)

example of this reconstructive method for modeling microstructure. In the RMC model, experimental SAXS and SANS data are used to generate a "target" pair distribution function $g_t(r)$. Carbon is introduced into a cubic, periodic simulation cell as rigid, sp^2-bonded graphene platelets. These platelets are derived by random construction, whereby sixfold aromatic rings are added to the exposed edges of a growing platelet. The number of aromatic rings for each platelet is chosen from a Gaussian distribution, and thus both the size and shapes of the graphene platelets are varying. After randomly positioning and orienting the platelets in the simula-

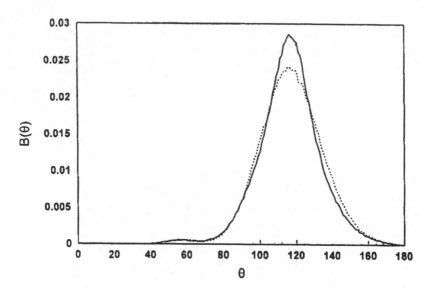

FIG. 77 Bond angle distribution for v25 sample (solid line) and v10 sample (dashed line). (From Ref. 829.)

tion cell, a series of RMC moves similar to traditional Monte Carlo steps are attempted. An RMC move is accepted whenever the *global error* in $g_s(r)$, the simulated pair distribution function, is improved. The authors define the global error by the square deviation function:

$$\chi^2 = \sum_{i=1}^{n} [g_s(r_i) - g_t(r_i)]^2 \tag{24}$$

The index $(i = 1, n)$ denotes the discretization of the radial distance r. If χ^2, measured after the attempted move, is smaller than before the move (i.e., the global error in $g_s(r)$ is reduced upon making the move), the RMC move is accepted.

Thomson and Gubbins included three varieties of moves. The first, *translation/reorientation*, involved randomly choosing a platelet and either displacing its center of mass, or rotating in some arbitrary direction about its center of mass. This type of move is analogous to traditional Monte Carlo moves in fluid simulations. The second type, *ring creation/annihilation*, involved randomly choosing an edge aromatic ring from a randomly determined platelet and either adding a new ring (creation) or removing that ring (annihilation). The purpose of this second RMC move type was to allow the model to redistribute aromatic rings among the platelets in the simulation cell. This, in effect, provided a mechanism for altering the shape and size of platelets.

To account for deviations in carbon density that may result from the ring modification steps, a third type of move was included termed *plate creation/ annihilation*. In generating a porous carbon model structure using the RMC method, a target carbon density is chosen. If, during the course of the RMC run, the simulated carbon density deviation is too high (low) due to excessive ring creation (annihilation) steps, then plate annihilation (creation) steps are attempted. During a plate annihilation move, an entire platelet, chosen at random, is removed from the simulation cell based on the criterion of reduced global error. Similarly, during a plate creation move an existing platelet is duplicated and randomly inserted into the simulation structure—again subject to the global error reduction criterion. These move combinations allow for fairly large errors in the initial structure. For example, if an actual carbon composed of relatively few yet large platelets was erroneously represented by an initial simulated structure containing many small platelets, the combination of ring creation and plate annihilation steps throughout the course of the RMC run would correct the structure and maintain the target carbon density.

Figure 78 shows the microstructure model produced by RMC for an experimental, KOH-activated mesocarbon microbead. For this particular carbon, a simulation box measuring 10.0 nm on a side was used with a target density of 0.97 g/cm^3. Figure 79 shows the initial and converged $g(r)$ compared to the experimental target (solid line). We note that beyond a radial distance of ~0.5 nm, near perfect matching is achieved in the pair distribution function. However, deviations occur at values of $r < 0.5$ nm. The authors [830] offered two possible reasons for this discrepancy: (1) inaccuracies in the large-angle $S(q)$ data, and (2) overly rigid constraints on the RMC platelet shape. It is expected that a better fit would have been achieved if defects in the form of nonaromatic rings and heteroatoms were included in the physical structure of the graphene platelets. Nevertheless, the RMC method offers a fairly good structural model of the micropore environment for use in characterization and adsorption analysis. Figure 80 shows the pore size distribution obtained from the structural model. From this plot, pore size peaks at ~0.48 nm, ~0.70 nm, and ~0.96 nm are evident. The first peak can be identified as the intracrystallite platelet spacing (in which no adsorption occurs), while the main peak at ~0.96 nm represents the most predominant pore size contribution—corresponding to an overlapping platelet configuration where adsorption occurs between two platelets separated by a third platelet.

While the RMC model of Thomson and Gubbins is reasonable for many graphitizable carbons, the use of graphene microcrystals as the basic units fails to allow for ring defects that are important in most activated carbons. Pikunic et al. [835] presented an improved molecular model also based on the original RMC technique developed by McGreevy and Pusztai [821]. In this approach, they use the experimental $S(q)$ as the target function and the following constraints: (1) the bond angle distribution is centered at 120° and (2) the distribution of the number

FIG. 78 Structural representation of the converged RMC [a-MCMB] carbon microbead structure. The spheres represent carbon atoms, which are shown at a scale much less than their van der Waals radii for reasons of clarity. (From Ref. 830.)

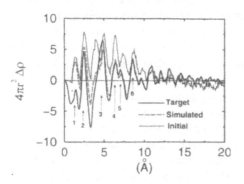

FIG. 79 C-C radial distribution functions for the [a-MCMB] carbon microbead structure. The experimental RDF (solid line); the simulated, converged RDF (long-dashed line); and the initial simulated RDF (dotted line) are shown. (From Ref. 830.)

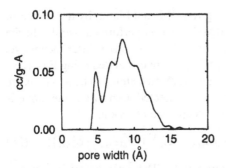

FIG. 80 Pore size distribution of the converged RMC [a-MCMB] structure. Units are in cubic centimeters of micropore volume per gram of carbon per angstrom. (From Ref. 830.)

of neighbors is centered at 3. Any two atoms are considered neighbors if the distance between their centers is within a minimal and a maximal bond length. These limits are arbitrarily set. This approach is different from the one presented by Thomson and Gubbins [830] in the sense that the interatomic distances and the bond angles are not restricted to specific values. The advantage of this method is that it allows modeling the local defects observed in real porous materials, such as curvature produced by nonaromatic rings and roughness. However, the definition of a graphene platelet is not trivial in this case because the curved plates are cross-linked, and therefore each atom must be moved independently.

One can think of this approach as a simultaneous minimization of three quantities.

$$\chi^2 = \sum_{i=1}^{n_{exp}} [S_{sim}(q_i) - S_{exp}(q_i)]^2 \tag{25}$$

$$\psi^2 = \sum_{i=1}^{n_\theta} \left[\theta_i - \frac{2\pi}{3} \right]^2 \tag{26}$$

$$\delta^2 = 1 - \frac{N_3}{N} \tag{27}$$

where $S_{sim}(q_i)$ is the simulated $S(q)$ and $S_{exp}(q_i)$ is the experimental $S(q)$ evaluated at q_i. The angles θ_i are the different bond angles in radians, n_θ is the total number of bond angles, N_3 is the number of atoms with three neighbors, and N is the total number of atoms in the simulation box.

The simulation is begun by generating an initial configuration with a random structure in a cubic simulation box, with periodic boundary conditions and a target density ρ to ensure that the resulting model is not biased by the choice of initial configuration. Only one type of RMC move is performed, similar to the Metropolis algorithm for MC simulations in the NVT ensemble. After each move, the three quantities to be minimized—χ^2, ψ^2, δ^2—are calculated using Eqs. (25), (26), and (27), respectively. If the move causes any two atoms to be at a distance less than the minimal bond length, the move is rejected. Otherwise, the move is accepted/rejected based on the following acceptance criterion:

$$P_{acc} = \min[1, \exp\{-[P_\chi(\chi^2_{new} - \chi^2_{old}) + P_\psi(\delta^2_{new} - \psi^2_{old}) + P_\delta(\delta^2_{new} - \delta^2_{old})]\}] \quad (28)$$

where P_χ, P_ψ, and P_δ are weighting parameters. The ratio between any two weighting parameters determines the relative importance of the structural properties. When the three weighting parameters are set to infinity and the maximal and minimal bond lengths are set to 0.142 nm, this method is equivalent to the basic carbon unit approach.

In this work, Pikunic et al. [831] modeled the glassy carbon formed at a heat treatment temperature of 2773 K presented in the work by O'Malley et al. [829]. The RMC simulations were performed in a box of 3 nm with a density of 1.57 g/ml. The simulated and the experimental $S(q)$ values are presented in Fig. 81. The structural representation is shown in Fig. 82. The simulated and the target $S(q)$ are in very good agreement, but there is some disagreement at small q. The structural representation reveals that the plates in this model contain a significant

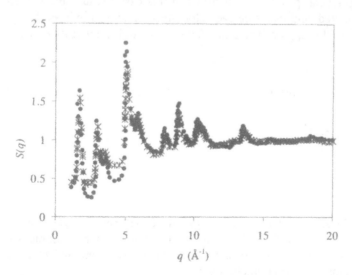

FIG. 81 Structure factor of glassy carbon. Experimental (dots) and RMC model (stars). (From Ref. 831.)

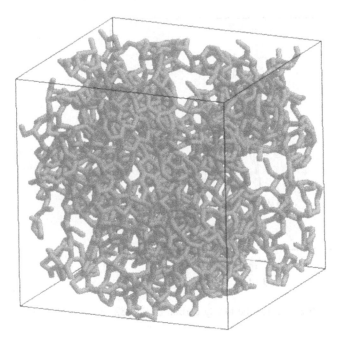

FIG. 82 Structural representation of glassy carbon. (From Ref. 831.)

number of rings of more than and less than six carbon atoms. These types of defects create the curvature that we observe in all the carbon plates. The bond angle distribution is presented in Fig. 83. The distribution is centered at 120°. O'Malley and coworkers found a small peak in the bond angle distribution at 60° [829], which revealed the presence of threefold rings. We observe that this peak does not appear when we include the angle constraint. The addition of this angle constraint also causes better agreement between the simulated and the experimental $S(q)$ in the q region corresponding to the first and second coordination shell.

The benefit of the RMC model is its incorporation of experimental structural data into the model protocol. This allows for versatility in the types of carbon studied and offers flexibility in the type of experimental data used. The correlation functions used as input to the RMC method are not restricted to pair distribution functions. In fact, any correlation function that can be simulated from a structural representation can be included in the data set. For example, fringe-lattice TEM micrograph data in the form of void-void or matrix-matrix correlation functions, or as chord length distribution functions, have been suggested. These could be especially important in describing mesopores, which is difficult to achieve in

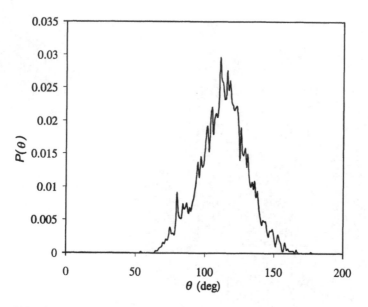

FIG. 83   Bond angle distribution for the glassy carbon of Fig. 82. (From Ref. 831.)

RMC methods incorporating atomic pair correlation data alone; pair distribution functions are relatively insensitive to large open-pore regions.

The RMC method is particularly sensitive to the choice of target density. Porous carbons are heterogeneous in nature and contain regions of vastly differing local density. The macroscopic dimensions of these density variations compared with the small simulation cell sizes necessary for the RMC model pose an interesting problem for the structural modeler. One has to have accurate density measurements in the regions where SAXS and SANS data are obtained for the model to make physical sense. This is because the RMC method is underconstrained. One is actually searching for the "best fit" structure based on specific constraints obtained from external sources, i.e., presumed constraints on the shape and local bonding of the graphene platelets, and measurements of local carbon density. It is possible to get converged structures for any reasonable value of the target density for any given target $g_i(r)$. Thus, a great deal of care must be taken in defining the system of interest and obtaining the parameters for input into the RMC method.

4. Carbon Aerogels

Aerogels, formed through sol-gel polymerization followed by drying under supercritical conditions (Section II.A.8), are composed of microscopic particles that are interlinked in a random network [834,835]. In carbon aerogels these particles are usually roughly spherical and from 5 to 10 nm in diameter (see Figs. 26 and 27). Such materials have often been modeled as a rigid matrix of solid spheres ("cannonball" solid), usually randomly arranged. Some of the possible variations on such a model are shown in Fig. 84 [836]. Models (a) and (b) differ at higher porosities in that some unconnected spheres can occur in (a), as shown by the sphere labeled 1, but not in (b). While model (b) is probably the most realistic representation of this kind of disordered media, models (c) and (d) are often easier to study theoretically. MacElroy and coworkers [836–838] developed models of this type for silica xerogels with solid spheres having a diameter of 2.4–2.8 nm. Silica spheres were generated using Monte Carlo simulations and a quench procedure, coupled with intermolecular interactions that included both pair and three-body potentials with parameters proposed by Feuston and Garofalini [839]. Connected matrix structures, such as (b) and (d) in Fig. 84, were formed by starting with a randomly close-packed assembly of hard spheres, with porosity of 0.367; such an assembly is assumed to be fully connected. Spheres were then randomly removed to get the desired porosity, with subsequent scanning to check that remaining spheres are all connected. Random structures such as (a) and (c) in Fig. 84 were generated by carrying out a Monte Carlo run to generate a random configuration for hard spheres. The desired porosity is obtained by setting an appropriate lattice spacing in the initial fcc lattice. MacElroy and coworkers used these structures to study adsorption and diffusion of methane in a silica xerogel. They found good agreement with experimental data using the more realistic connected model, (d), of Fig. 84. Connectivity of the spheres forming the matrix was found to have a pronounced effect on the diffusion rates.

In the case of carbon aerogels, two pore-sized regimes are present: (1) mesopores (typically 3–15 nm in width) that occur between the carbon particles, and (2) micropores (usually 0.7–1.5 nm in width) within the individual carbon particles (see Fig. 27). Gavalda et al. [840,841] have presented a model for carbon aerogels in which the microporous carbon particles are spherical and connected in an overlapping network (similar to the model of Fig. 84b). TEM images of the aerogel studied by Gavalda et al. showed that the carbon spheres had an average diameter of 6 nm and can overlap by up to 40% of the sphere diameter. Preparation of the model followed a similar Monte Carlo procedure to that used by Park and MacElroy [836] and requires two steps. First, a random close-packed structure of overlapping spheres in a cubic box is generated; spheres are then randomly removed to attain the targeted porosity (0.55 in the aerogel studied).

The random close-packed overlapping system is generated using a procedure similar to that described by Mason [842]. Initially a number of points are generated at random inside the simulation box. These points are then considered to be small spheres of a chosen radius. If two spheres overlap by more than the permitted amount (40%), they are moved apart along the line of centers until the targeted overlapping is reached. The spheres are then increased in size by a chosen incre-

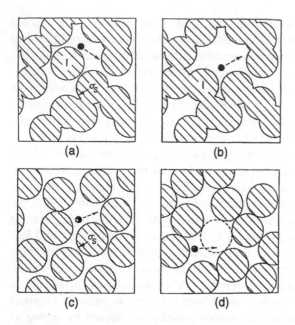

FIG. 84 Some models for an aerogel: (a) random overlapping spheres; (b) randomly connected overlapping spheres; (c) random nonoverlapping spheres; (d) randomly connected nonoverlapping spheres. Shaded regions are the solid matrix; small solid circles represent a fluid molecule moving through the void space. (From Ref. 836.)

ment and the process repeated. The close-packed structure obtained has a porosity of 0.2 (Fig. 85a). In the second step, the removal of spheres is restricted by connectivity, i.e., the remaining spheres must all be connected in a single cluster. A cluster-labeling algorithm similar to that of Hoshen and Kopelman [843] is used for this purpose. The number of spheres to be removed is determined by the porosity desired. The computation of porosity involves a Monte Carlo estimation of the volume of void space accessible to a test particle modeled on nitrogen, divided by the total volume of the system. The final structure is shown in Fig. 85b. The structural characteristics (mesopore surface area, mesopore volume, pore size distribution) of the model material can be determined by Monte Carlo methods [730] and were close to the values estimated from experiment. It was found that the structural characteristics were somewhat dependent on the simulation box size. For the 6-nm carbon particles studied by Gavalda et al., box lengths up to 72 nm were studied. A minimal box length for a realistic representation of the aerogel was found to be 36 nm. The surface area and pore size distributions

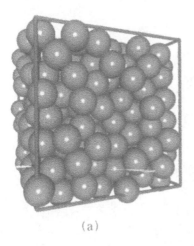
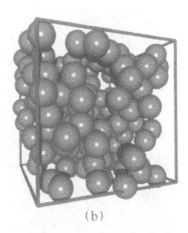

(a)   (b)

FIG. 85 Carbon aerogel model structure: (a) close-packed configuration; (b) final configuration. (From Ref. 841.)

were particularly sensitive to box size, and results for these properties varied somewhat as the box length was decreased from 72 to 36 nm.

The mesoporous model produced by the above procedure does not take into account the micropore structure of the carbon particles. To generate this structure the RMC technique [830] was used (see Section IV.B.3 above). Starting from a random arrangement of microcrystals composed of carbon basal plane fragments, Monte Carlo-type moves eventually result in a structure that matches experimental X-ray scattering data for the aerogel. Three different types of Monte Carlo moves were performed: (1) plate translation/reorientation, (2) plate creation/annihilation, and (3) ring creation/annihilation. The final micropore structure obtained by this RMC procedure is shown in Fig. 86.

Gavalda et al. [841] have used this model to study nitrogen adsorption in the aerogel at 77 K, using massively parallel GCMC simulations. Because of the large box sizes needed to provide a satisfactory representation of the aerogel, these simulations typically involve millions of molecules and require about 2 billion Monte Carlo moves for each point. The results are shown in Fig. 87, together with the experimental data [844]. Good agreement with experiment is obtained for reduced pressures up to about 0.8. However, the model predicts capillary condensation at somewhat lower relative pressures than those observed experimentally. There are several possible reasons for this discrepancy. The model does not account for variations in carbon particle size but treats the particles as being all of the same diameter of 6 nm. In addition, the experimental

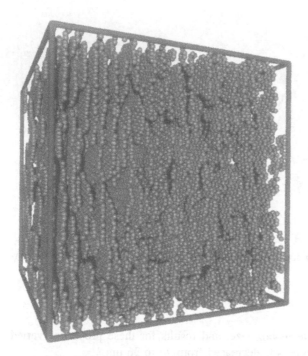

FIG. 86 Micropore structure of carbon aerogel, as determined by reverse Monte Carlo. Each sphere represents a carbon atom. (From Ref. 841.)

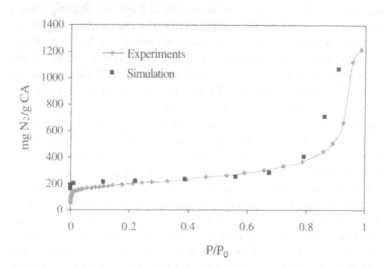

FIG. 87 Nitrogen adsorption isotherms at 77 K for a carbon aerogel. Squares are GCMC simulation results. Diamonds and the curve are experimental data. (From Ref. 841.)

estimate of porosity is subject to substantial uncertainty because it relies on estimates of the density of the carbon spheres, which are themselves uncertain.

5. Chemically Heterogeneous Surfaces

Most studies of the influence of chemical heterogeneity have been for slit-shaped pores, and these can be divided into two types. In the first, the pore surface is divided into strips of strongly and weakly attractive regions [845–854]. In the second, individual functional groups are attached to the planar surface [611,855–862]. For the case of chemically striped surfaces, the behavior depends strongly on the ratio d/H, where d is the width of the stripe and H is the pore width. For individual groups on the surface the corresponding variable is n/H, where n is the site density on the surface. When d/H or n/H are sufficiently small, the behavior is qualitatively similar to that for homogeneous walls, since the effects of surface heterogeneity are not felt far from the walls. When d/H or n/H are large, the behavior is similar to that of a collection of independent pores of various widths or fluid-adsorbate interactions. Intermediate cases between these two extremes are of particular interest. In addition to these studies for slit pores, Segarra and Glandt [799] have studied a model of water on activated carbons in which the carbon is modeled as made up of randomly oriented platelets of graphite with a dipole distributed uniformly over the edge of the platelets to mimic the activation (see Section IV.B.3.).

For striped surfaces, phase transitions in an adsorbed phase tend to occur at different conditions for adsorbate over strongly and weakly attractive regions of the surface [e.g., 847–854]. Capillary condensation often occurs as two transitions at a given temperature, one over the more attractive surface regions to form a "bridge" phase at a lower pressure, and the other over the less attractive regions (to form a dense liquid-like phase throughout the pores) at a higher pressure.

Surfaces with individual functional groups have been studied by lattice gas models [855]. GCMC simulations have been reported for water [858–862], methane [862], and for associating chain molecules [856,857,859] adsorbed on model activated carbons composed of slit pores in which the surfaces are decorated with functional groups. In the case of water and associating chains, these groups can hydrogen-bond to the adsorbate molecules and have been represented by either site-site potentials of the OPLS type [866] (Lennard-Jones sites and point charges) or by square well sites. For water it is found that if the site density is low enough the adsorption is similar to that in graphitic pores [611,864]. That is, almost no adsorption occurs at low to moderate relative pressures, but this is followed at higher pressures by capillary condensation. As the surface density of sites increases, more adsorption occurs at low pressures and the pore filling moves to lower pressures. At these higher site densities the pore filling ceases to be via a sharp phase transition, but occurs continuously. Examination of snapshots of the adsorption at increasing pressures shows that water molecules first adsorb singly on the surface sites, and these adsorbed molecules form adsorption sites for additional water molecules to H-bond to. Thus, water clusters are formed around surface sites, and these grow until eventually they link up with other clusters on the same wall, or with clusters on the opposing wall in narrow pores. This filling mechanism is quite different from that for simple fluids on homogeneous pore walls, where the filling occurs a layer at a time followed by capillary condensation.

Recently, a more realistic carbon structure has been used to study the adsorption of water [865,866]. The RMC method was used to generate a model of a

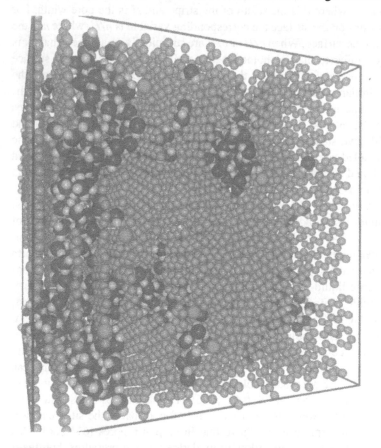

FIG. 88 Water adsorbed in a model of a microporous carbon produced by reverse Monte Carlo, with =O sites added randomly to the edges of the microcrystals. In this model the surface site density is $n = 0.65$ site/nm^2. Oxygen atoms of water molecules, carbon atoms, and surface sites are represented by black, small gray, and large gray spheres, respectively. (From Ref. 866.)

microporous graphitizable carbon composed of rigid, basal plane microcrystals, whose microscopic structure matches X-ray scattering data. Oxygenated surface sites were randomly placed at the edges of the microcrystals, with densities ranging from $n = 0.25$ to 2.5 site/nm². The surface sites were represented as oxygen sites, $=O$, modeled as a single Lennard-Jones sphere with a point charge at its center, with potential parameters for the OPLS model [863] and a $C=O$ bond length of 0.1214 nm. This model has been used to study the adsorption of water as a function of pressure and surface site density. Three-dimensional water clusters similar to those observed in the slit-pore model studies were also found in these models. Furthermore, these clusters were found to significantly reduce the volume accessible to gas mixtures entering the porous carbon. The results agree qualitatively with the behavior observed experimentally. An example of this structure is shown in Fig. 88.

C. Modeling Carbonization and Physical Activation

1. Semiempirical and Ab Initio Approaches

In generating models of porous carbons, it is recognized that approximations must be made that limit to some extent the "chemical reality" of the carbon structure. Specifically, the representation of the chemical bonds that make up the structural building blocks are often of necessity ignored or approximated. Similarly, interactions of adsorbed molecules like methane, carbon dioxide, and nitrogen are typically handled by employing empirical, pair-wise additive interaction terms. While these simple expressions may provide phenomenological insight into carbon-host interactions, they are nevertheless approximate.

The chemical complexity of the carbon structure and its formation from raw carbonaceous materials provide interesting challenges in the formation of structural models based on *mimetic simulations*. The reason for these challenges is that formal analysis of bond formation and dissociation involves aspects of quantum chemistry. The current state of *ab initio* (i.e., first principles) computation unfortunately precludes system sizes on the order of that required to generate reasonable representations of the carbon microstructure. In addition, the nature of porous carbon synthesis is rather complex even for traditional chemical analysis. Despite these shortcomings, semiempirical versions of *ab initio* computations have helped our understanding of porous carbon synthesis chemistry as well as providing tools for analysis of microstructure.

Ab initio or first-principles methods seek to describe the electrons of a chemical system through solution of the *many-body Schrödinger equation* (Hamiltonian). One imagines a collection of positively charged nuclei fixed in space. One then solves for the ground-state wavefunction of the electrons interacting with the collective Coulomb potential of the nuclei. From this solution, the electronic structure of the chemical system is described completely. Electron density, internuclear forces, stress, and electronic energy can all be derived from elementary analysis of the electron wavefunction.

The *many-body wavefunction* is the solution of lowest eigenvalue of the many-body Schrödinger equation:

$$\left(\sum_{i=1}^{N} \left(-\frac{1}{2} \vec{\nabla}_i^2 + V_{ext}(\vec{r}_i) \right) + \sum_{i=1}^{N-1} \sum_{j>i} \frac{e^2}{|\vec{r}_i - \vec{r}_j|} \right)$$

$$\Psi(\vec{r}_1, \vec{r}_2, \ldots, \vec{r}_N) = E\Psi(\vec{r}_1, \vec{r}_2, \ldots, \vec{r}_N)$$

where V_{ext} is the collective Coulomb potential of the nuclear field and N is the total number of electron states (not including spin degenerate states). Finding an analytical solution to this problem for relevant system sizes is severely complicated by the double-sum term (representing the electron-electron interactions). Consequently, approximations are employed that first limit the form of the many-body wavefunction and then attempt to correct for those limitations. For a detailed description of these methods and their application, see the excellent review by Sauer [867].

The computational expense associated with *ab initio* and semiempirical methods means there are very few reports of their use to study precursor carbonization or carbon activation, and they have not been used in any mimetic approach. Perhaps the only quantum level study of the carbonization process, at least in its broadest sense, is that of Pappano et al. [868], who used a semiempirical method to study the polymerization of phenanthrene and anthracene. Interestingly, the polymerized phenanthrene system yielded by the simulations contained holes and curvature induced by the presence of five-membered rings, while the polymerized anthracene system contained little curvature and was not dissimilar to a sheet of paper riven by tears. The gasification of graphite by CO_2, H_2O, and O_2, which is of relevance to activation, has been extensively considered by Chen and Yang in a series of studies at both the semiempirical [869] and, more recently, *ab initio* [870] levels. They have considered both the uncatalyzed systems and those catalyzed by alkali and alkaline metals, indicating under what conditions catalytic effects are likely to prevail and suggesting reaction mechanisms. Ma and coworkers [871] have used a semiempirical method to study the inhibitive effects of boron, while Kyotani and Tomita [872] have used an *ab initio* method to consider the reaction of carbon with NO and N_2O.

Although *ab initio* and semiemperical methods have yielded significant insights into the carbon gasification process, their computational expense means consideration is limited to models that are approximately 10 rings or less. As this is far smaller than is necessary for the study of more realistic carbon structures, approximations in the form of the Hamiltonian (i.e., Schrödinger) equation must be made. The tight-binding method is one such approximation that has been successfully applied to amorphous and porous carbon systems. In the tight-binding approximation, the form of the many-body wavefunction is expressed as a product of individual atomic orbitals, localized on the atomic centers, and chosen to have the proper symmetry properties for the molecular system. The tight-binding Hamiltonian is then constructed by including contributions to the Schrödinger equation that are dependent only on any two given atomic orbitals. In this fashion, these terms are reduced to parameters that depend only on interatomic separation and the specific orbital symmetries. See Goringe et al. for an excellent review on the topic [873].

The tight-binding model was used by Charlier et al. [874] to study distorted stacking of graphene layers, termed *pregraphitic* or *turbostratic* carbon. The electronic structure of crystalline graphite is well known. It can be described adequately as individual graphene layers interacting weakly though π bonds formed by adjacent p_z orbitals; the p_z orbitals are not involved in sp^2 hybridization. It is expected that the electronic structure of turbostratic carbon should be similar to that of graphite for the majority of the energy spectrum. Deep below the Fermi level, the states are dominated by the in-plane interactions. However, the states right at the Fermi level describe the π and π^* (bonding and antibonding) splitting due to the weak interplate interactions.

The turbostratic structure was obtained by generating an amorphous cluster of graphene plates. As each plate was added in succession to the structure, it was randomly translated and rotated to produce the disordered stacking arrangement. The clusters contained 24 graphene layers and 35,736 carbon atoms. Calculations of the density of states (number of electrons/eV-atom per spin) showed that the disordered packing more closely resembled the Bernal stacking of graphite (ABAB) in density and shape. The densities of states were compared at the Fermi level, where their importance for excitations and electron transport properties are the greatest. Other comparisons included rhombohedral graphene (ABCABC stacking) and a hypothetical stacking (AAAA). The tight-binding method has also been applied to amorphous carbons—sometimes referred to as diamond-like carbon—again with the interest in studying the electronic structure [875,876]. However. Wang et al. [877] used the method as a means of establishing a structural model of porous carbon. The authors used tight-binding molecular dynamics (TBMD) to simulate a high-temperature quench of liquid carbon at various densities. The molecular dynamics scheme propagates the carbon atoms based on the equation of motion. Forces are calculated through a combination of the tight-binding orbitals and the Coulomb repulsion of the carbon nuclei. Initial configurations of carbon were quenched from 6000–10,000 K to 0 K gradually throughout the simulation. Structurally relaxed configurations were then obtained by refining the 0 K configurations into local minimal energy states. Finally the structures were reannealed by heating to 2500 K and then cooled back to 0 K.

An interesting observation in this work is that different bonding topologies resulted from varying the initial carbon density. High carbon density resulted in *diamond-like carbon* characterized by a large fraction of sp^3 bonding sites (72%). Low carbon density resulted in an amorphous carbon structure dominated by three- and twofold coordinated atoms. This structure was highly curved and resembled distorted graphitic sheets containing many nonaromatic ring members. Figure 89 shows a supercell of the low-density configuration (1.25 g/cm³). One can clearly see the curved graphene sheets containing nonaromatic five-, six-, and seven-fold carbon rings.

A particularly interesting application of these types of models was proposed by Rosato et al. [878], which in effect combines the TBMD and RMC methods. The authors first used the RMC method to generate amorphous carbon configurations based on experimental radial distribution functions. These were subjected

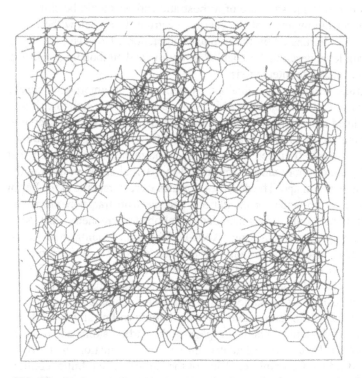

FIG. 89 The network of the amorphous carbon with the average density of 1.25 g/cm³. Periodic boundary conditions are used to extend the network beyond the simulation unit cell. (From Ref. 877.)

to the constraints on the fraction of different coordinated carbon sites in the model. The converged structures then underwent relaxation through the TBMD procedure. The idea is to get a configuration quickly, through RMC, which is close enough physically to the structure of interest. Then the configuration is relaxed through more rigorous means (the TBMD procedure), incorporating local bonding and interactions into the framework. This method is desirable in that it adopts the chemistry of amorphous carbon structures at a rigorous level yet is flexible enough—through the RMC prescription—to be applicable to a wide range of experimental carbons. Long-range structural features, which can be captured experimentally through appropriate correlation functions, can be included through the RMC portion of the procedure. Short-range bonding configurations, best handled through *ab initio* techniques, are then optimized to provide a realistic model of porous carbons on many levels.

2. Stochastic Approaches

The high computational expense of the deterministic approaches outlined in the previous section means that they are unlikely to form the basis of any mimetic approach for the foreseeable future. Some efforts have, therefore, been directed to the development of stochastic-based mimetic models. Based on the work of Stein and Brown [879], Kyotani et al. [880] have modeled the gasification of polyaromatic molecules using a Monte Carlo process whereby the probability of a carbon reacting is determined using Hückel molecular orbital (HMO) theory; the relative simplicity of HMO theory means that extension to more complex structures appears possible within current computational capabilities. Although principally used to predict the tar evolution profile during coal devolatilization, the DVC (depolymerization, vaporization, and cross-linking) component of the FG-DVC model (functional group: DVC) of Solomon and coworkers (see review of Solomon and Fletcher [881] for details of this and other similar models) could potentially be used to follow the precursor-char transition. In this case, the precursor is modeled by sites representing irreducible molecular fragments of varying molecular weight and chemistry connected by labile links that may be broken with a probability defined by their reactivity. The characteristics of the sites and links are determined from experimental characterizations of the precursor, including those obtained from NMR of the precursor and mass spectra of the precursor pyrolysis products. Jones et al. [882] have built structures along the coal-char pathway that match experimental data and those extrapolated from the precursor by the DVC method by breaking the labile bonds of a molecular structure that matches experimental data of the precursor, and allowing condensation reactions to occur. Although published details are scarce at this stage, it appears that Mathews et al. [883] may have also modeled devolatilization of a coal by similarly modifying a model molecular structure of coal.

V. CONCLUSIONS

The structural models available for porous carbons can be roughly divided into two classes: simple geometrical models, such as collections of slit- or wedge-shaped pores, and more complex models in which more realistic features, such as pore connectivity, tortuosity, curved and defective carbon sheets, are included. The simple geometrical models are easy to apply and can give a good account of adsorption when the pore size distribution is fitted to experimental data. Such models are now incorporated into the software of most sorptometers, and are used to estimate surface areas and pore size distributions. However, such models omit many important features of porous carbon structure including pore connectivity, tortuosity, variations in pore shape, chemical heterogeneity of the surfaces, and so forth. Such models may give poor results even for adsorption if extrapolated to temperatures or adsorbate gases far from the region of fit. They are particularly poor in representing diffusion in carbons, where connectivity, tortuosity, and surface heterogeneity have a large influence on the diffusive flux. Diffusion rates calculated using slit-pore models can be in error by an order of magnitude or more.

In the last few years, several more realistic models have been proposed in an attempt to include connectivity, variations in pore morphology, defective ring structures and curved carbon plates, and so on. These more complex models include the virtual carbon model, the chemically constrained model, and RMC models. None of these models have been fully developed or tested, but they offer the prospect of considerably more sophisticated and accurate modeling of carbons. What is needed are carefully designed efforts to test and refine these models through collaborative research programs involving complementary experimental and modeling studies. Eventually, it should be possible to replace the simple geometrical models by the more complex models in practical applications, such as predictions of adsorption, separations, diffusion rates, and so on.

The existing structural models are overwhelmingly of the reconstructive type, in which the model is constructed based on experimental structural data. This is a result of the complex and poorly understood synthesis of the carbons. Mimetic simulation methods, in which the synthesis is modeled using molecular or *ab initio* simulations, has been successfully used for some other porous materials, e.g., porous glasses [730–732,884] and MCM-41 [885]. Such approaches are desirable because they can produce unique and physically realistic structures. Moreover, they offer insight into the synthetic route itself and may suggest ways to improve it. Some attempts to model a part of the synthetic process using mimetic *ab initio* methods have been made (see Section IV.C). At first sight it would seem hopeless to attempt mimetic methods to simulate the entire synthetic process for activated carbons. However, reasonably successful intermolecular potentials exist for carbon [e.g., 753]. These cannot be expected to produce realistic structures for activated carbons via direct simulations. The synthetic process involves many chemical reactions, the details of which are largely unknown, and the final carbon structures are not equilibrium ones. However, it may prove possible to improve the models by incorporating the potential in the reconstruction in some way.

While the principal stumbling block remains the development of more realistic models, improvements in experimental techniques are also needed. In the measurements of structure by diffraction or TEM, higher resolution and accuracy is desirable to provide a clearer picture of the atomic and surface structure. In the case of TEM measurements, the development of methods to obtain three-dimensional structures, as opposed to the two-dimensional thin-film structures currently possible, would provide a major advance. In the surface chemistry studies, further resolution of both the location and species of surface groups is needed.

ACKNOWLEDGMENTS

We thank the National Science Foundation (grant no. CTS-9908535), Department of Energy (grant no. DE-FG02-88ER13974), and the Engineering and Physical Sciences Research Council (UK) for support of this research.

REFERENCES AND NOTES

1. M Smísek, S Cerný. Active Carbon: Manufacture Properties and Applications. Amsterdam: Elsevier, 1970, Chapter 5.
2. RC Bansal, JB Donnet, F Stoeckli. Active Carbon. New York: Marcel Dekker, 1988, Chapter 6.
3. A Jankowska, Swiatkowski, J Choma. Active Carbon. Chichester: Ellis Horwood, 1991.
4. TJ Mays. In: TD Burchell, ed. Carbon Materials for Advanced Technologies. Amsterdam: Pergamon, 1999, Chapter 3.
5. GM Roy. Activated Carbon Applications in the Food and Pharmaceutical Industries. Lancaster, PA: Technomic Publishing, 1995.
6. F Erb, D Gairin, N Leroux, J Toxicol Clin Exp 9:235, 1989
7. A Bailey. In: J W Patrick, ed. Porosity in Carbons. London: Edward Arnold, 1995, Chapter 8.
8. P Chaimbault, K Petritis, C Elfakir, M Dreux. J Chromatogr A 870:245, 2000.
9. S Sircar, TC Golden, MB Rao. Carbon 34:1, 1996.
10. CB Eom, AF Hebard, LE Trimble, GK Celler, RC Haddon. Science 259: 1887–1890, 1993.
11. TE Lipman, MA DeLucchi, Int J Vehicle Design, 17: 562, 1996.
12. TL Cook, C Komodromos, DF Quinn, S Ragan. In: TD Burchell, ed. Carbon Materials for Advanced Technologies, Amsterdam: Pergamon, 1999, Chapter 9.
13. GE Gadd, PJ Evans, S Kennedy, M James, M Elcombe, D Cassidy, S Moricca, J Holmes, N Webb, A Dixon, P Prasad. Fullerene Sci Technol 7:1043–1143, 1999.

14. H Jüntgen, H Kühl. In: PA Thrower, ed. Chemistry and Physics of Carbon, Vol. 22. New York: Marcel Dekker, 1992, p. 145.
15. F Rodríguez-Reinoso. Carbon 36:159, 1998.
16. BK Arumugam. PC Wankat, Adsorption 4:345, 1998.
17. E Auer, A Freund, J Pietsch, T Tacke. Appl Catal A 173:259–271, 1998.
18. M Nemati, C Webb, J Chem Technol Biotech 74:562–570, 1999.
19. B Coq, JM Planeix, V Brotons. Appl Catal A 173:175–183, 1998.
20. M Sharon, K Mukhopadhay, K Yase, S Iijima, Y Ando, X Zhao. Carbon 36:507, 1998.
21. T Zheng, JR Dahn. In: TD Burchell, ed. Carbon Materials for Advanced Technologies. Amsterdam: Pergamon, 1999, Chapter 11.
22. K Tokumitsu, H Fujimoto, A Mabuchi, T Kasuh, Carbon 37:1599, 1999.
23. RW Pekala, JC Farmer, CT Alviso, TD Tran, ST Mayer, JM Miller, B Dunn, J Non-Cryst Solids 225:74, 1998.
24. C Niu, EK Sichel, R Hoch, D Moy, H Tennent. Appl Phys Lett 70:1480, 1997.
25. J Fricke, A Emmerling. J Sol-Gel Sci Technol 13:299, 1998.
26. RE Critoph. In: TD Burchell, ed. Carbon Materials for Advanced Technologies. Amsterdam: Pergamon, 1999, Chapter 10.
27. I Tanahashi, H Nakama, A Nishino, J Takeda. Bull Chem Soc Jpn 64:2220, 1991.
28. PJF Harris. Int Mater Rev 42:206–218, 1997.
29. F Rodríguez-Reinoso, A Linares-Solano. In: PA Thrower, ed. Chemistry and Physics of Carbon, Vol. 21. New York: Marcel Dekker, 1989, pp. 1–146.
30. A Oberlin. In: PA Thrower, ed. Chemistry and Physics of Carbon, Vol 22. New York: Marcel Dekker, 1989, pp. 1–143.
31. E Hoinkis. In: PA Thrower, ed. Chemistry and Physics of Carbon, Vol. 25. New York: Marcel Dekker, 1987, pp. 169–174.
32. A Oberlin, S Bonnamy, PG Rouxhet. In: PA Thrower, L R Radovic, eds. Chemistry and Physics of Carbon, Vol. 26. New York: Marcel Dekker, 1999, pp. 1–148.
33. JW Patrick, ed. Porosity in Carbons. New York: Halsted Press, 1995, Chapters 1,2.
34. TD Burchell, ed. Carbon Materials for Advanced Technologies. Oxford: Pergamon, 1999, Chapters 1–6.
35. TW Ebbesen, ed. Carbon Nanotubes. Boca Raton: CRC Press, 1999, Chapters 1–3.
36. F Rouquerol, J Rouquerol, KS W Sing. Adsorption by Powders and Porous Solids. London: Academic Press, 1999, Section 91.
37. E Fitzer, KH Köchling, HP Boehm, H Marsh. Pure Appl Chem 67:473–506, 1995.
38. RB Heimann, SE Evsyukov, Y Koga. Carbon 35:1654–1658, 1997.
39. Polytypes are allotropes that have identical unit cell parameters in two directions with the third being different [38]. In the case of graphite, the difference occurs in the c-direction.
40. H Shi, J Baker, MY Saidi, R Koksbang, L Morris. J Power Sources 68:291–295, 1997.
41. WN Reynolds. Physical Properties of Graphite. Amsterdam: Elsevier, 1968, Chapter 1.
42. S Amelinckx, P Delavignette, M Heerschap. In: PL Walker Jr, ed. Chemistry and Physics of Carbon, Vol. 1. New York: Marcel Dekker, 1965, pp. 1–71.

43. A Grenall, A Sosin. Proc 4th Conf Carbon:371–401, 1960.
44. FP Bundy, HM Strong RH Wentorf Jr. In: PL Walker Jr, PA Thrower, eds. Chemistry and Physics of Carbon, Vol. 10. New York: Marcel Dekker, 1973, pp. 213–263.
45. KE Spear, AW Phelps, WB White. J Mater Res 5:2277–2285, 1990.
46. AW Phelps, W Howard, DK Smith. J Mater Res 8:2835–2839, 1993.
47. R Kapil, BR Mehta, VD Vankar. Thin Solid Films 312:106–110, 1998.
48. DM Bibby. In: PA Thrower, ed. Chemistry and Physics of Carbon, Vol. 18. New York: Marcel Dekker, 1982, pp. 1–91.
49. YuP Kudryavtsev, RB Heimann, S Evsyukov. J Mater Sci 31:5557–5571, 1996.
50. YuP Kudryavtsev, S Evsyukov, M Guseva, V Babaev, V Khvostov. In: PA Thrower, ed. Chemistry and Physics of Carbon, Vol. 25. New York: Marcel Dekker, 1997, pp. 1–69.
51. RB Heimann, SE Evsyukov, L Kavan, eds. Carbyne and Carbynoid Structures. Dordrecht: Kluwer Academic, 1999.
52. YuP Kudryavtsev. In: SE Evsyukov and L Kavan, eds. Carbyne and Carbynoid Structures Dordrecht: Kluwer Academic, 1999, pp. 1–6
53. RB Heimann. In: SE Evsyukov, L Kavan, eds. Carbyne and Carbynoid Structures. Dordrecht: Kluwer Academic, 1999, pp. 7–15
54. VI Kasatochkin, VV Korshak, YuP Kudryavtsev, AM Sladkov, LE Sterenberg. Carbon 11:70–72, 1973.
55. SI Tanuma, A Palnichenko. J Mater Res 10:1120–1125, 1995.
56. SE Evsyukov, L Kavan, eds. Carbyne and Carbyne Structures. Dordrect: Kluwer Academic, 1999, Chapter 4.
57. HW Kroto, DRM Walton. Phil Trans R Soc Lond A 343:103–112, 1993.
58. RB Heimann, SE Evsyukov, L Kavan, eds. Carbyne and Carbynoid Structures. Dordrecht: Kluwer Academic, 1999, pp. 235–268.
59. YuP Kudryavtsev, SE Evsyukov, VG Babaev, MB Guseva, VV Khvostov, LM Krechko. Carbon 30:213–221, 1992.
60. RB Heimann, J Kleiman, NM Salansky. Carbon 22:147–155, 1984.
61. M Springborg, L Kavan. Chem Phys 168:249–258, 1992.
62. RB Heimann. Carbon 35:1669–1671, 1992.
63. HW Kroto, JR Heath, SC O'Brien, RF Curl, RE Smalley. Nature 318:162–163, 1985.
64. W Krätschmer, LD Lamb, K Fostiropoulos, DR Huffman. Nature 347:354–358, 1990.
65. F Wudl, RE Smalley, AB Smith, R Taylor, E Wasserman, EW Godly. Pure Appl Chem 69:1412–1434, 1997.
66. PW Fowler. Phil Trans R Soc London A 343:39–52, 1993.
67. PW Fowler. Contemp Phys 37:235–247, 1996.
68. A search of the ISI on-line database indicated that only a handful of publications have used this term to date.
69. F Diederich, RL Whetten. Acc Chem Res 25:119–126, 1992.
70. RF Curl, RE Smalley. Sci Am 265:54, 1991.
71. GP Moss, PAS Smith, D Tavernier. Pure Appl Chem 67:1307–1375, 1995.

72. D Dubois, KM Kadish, S Flanagan, LJ Wilson. J Am Chem Soc 113:7773–7774, 1991.
73. B Kubler, E Millon, JJ Gaumet, JF Muller. Fullerene Sci Technol 4:1247–1261, 1996.
74. H Richter, SC Emberson, A Fonseca. Rev Inst Fr Petrol 49:413–419, 1994.
75. MS Dresselhaus, PC Eklund, G Dresselhaus. In: TD Burchell, ed. Carbon Materials for Advanced Technology. Amsterdam: Pergamon, 1999, Chapter 2.
76. C Piskoti, J Yager, A Zettl. Nature 393:771–774, 1998.
77. S Maruyama, MY Lee, RE Haufler, Y Chai, RE Smalley. Z Phys D 19:409–412, 1991.
78. S Maruyama, LR Anderson, RE Smalley. Rev Sci Instrum 61:3686–3693, 1990.
79. J Ahrens, M Bachmann, Th Baum, J Griesheimer, R Kovacs, P Weilmünster, KH Homann. Int J Mass Spectrom Ion Proc 138:133–148, 1994.
80. F Beer, A Gügel, K Martin, J Räder, K Müllen. J Mater Chem 7:1327–1330, 1997.
81. G von Helden, N Gott, MT Bowers. Nature 363:60–63, 1993.
82. H Prinzbach, A Weiler, P Landenberger, F Wahl, J Worth, LT Scott, M Gelmont, D Olevano, B v Issendorff. Nature 407:60–63, 2000.
83. RE Smalley. Acc Chem Res 25:98–105, 1992.
84. A Maiti, CJ Brabec, J Bernholc. Phys Rev Lett 70:3023–3026, 1993.
85. GE Scuseria. Chem Phys Lett 243:193–198, 1995.
86. S Itoh, P Ordejón, DA Drabold, RM Martin. Phys Rev B 53:2132–2140, 1996.
87. D York, JP Lu, W Yang. Phys Rev B 49:8526–8528, 1994.
88. JP Lu, W Yang. Phys Rev B 49:11421–11424, 1994.
89. BC Wang, L Chen, HW WanG, YM Chou. Synth Metals 103:2448–2449, 1999.
90. YY Astakhova, GA Vinogradov. J Mol Struct Theochem 430:259–268, 1998.
91. J Aihara. Bull Chem Soc Jpn 72:7–11, 1999.
92. B Sundqvist. Adv Phys 48:1–134, 1999.
93. S Wasa, K Suito, M Kobayashi, A Onodera. Solid State Commun 114:209–213, 2000.
94. SM Bennington, N Kitamura, MG Cain, MH Lewis, M Arai. Physica B:263–264, 632–635, 1999.
95. E Sandré, F Cyrot-Lackmann. Solid State Commun 90:431–434, 1994.
96. B Wei, J Liang, Z Gao, J Zhang, Y Zhu, Y Li, D Wu. J Mater Process Technol 63:573–578, 1997.
97. VD Blank, BA Kulnitskiy, YeV Tatyanin, OM Zhigalina. Carbon 37:549–554, 1999.
98. JE Fischer, PA Heiney, DE Luzzi, DE Cox. ACS Symp Series 481:55–69, 1992.
99. CNR Rao, R Seshadri, A Govindaraj, R Sen. Mater Sci Eng R15:209–262, 1995.
100. JH Weaver, DM Poirier. Solid State Phys 48:1–107, 1994.
101. Y Saito, T Yoshikawa, N Fujimoto, H Shinohara. Phys Rev B 48:9182–9185, 1993.
102. H Rietschel. Fullerene Sci Technol 4:743–755, 1996.
103. S Margadonna, CM Brown, TJS Dennis, A Lappas, P Pattison, K Prassides, H Shinohara. Chem Mater 10:1742–1744, 1998.
104. Y Saito, N Fujimoto, K Kikuchi, Y Achiba. Phys Rev B 49:14794–14797, 1994.

105. JF Armbruster, HA Romberg, P Schweiss, P Adelmann, M Knupfer, J Fink, RH Michel, J Rockenberger, F Hennrich, H Schreiber, MM Kappes. Z Phys B 95:469–474, 1994.
106. H Kawada, Y Fujii, H Nakao, Y Murakami, T Watanuki, H Suematsu, K Kikuchi, Y Achiba, I Ikemoto. Phys Rev B 51:8723–8730, 1995.
107. Y Maniwa, K Kume, K Kikuchi, K Saito, I Ikemoto, S Suzuki, Y Achiba. Phys Rev B 53:14196–14199, 1996.
108. H Nakao, Y Fujii, T Watanuki, K Ishii, H Suematsu, H Kawada, Y Murakami, K Kikuchi, Y Achiba, Y Maniwa. J Phys Soc Jpn 67:4117–4123, 1998.
109. M Côté, JC Grossman, SG Louie, ML Cohen. Bull Am Phys Soc 42:270, 1997.
110. Although nonequilibrium conditions are also known to affect specifics of the phase transition behavior (eg Epanchintsev et al., J Phys Chem Solids 58:1785–1788 1997), the qualitative features of this diagram still stand.
111. WFI David, RM Ibberson, TJS Dennis, JP Hare, K Prassides. Europhys Lett 18: 219–225, 1992.
112. This is an assumption, but it is acceptable in general [92].
113. JE Fischer, PA Heiney, AR McGhie, WJ Romanow, AM Denenstein, JP McCauley, AB Smith. Science 252:1288–1290, 1991.
114. GA Samara, JE Schirber, B Morosin, LV Hansen, D Loy, AP Sylwester. Phys Rev Lett 67:3136–3139, 1991.
115. JA Wolk, PJ Horoyski, MLW Thewalt. Phys Rev Lett 74:3483–3486, 1995.
116. NP Lalla, ON Srivastava. Prog Crystal Growth Charact 34:53–80, 1997.
117. YZ Li, JC Patrin, M Chander, JH Weaver, LPF Chibante, RE Smalley. Science 252:547–548, 1991.
118. D Bernaerts, G Van Tendeloo, S Amelinckx, K Hevesi, G Gensterblum, LM Yu, JJ Pireaux, F Grey, J Bohr. J Appl Phys 80:3310–3318, 1996.
119. NI Gorbenko, EN Zubarev, SA Medvedev, AT Pugachev, NP Churakova. Low Temp Phys 25:79–80 1999.
120. VY Gusev, S Ruetsch, LA Popeko, IE Popeko. J Phys Chem B 103:6498–6503, 1999.
121. P Milani, E Barborini, P Piseri, CE Bottani, AC Ferrari, A Li Bassi. Eur Phys J D. 9:63–68, 1999.
122. PA Heiney. J Phys Chem Solids 53:1333–1352, 1992.
123. H Werner, D Bublak, U Göbel, B Henschke, W Bensch, R Schlögl. Angew Chem Int Ed Engl 31:868–870, 1992.
124. Z Belahmer, P Bernier, L Firlej, JM Lambert, M Ribet. Phys Rev B 47:15980–15983, 1993.
125. M Gu, Y Wang, TB Tang, W Zhang, C Hu, F Yan, D Feng. Phys Lett A 223:273–279, 1996.
126. CS Sundar, Y Hariharan, A Bharathi, M Premila, VS Sastry, GVN Rao, J Janaki, DV Natarajan, KV Devadhasan, TS Radhakrishnan, N Subramanian, PCh Sahu, M Yousuf, S Ruju, VS Raghunathan, MC Valsakumar. Prog Crystal Growth Charact 34:11–23 1997.
127. A Zahab, L Firlej, P Bernier, R Aznar, A Rassat, C Fabre. Solid State Commun 84:429–433, 1992.

128. T Hashizume, K Motai, XD Wang, H Shinohara, Y Saito, Y Maruyama, K Ohno, Y Kawazoe, Y Nishina, HW Pickering, Y Kuk, T Sakurai. Phys Rev Lett 71:2959–2962, 1993.

129. PC Eklund, AM Rao, P Zhou, Y Wang, JM Holden. Thin Solid Films 257:185–203, 1995.

130. AM Rao, PC Eklund. Mater Sci Forum 232:173–206, 1996.

131. VD Blank, SG Buga, GA Dubitsky, NR Serebryanaya, MYu Popov, B Sundqvist. Carbon 36:319–343, 1998.

132. M Núñez-Regueiro, L Marques, JL Hodeau, O Béthoux, M Perroux. Phys Rev Lett. 74:278–281, 1995.

133. RC Haddon, AF Hebard, MJ Rosseinsky, DW Murphy, SJ Duclos, KB Lyons, B Miller, JM Rosamilia, RM Fleming, AR Kortan, SH Glarum, AV Makhija, AJ Muller, RH Eick, SM Zahurak, R Tycko, G Dabbagh, FA Thiel. Nature 350:320–322, 1991.

134. MJ Rosseinsky. J Mater Chem 10:1497–1513, 1995.

135. MJ Rosseinsky. Chem Mater 10:2665–2685, 1998.

136. Y Chai, T Guo, C Jin, RE Haufler, LPF Chibante, J Fure, L Wang, JM Alford, RE Smalley. J Phys Chem 95:7564–7568, 1991.

137. M Saunders, HA Jiménez-Vázquez, RJ Cross, S Mroczkowski, ML Gross, DE Giblin, RJ Poreda. J Am Chem Soc 116:2193–2194, 1994.

138. S Stevenson, G Rice, T Glass, K Harich, F Cromer, MR Jordan, J Craft, E Hadju, R Bible, MM Olmstead, K Maitra, AJ Fisher, AL Balch, HC Dorn. Nature 410:55–57, 1999.

139. DS Bethune, RD Johnson, JR Salem, MS de Vries, CS Yannoni. Nature 366:123–128, 1993.

140. H Shinohara. Rep Prog Phys 63:843–892, 2000.

141. S Liu, S Sun. J Organometal Chem 599:74–86, 2000.

142. R Taylor, ed. The Chemistry of Fullerenes. Singapore: World Scientific, 1995.

143. CR Reed, RD Bolskar. Chem Rev 100:1075–1120, 2000.

144. U Zimmermann, N Malinowski, A Burkhardt, TP Martin. Carbon 33:995–1006, 1995.

145. U Zimmermann, N Malinowski, U Näher, S Frank, TP Martin. Phys Rev Lett 72:3542–3545, 1994.

146. F Tast, N Malinowski, S Frank, M Heinebrodt, IML Billas, TP Martin. Z Phys D 40:351–354, 1997.

147. F Tast, N Malinowski, M Heinebrodt, IML Billas, TP Martin. J Chem Phys 106:9372–9375, 1997.

148. TR Ohno, Y Chen, SE Harvey, GH Kroll, PJ Benning, JH Weaver, LPF Chibante, RE Smalley. Phys Rev B 47:2389–2393, 1993.

149. A Hirsch. J Phys Chem Solids 58:1729–1740, 1997.

150. F Wudl. Acc Chem Res 25:157–161, 1992.

151. F Diederich, L Isaacs, D Philp. Chem Soc Rev 23:243–255, 1994.

152. PJ Fagan, JC Calabrese, B Malone. Acc Chem Res 25:134–142, 1992.

153. AL Balch, MM Olmstead. Chem Rev 98:2123–2165, 1998.

154. AHH Stephens. MLH Green. Adv Inorg Chem 44:1–43, 1997.

155. JR Browser. Adv Organometal Chem. 36:57–94, 1994.

156. M Prato. J Mater Chem 7:1097–1109, 1997.
157. F Diederich, M Gómez-López. Chem Soc Rev 28:263–277, 1999.
158. F Diederich, Y Rubin. Angew Chem Int Ed Engl 31:1101–1123, 1992.
159. UHF Bunz, Y Rubin, Y Tobe. Chem Soc Rev 28:107–119, 1999.
160. A Hirsch. Adv Mater 5:859–861, 1993.
161. KE Geckeler, S Samal. Polym Int 48:743, 1999.
162. JC Hummelen, B Knight, J Pavlovich, R González. F Wudl. Science 269:554–1556, 1995.
163. B Nuber, A Hirsch. J Chem Soc Chem Commun 1421–1422, 1996.
164. JC Hummelen, C Lund-Bellavia, F Wudl. Top Curr Chem 199:93–134, 1999.
165. U Reuther, A Hirsch. Carbon 38:1539–1549, 2000.
166. T Guo, C Jin, RE Smalley. J Phys Chem 95:4948–4950, 1991.
167. MM Alvarez, EG Gillan, K Holczer, RB Kaner, KS Min, RL Whetten. J Phys Chem 95:10561–10563, 1991.
168. DE Clemmer, JM Hunter, KB Shelimov, MF Jarrold. Nature 372:248–250, 1994.
169. T Kimura, T Sugai, H Shinohara. Chem Phys Lett 256:269–273, 1996.
170. MS Dresselhaus, G Dresselhaus, R Saito. Carbon 33:883–891, 1995.
171. R Saito, G Dresselhaus, MS Dresselhaus. Physical Properties of Carbon Nanotubes. London: Imperial College Press, 1998, Appendix.
172. S Iijima. Nature 354:56–58, 1991.
173. M Liu, JM Cowley. Carbon 32:393–403, 1994.
174. D Bernaerts, MO de Beeck, S Amelinckx, J Van Landuyt, G Van Tendeloo. Phil Mag A 74:723–740, 1996.
175. A Hassanien, M Tokumoto, S Ohshima, Y Kuriki, F Ikazaki, K Uchida, M Yumura. Appl Phys Lett 75:2755–2757, 1999.
176. RR Bacsa, Ch Laurent, A Peigney, WS Bacsa, Th Vaugien, A Rousset. Chem Phys Lett 323:566–571, 2000.
177. M Terrones, N Grobert, J Olivares, JP Zhang, H Terrones, K Kordatos, WK Hsu, JP Hare, PD Townsend, K Prassides, AK Cheetham, HW Kroto, DRM Walton. Nature 388:52–55, 1997.
178. CJ Lee, JH Park, J Park. Chem Phys Lett 323:560–565, 2000.
179. AY Kasumov, H Bouchiat, B Reulet, O Stephan, II Khodos, Yu B Gorbaatov, C Colliex. Europhys Lett 43:89–94, 1998.
180. AP Burden, SRP Silva. Appl Phys Lett 73:3082–3084, 1998.
181. ZW Pan, SS Xie, BH Chang, CY Wang, L Lu, W Liu, WY Zhou, WZ Li, LX Qian. Nature 394:631–632, 1998.
182. C-H Kiang, M Endo, PM Ajayan, G Dresselhaus, MS Dresselhaus. Phys Rev Lett 81:1869–1872, 1998.
183. (a) S Iijima, PM Ajayan, T Ichihashi. Phys Rev Lett 69:3100–3103, 1992. (b) S Iijima. Mater Sci Eng B19:172–180, 1993.
184. Y Saito, T Yoshikawa. J Crystal Growth 134:154–156, 1993.
185. M Liu, JM Cowley. Mater Sci Eng A185:131–140, 1994.
186. TW Ebbesen, T Takada. Carbon 33:973–978, 1995.
187. T Guo, P Nikolaev, AG Rinzler, D Tománek, DT Colbert, RE Smalley. J Phys Chem 99:10694–10697, 1995.

188. NA Kiselev, J Sloan, DN Zakharov, EF Kukovitskii, JL Hutchison, J Hammer, AS Kotosonov. Carbon 36:1149–1157, 1998.

189. W Li, S Xie, W Liu, R Zhao, Y Zhang, W Zhou, G Wang. J Mater Sci 34:2745–2749, 1999.

190. K Kaneto, M Tsuruta, G Sakai, WY Cho, Y Ando. Synth Metals 103:2543–2546, 1999.

191. R Andrews, D Jacques, AM Rao, F Derbyshire, D Qian, X Fan, EC Dickey, J Chen. Chem Phys Lett 303:467–474, 1999.

192. A Thess, R Lee, P Nikolaev, H Dai, P Petit, J Robert, C Xu, YH Lee, SG Kim, AG Rinzler, DT Colbert, GE Scuseria, D Tománek, JE Fischer, RE Smalley. Science 273:483–487, 1996.

193. C Journet, WK Maser, P Bernier, A Loiseau, M Lamy de la Chapelle, S Lefrant, P Deniard, R Lee, JE Fischer. Nature 388:756–758, 1997.

194. HM Cheng, F Li, X Sun, SDM Brown, MA Pimenta, A Marucci, G Dresselhaus, MS Dresselhaus. Chem Phys Lett 289:602–610, 1998.

195. H Dai, AG Rinzler, P Nikolaev, A Thess, DT Colbert, RE Smalley. Chem Phys Lett 260:471–475, 1996.

196. J Kong, HT Soh, AM Cassell, CF Quate, H Dai. Nature 395:878–881, 1998.

197. A Hassanien, M Tokumoto, Y Kumazawa, H Kataura, Y Maniwa, S Suzuki, Y Achiba. Appl Phys Lett 73:3839–3841, 1998.

198. Y Ando, X Zhao, K Hirahara, K Suenaga, S Bandow, S Iijima. Chem Phys Lett 323:580–585, 2000.

199. J Lefebvre, R Antonov, AT Johnson. Appl Phys A 67:71–74, 1998.

200. W Clauss, DJ Bergeron, AT Johnson. Phys Rev B 58: R4266–R4269, 1998.

201. S Iijima and T Ichihashi. Nature 363:603–605, 1993.

202. X Lin, XK Wang, VP Dravid, RPH Chang, JB Ketterson. Appl Phys Lett 64:181–183, 1994.

203. Ph Lambin, V Meunier, L Henrard, AA Lucas. Carbon 38:1713–1721, 2000.

204. VT Binh, P Vincent, F Feschet, JM Bonard. J Appl Phys 88:3385–3391, 2000.

205. DB Mawhinney, V Naumenko, A Kuznetsova, JT Yates, J Liu, RE Smalley. Chem Phys Lett 324:213–216, 2000.

206. BI Dunlap. Phys Rev B 46:1933–1936, 1992.

207. S Itoh, S Ihara. Phys Rev B 49:13970–13974, 1994.

208. LA Chernozatonskii. Phys Lett A 166:55–60, 1992; 170:37–40, 1992.

209. S Ihara, S Itoh, J Kitakami. Phys Rev B 48:5643–5647, 1993.

210. XF Zhang, Z Zhang. Phys Rev B 52:5313–5317, 1995.

211. D Bernaerts, XB Zhang, XF Zhang, S Amelinckx, G Van Tendeloo, J Van Landuyt, V Ivanov, JB Nagy. Phil Mag A 71:605–630, 1995.

212. A Volodin, M Ahlskog, E Seynaeve, C Van Haesendonck, A Fonseca, JB Nagy. Phys Rev Lett 84:3342–3345, 2000.

213. LP Bíró, SD Lazarescu, PA Thiry, A Fonseca, JB Nagy, AA Lucas, Ph Lambin. Europhys Lett 50:494–500, 2000.

214. J Liu, H Dai, JH Hafner, DT Colbert, RE Smalley, SJ Tans, C Dekker. Nature 385: 780–781, 1997.

215. R Martel, HR Shea, P Avouris. Nature 398:299, 1999.

216. M Ahlskog, E Seynaeve, RJM Vullers, C Van Haesendonck, A Fonseca, K Hernadi, JB Nagy. Chem Phys Lett 300:202–206, 1999.
217. J Han. Chem Phys Lett 282:187–191, 1998.
218. Note that Huang et al. (J Mater Chem 9:1221–1222, 1999) have, unfortunately, also used the term nanotube crop circle to describe a very different system of nanotubes, where they occur as a ring of nanotubes normal to the substrate on which they were grown.
219. D Ugarte. Nature 359:707–709, 1992.
220. D Ugarte. Carbon 33:989–993, 1995.
221. T Cabioc'h, JC Girard, M Jaouen, MF Denanot, G Hug. Europhys Lett 38:471–475, 1997.
222. T Cabioc'h, E Thune, M Jaouen. Chem Phys Lett 320:202–205, 2000.
223. AC Tang, FQ Huang. Phys Rev B 52:17435–17438, 1995.
224. JL Morán-López, KH Bennemann, M Carrera-Trujillo, J Dorantes-Dávila. Solid State Commun 89:977–981, 1994.
225. R Pis Diaz, MP Iniguez. Int J Quantum Chem 56:689–696, 1995.
226. MI Heggie, M Terrones, BR Eggen, G Jungnickel, R Jones, CD Latham, PR Briddon, H Terrones. Phys Rev B 57:13339–13342, 1998.
227. MS Zwanger, F Banhart, A Seeger. J Cryst Growth 163:445–454, 1996.
228. T Cabioc'h, JP Rivière, J Delafond. J Mater Sci 30:4787–4792, 1995.
229. T Oku, T Hirano, M Kuno, T Kusunose, K Niihara, K Suganuma. Mater Sci Eng B 74:206–217, 2000.
230. R Selvan, R Unnikrishnan, S Ganapathy, T Pradeep. Chem Phys Lett 316:205–210, 2000.
231. F Banhart, PM Ajayan. Nature 382:433–435, 1996.
232. MS Zwanger, F Banhart. Phil Mag B 72:149–1782, 1995.
233. The vast majority of simulation studies support the notion of largely spherical carbon onions; Maiti et al. [84] is one of the few studies suggesting that carbon onions are polyhedral.
234. DJ Srolovitz, SA Safran, M Homyonfer, R Tenne. Phys Rev Lett 74:1779–1782, 1995.
235. H Terrones, M Terrones. J Phys Chem Solids 58:1789–1796, 1997.
236. A Jankowska, Swiatkowski, J Choma. Active Carbon. Chichester: Ellis Horwood, 1991.
237. J Biscoe, BE Warren. J Appl Phys 13:364–371, 1942.
238. RE Franklin. Proc R Soc London 209:196–218, 1951.
239. RC Bansal, JB Donnet, F Stoeckli. Active Carbon. New York: Marcel Dekker, 1988, Chapter 2.
240. LL Ban. Surf Defect Prop Solids 1:54–94, 1972.
241. GR Millward, DA Jefferson. In: PL Walker Jr, PA Thrower, eds. Chemistry and Physics of Carbon, Vol. 14. New York: Marcel Dekker, 1978, pp. 1–82.
242. JR Fryer. Carbon 19:431–439, 1979.
243. SC Bennett, DJ Johnson. Carbon 17:25–39, 1979.
244. H Marsh, D Crawford. Carbon 22:413–422, 1984.
245. KJ Masters, B McEnaney. Carbon 22:595–601, 1984.

246. A Oberlin. In: PA Thrower, ed. Chemistry and Physics of Carbon, Vol. 22. New York: Marcel Dekker, 1989, pp. 1–143.

247. X Bourrat, A Oberlin, JC Escalier. Fuel 65:1490–1500, 1986.

248. A Oberlin, M Villey, A Combaz. Carbon 18:347–353, 1980.

249. A Oberlin, M Oberlin. J Micro 132:353–363, 1983.

250. HH Lowry, RM Bozorth. J Chem Phys 32:1524–1527, 1928.

251. As a result of the very first powder X-ray diffraction experiments on carbons, Debye and Scherrer (Phys Zeitschr 18:291–301, 1917) indicated that microporous carbons might be constituted from extremely small graphite crystallites. No statement was made regarding the turbostratic nature of the crystallite, which had to await the work of Biscoe and Warren [237].

252. A Oberlin. Carbon 22:521–541, 1984.

253. FG Emmerich. Carbon 33:1709–1715, 1995.

254. PJF Harris, SC Tsang. Philos Mag A 76:667–677, 1997.

255. PJF Harris, A Burian, S Duber. Philos Mag Lett 80:381–386, 2000.

256. HC Foley. Microporous Mater 4:407–433, 1995.

257. T Kyotani. Carbon 38:269–286, 2000.

258. JB Donnet, A Voet. Carbon Black. New York: Marcel Dekker, 1976, Chapter 1.

259. CL Mantell. Carbon and Graphite Handbook. New York: Interscience, 1968, Chapter 6.

260. RD Heidenreich, NM Hess, LL Ban. J Appl Crystallogr 1:1–19, 1968.

261. JB Donnet, A Voet. Carbon Black. New York: Marcel Dekker, 1976.

262. CJ Brinker, GW Scherer. Sol-Gel Science. San Diego: Academic Press, 1990.

263. N Hüsing, U Schubert. Angew Chem Int Ed. 37:22–45, 1998.

264. GC Ruben, RW Pekala, TM Tillotson, LW Hrubesh. J Mater Sci 27:4341–4349, 1992.

265. GC Ruben, RW Pekala. J Non-Cryst Solids 186:219–231, 1995.

266. MH Nguyen, LH Dao. J Non-Cryst Solids 225:51–57, 1998.

267. H Tamon, H Ishizaka, T Araki, M Okazaki, Carbon 36:1257–1262, 1998.

268. V Bock, A Emmerling, J Fricke. J Non-Cryst Solids 225:69–73, 1998.

269. GAM Reynolds, AWP Fung, ZH Wang, MS Dresselhaus, RW Pekala. J Non-Cryst Solids 188:27–33, 1995.

270. RW Pekala, CT Alviso, X Lu, J Gross, J Fricke. J Non-Cryst Solids 188:34–40, 1995.

271. B Mathieu, S Blacher, R Pirard, B Sahouli, F Brouers. J Non-Cryst Solids 212: 250–261, 1997.

272. R Kocklenberg, B Mathieu, S Blacher, R Pirard, JP Pirard, R Sobry, G Van den Bossche. J Non-Cryst Solids 225:8–13, 1998.

273. C Lin, JA Ritter. Carbon 38:849–861, 2000.

274. DW Schaefer. Science 243:1023–1027, 1989.

275. WmR Even, RW Crocker, MC Hunter, NC Yang, TJ Headly. J Non-Cryst Solids 186:191–199, 1995.

276. RW Pekala, ST Mayer, JL Kaschmitter, FM Kong. In: YA Attia, ed. Sol-Gel Processing and Applications. New York: Plenum Press, 1994, pp. 369–377.

277. AWP Fung, ZH Wang, K Lu, MS Dresselhaus, RW Pekala. J Mater Res 8:1875–1885, 1993.

278. RW Pekala, CT Alviso. Mater Res Soc Symp Proc 270:3, 1992.

279. ST Mayer, RW Pekala, JL Kaschmitter. J Electrochem Soc 140:446, 1993.

280. U Fischer, R Saliger, V Bock, R Petricevic, J Fricke. J Porous Mater 4:281, 1997.

281. X Lu, MC Arduini-Schuster, J Kuhn, O Nilsson, J Fricke, RW Pekala. Science 255:971, 1992.

282. J Gross, J Fricke, LW Hrubesh. J Acoust Soc Am 91:2004, 1992.

283. RW Pekala, JC Farmer, CT Alviso, TD Tran, ST Mayer, JM Miller, B Dunn. J Non-Cryst Solids 225:74–80, 1998.

284. H Tamon, H Ishizaka, T Yamamoto, T. Suzuki. Carbon 37:2049–2055, 1999.

285. AL Mackay, H Terrones, Nature 352:762, 1991.

286. TJ Lenosky, X Gonze, M Teter, V Elser. Nature 355:333–335, 1992.

287. D Vanderbilt, J Tersoff. Phys Rev Lett 68:511–513, 1992.

288. M O'Keeffe, GB Adams, OF Sankey. Phys Rev Lett 68:2325–2328, 1992.

289. SJ Townsend, TJ Lenosky, DA Muller, CS Nicholas, V Elser. Phys Rev Lett 69: 921–924, 1992.

290. H Terrones, M Terrones. Phys Rev B 55:9969–9974, 1997.

291. SC Tsang, PJF Harris, JB Claridge, MLH Green. J Chem Soc Chem Commun 1519–1522, 1997.

292. LN Bourgeois, LA Bursill. Phil Mag A 76:753–768, 1997.

293. D Donadio, L Colombo, P Milani, G Benedek. Phys Rev Lett 83:776–779, 1999.

294. MM Haley, SC Brand, JJ Pak. Angew Chem Int Ed Engl 36:836–838, 1997.

295. F Diederich. Nature 369:199–207, 1994.

296. AT Balaban, CC Rentia, E Ciupitu. Rev Roum Chim 13:231, 1968.

297. RB Heimann. In: SE Evsyukov, L Kavan, eds. Carbyne and Carbynoid Structures. Dordrecht: 1999, pp. 139–148

298. CM Lieber, CC Chen. Solid State Phys 48:109–148, 1994.

299. C Journet, P Bernier. Appl Phys A 67:1–9, 1998.

300. M Razvigorova, T Budinova, N Petrov, V Minkova. Wat Res 32:2135–2139, 1998.

301. G Kovacik, B Wong, E Furimsky. Fuel Process Technol 41:89–99, 1995.

302. JS Qiu, Y Zhou, ZG Yang, DK Wang, SC Guo, SC Tsang, PJF Harris. Fuel 79: 1303–1308, 2000.

303. LSK Pang, AM Vassallo, MA Wilson. Energy Fuels 6:176–179, 1992.

304. M Jagtoyen, M Thwaites, J Stencel, B McEnaney, F Derbyshire. Carbon 30:1089–1096, 1992.

305. YC Chiang, PC Chiang, EE Chang. Chemosphere 37:237–247, 1998.

306. KA Williams, M Tachibana, JL Allen, L Grigorian, SC Cheng, SL Fang, GU Sumanasekera, AL Loper, JH Williams, PC Eklund. Chem Phys Lett 310:31–37, 1999.

307. AK Dalai, J Zaman, ES Hall, EL Tollefson. Fuel 75:227–237, 1996.

308. OC Kopp, EL Fuller, CR Sparks, MR Rogers, ML McKinney. Carbon 35:1765–1779, 1997.

309. JS Qiu, Y Zhou, LN Wang, SC Tsang. Carbon 36:465–467, 1998.

310. C Toles, S Rimmer, JC Hower. Carbon 34:1419–1426, 1996.

311. XQ Lu, DR Smith, DW Johnson, HR Rose, MA Wilson. Carbon 34:1145–1162, 1996.

312. E Tütem, R Apak, CF Ünal. Wat Res 32:2315–2324, 1998.

313. K Fisher, C Largeau, S Derenne. Org Geochem 24:715–723, 1996.
314. HC Messman. ACS Symp Series 21:77–84, 1976.
315. DD Edie. Carbon 36:345–362, 1998.
316. M Suzuki. Carbon 32:577–586, 1994.
317. R Zhang, Y Achiba, KJ Fisher, GE Gadd, FG Hopwood, T Ishigaki, DR Smith, S Suzuki, GD Willett. J Phys Chem 103:9450–9458, 1999.
318. J Osterodt, A Zett, F Vögtle. Tetrahedron 52:4949–4962, 1996.
319. KH Homann. Angew Chem Int Ed 37:2434–2451, 1998.
320. M Ozawa, P Deota, E Osawa. Fullerene Sci Technol 7:387–409, 1999.
321. A Sokolowska, A Olszyna. In: SE Evsyukov, L Kavan, eds. Carbyne and Carbynoid Structures. Dordrect: Kluwer Academic, 1999, pp. 117–131.
322. YuP Kudryavtsev. In: RB Heimann, SE Evsyukov, L Kavan, eds. Carbyne and Carbynoid Structures. Dordrecht: Kluwer Academic, 1999, pp. 39–45.
323. IA Udod. In: RB Heimann, SE Evsyukov, L Kavan, eds. Carbyne and Carbynoid Structures. Dordrecht: Kluwer Academic, 1999, pp. 93–116.
324. M Yumura. In: The Science and Technology of Carbon Nanotubes. Amsterdam: Elsevier, 1999, Chapter 2.
325. R Arriagada, R García, P Reyes. J Chem Tech Biotechnol 60:427–435, 1994.
326. M Jagtoyen, F Derbyshire. Carbon 36:1085–1097, 1998.
327. Z Hu, MP Srinivasan. Micropor Mesopor Mater 27:11–18, 1999.
328. K Gergova, N Petrov, S Eser. Carbon: 693–702, 1994.
329. W Heschel, E Klose. Fuel 74:1786–1791, 1995.
330. S Ogasawara, M Kuroda, N Wakao. Ind Eng Chem Res 26:2556–2557, 1987.
331. JL Allen, JL Gatz, PC Eklund. Carbon 1485–1489, 1999.
332. GQ Lu, DD Lau. Gas Sep Purif 10:103–111, 1996.
333. MJ Martin, MD Balaguer, M Rigola. Environ Technol 17:667–672, 1996.
334. BR Kim, EM Kalis, IT Salmeen, CW Kruse, I Demir, SL Carlson, M Rostam-Abadi. J Environ Eng ASCE 532–539, 1996.
335. H Lutz, GA Romeiro, RN Damasceno, M Kutubuddin, E Bayer. Bioresource Technol 74:103–107, 2000.
336. J Hayashi, A Kazehaya, K Muroyama, AP Watkinson. Carbon 38:1873–1878, 2000.
337. M Acharya, BA Raich, HC Foley, HP Harold, JJ Lerou. Ind Eng Chem Res 36: 2924–2930, 1997.
338. MB Rao, S Sircar. J Membrane Sci 85:253–264, 1993.
339. TA Centeno, AB Fuertes. Carbon 38:1067–1073, 2000.
340. K Kusakabe, M Yamamoto, S Morooka. J Membrane Sci 149:59–67, 1998.
341. EEB Campbell, G Ulmer, B Hassalberger, HG Busmann, IV Hertel. J Chem Phys 93:6900–6907, 1990.
342. TA Centeno, AB Fuertes. J Membrane Sci 160:201–211, 1999.
343. H Kita, M Yoshino, K Tanaka, K Okamoto. J Chem Soc Chem Commun 1051–1052, 1997.
344. P Delhaes, F Carmona. In: P Walker, PA Thrower, eds. Chemistry and Physics of Carbon, Vol. 17. New York: Marcel Dekker, 1971, pp. 89–174.
345. T Wigmans. In A Capelle, F de Vooys, eds. Activated Carbon: A Fascinating Material, Amersfoort, The Netherlands: Norit NV, 1983, pp. 58–80.

346. F Derbyshire, M Jagtoyen, M Thwaites. In: JW Patrick, ed. Porosity in Carbons. London: Edward Arnold, 1995, pp. 229.
347. JC Gonzalez, MT Gonzalez, M Molina-Sabio, F Rodriguez-Reinoso, A Sepulveda-Escribano. Carbon 33:1175–1177, 1995.
348. F Stoeckli, E Daguerre, A Guillot. Carbon 37:2075–2077, 1999.
349. AC O'Sullivan. Cellulose 4:173–207, 1997.
350. EA Bayer, H Chanzy, R Lamed, Y Shoham. Curr Opin Struct Biol 8:548–557, 1998.
351. RM Brown. Pure Appl Chem 71:767–775, 1999.
352. F Derbyshire, A Marzec, HR Schulten, MA Wilson, A Davis, P Tekely, JJ Delpuech, A Jurkiewicz, CE Bronnimann, RA Wind, GE Maciel, R Narayan, K Bartle, C Snape. Fuel 68:1091–1106, 1989.
353. M Nishioka. Fuel 72:1719–1724, 1993.
354. K Qin, S Guo, S Li. Chinese Sci Bull 43:2025–2034, 1998.
355. SM Shevchenko. Croatica Chemica Acta 67:95–124, 1994.
356. JL Faulon, GA Carlson, PG Hatcher. Org Geochem 21:1169–1179, 1994.
357. SM Shevchenko, GW Bailey. J Mol Struct Theochem 364:197–208, 1996.
358. HM Colquhoun, DJ Williams. Acc Chem Res 33:189–198, 2000.
359. F Khorasheh, R Khaledi, MR Gray. Fuel 77:247–253, 1998.
360. JL Faulon, M Vandenbroucke, JM Drappier, F Behar, M Romero. Org Geochem 16:981–993, 1989.
361. JL Faulon. J Chem Inform Comput Sci 32:338–348, 1992.
362. JL Faulon. J Chem Inform Comput Sci 34:1204–1218, 1994.
363. JL Faulon. Chemtech 25:16–23, 1995.
364. JL Faulon. J Chem Inform Comput Sci 36:731–740, 1996.
365. M Nomura, K Matsubayashi, T Ida, S Murata. Fuel Process Technol 31:169–179, 1992.
366. T Ohkawa, T Sasai, N Komoda, S Murata, M Nomura. Energy Fuels 11:937–944, 1997.
367. S Murata, M Nomura, K Nakamura, H Kumagai, Y Sanada. Energy Fuels 7:469–472, 1993.
368. M Nomura, L Artok, S Murata, A Yamamoto, H Hama, H Gao, K Kidena. Energy Fuels 12:512–523, 1998.
369. PG Hatcher, JL Faulon, KA Wenzel, GD Cody. Energy Fuels 6:813–820, 1992.
370. JL Faulon, PG Hatcher, GA Carlson, KA Wenzel. Fuel Process Technol 34:277–293, 1993.
371. JL Faulon, PG Hatcher. Energy Fuels 8:402–407, 1994.
372. I Kowalewski, M Vandenbroucke, AY Huc, MJ Taylor, JL Faulon. Energy Fuels 10:97–107, 1996.
373. V Calemma, R Rausa, P D'Antona, L Montanari. Energy Fuels 12:422–428, 1998.
374. YF Zhang, F Li, KC Xie. Fuel Sci Technol Intl 13:957–972, 1995.
375. A Nabeel, MA Khan, S Husain, B Krishnamacharyulu, RN Rao, DK Sharma. Energy Sources 22:57–65, 2000.
376. The processes and technology associated with the manufacture of powdered, granular, and monolithic porous carbons from natural precursors have been reviewed many times, including: a) M Smísek, S Cerný. Active Carbon: Manufacture, Prop-

erties and Applications, Amsterdam: Elsevier, 1967, Chapter 2; b) JW Hassler. Activated Carbon, London: Leonard Hill Books, 1967, Chapter 8; c) A Yehaskel. Activated Carbon: Manufacture and Regeneration, Park Ridge, NJ: Noyes Data Corporation, 1978; d) J Wilson. Fuel 60:823–831, 1981); e) RC Bansal, J-B Donnet, F Stoeckli, Active Carbon. New York: Marcel Dekker, 1988, Chapter 1; f) H Jankowska, A Swiatkowski, J Choma, Active Carbon. Chichester: Ellis Horwood, 1991, Chapter 2; and g) F Derbyshire, M Jagtoyen, M Thwaites. In: JW Patrick, ed. Porosity in Carbons. London: Edward Arnold, 1995, pp. 233–242.

377. To our knowledge, there is no extensive review of the processes and technology used in the production of carbon gels, most likely due to the low levels of commercial production. However, references 263, 273, 276, and 284 outline aspects of preparation of carbon gels in the laboratory whereas the following article discusses the technology, markets, and costs associated with the commercialization of aerogels: G Carlson, D Lewis, K McKinley, J Richardson, T Tillotson. J Non-Cryst Solids 186:372–379, 1995.

378. The processes and technology associated with the manufacture of carbon fiber and activated carbon fiber have been reviewed from time to time, including references 315 and 316 and: a) A Yehaskel. Activated Carbon: Manufacture and Regeneration, Park Ridge, NJ: Noyes Data Corporation, 1978; b) M Sittig. Carbon and Graphite Fibers: Manufacture and Applications. Park Ridge, NJ: Noyes Data Corporation, 1980; c) W Watt, BV Perov, eds. Strong Fibers. Amsterdam: North-Holland, 1985, Chapters 8, 12, and 13; d) IN Ermolenko, IP Lyubliner, NV Gulko, Chemically Modified Carbon Fibres. Weinheim: VCH, 1990; and e) J-B Donnet, TK Wang, JCM Peng, S Rebouillat, eds. Carbon Fibres. New York: Marcel Dekker, 1998.

379. To our knowledge, there is currently no review of the processes and technology used in the production of porous carbon membranes, most likely due to the relatively low levels of commercialization. However, as carbon membranes are largely derived from the pyrolysis of organic membranes, of which there is a high level of commercialization, the reader is referred to the following reviews of the manufacture processes for organic membrane: a) M Mulder. Basic Principles of Membrane Technology. Dordrecht: Kluwer Academic, 1991, Chapter 3; b) I Pinnau, BD Freeman. ACS Symp Series 744:1–22, 2000; c) TS Chung. Polym Polym Compos 4:269–283, 1996; d) HP Hsieh. Inorganic Membranes for Separation and Reaction Amsterdam: Elsevier, 1996, pp. 70–71. References 337–340, 342, and 343 consider the preparation of specific types of microporous carbon membranes from organic membranes.

380. The vast patent literature for Europe, the United States, Japan, and many other countries can now be searched on-line http://epespacenetcom.

381. Z Hu, EF Vansant. Carbon 33:561–567, 1995.

382. M Molina-Sabio, F Rodríguez-Reinoso, F Caturla, MJ Sellés. Carbon 34:457–462, 1996.

383. V Gómez-Serrano, J Pastor-Villegas, A Perez-Florindo, CJ Durán-Valle, C Valenzuela-Calahorro. J Anal Appl Pyrol 36:71–80, 1996.

384. J Pastor-Villegas, CJ Durán-Valle, C Valenzuela-Calahorro, V Gómez-Serrano. Carbon 36:1251–1256, 1998.

385. H Marsh. Chem Soc Special Pub 32:133–174, 1978.

386. PL Walker, RL Taylor, JM Ranish. Carbon 29:411–421, 1991.

387. JA Moulijn, F Kapteijn. Carbon 33:1155–1165, 1995.

388. SG Chen, RT Yang. Energy Fuels 11:421–427, 1997.

389. SG Chen, RT Yang. J Phys Chem A 102:6348–6356, 1998.

390. F Rodríguez-Reinoso, M Molina-Sabio. Carbon 30:1111–1118, 1992.

391. M Jagtoyen, F Derbyshire. Carbon 31:1185–1192, 1993.

392. MS Solum, RJ Pugmire, M Jagtoyen, F Derbyshire. Carbon 33:1247–1254, 1995.

393. M Jagtoyen, F Derbyshire, S Rimmer, R Rathbone. Fuel 74:610–614, 1995.

394. H Benaddi, TJ Bandosz, J Jagiello, JA Schwarz, JN Rouzaud, D Legras, F Béguin. Carbon 38:669–674, 2000.

395. H Teng, TS Yeh, LY Hsu. Carbon 36:1387–1395, 1998.

396. LY Hsu, H Teng. Fuel Process Technol 64:155–166, 2000.

397. P Ehrburger, A Addoun, F Addoun, JB Donnet. Fuel 65:1447–1449, 1986.

398. YuO Begak, SD Kolosentsev, NF Fedorov. J Appl Chem USSR 65:1957–1959, 1992.

399. B Taraba, Fuel, 73:1679–1681, 1994.

400. HM Cheng, M Liu, ZH Shen, JZ Xi, H Sano, Y Uchiyama, K Kobayashi. Carbon 35:869–874, 1997.

401. H Wakayama, J Mizuno, Y Fukushima, K Nagano, T Fukunaga, U Mizutani. Carbon 37:947–952, 1999.

402. F Salver–Disma, JM Tarascon, C Clinard, JN Rouzaud. Carbon 37:1941–1959, 1999.

403. B Rubio, MT Izquierdo, E Segura. Carbon 37:1833–1841, 1999.

404. R Sakurovs. Fuel 76:615–621, 1997.

405. R Sakurovs. Energy Fuels 12:631–636, 1998.

406. R Sakurovs. Fuel 79:379–389, 2000.

407. DK Sharma, SK Singh. Energy Sources 17:485–493, 1995.

408. M Krzesinska. Energy Fuels 11:686–690, 1997.

409. P Tekely, D Nicole, JJ Delpuech, E Totino, JF Muller. Fuel Process Technol 15:225–231, 1987.

410. DV Franco, JM Gelan, HJ Martens, DJM Van der Zande. Fuel 70:811–817, 1991.

411. EW Hagaman, L Farrow, EG Galipo. Solid State Nucl Magn Reson 16:69–75, 2000.

412. JW Larsen, CS Pan, S Shawver. Energy Fuels 3:557–561, 1989.

413. J Kister, M Guiliano, G Mille, H Dou. Fuel 67:1076–1082, 1988.

414. OP Mahajan, PL Walker. Fuel 58:333–336, 1979.

415. RG Jenkins, SP Nandi, PL Walker. Fuel 52:288–293, 1973.

416. KJ Hüttinger, C Nattermann. Fuel 73:1682–1684, 1994.

417. MJ Muñoz-Guillena, MJ Illán-Gómez, JM Martín-Martínez, A Linares-Solano, C Salinas-Martínez de Lecea. Energy Fuels 6:9–15, 1992.

418. MV López-Ramón, C Moreno-Castilla, J Rivera-Utrilla, R Hidalgo-Alvarez. Carbon 31:815–819, 1993.

419. P Samaras, E Diamadopoulos, GP Sakellaropoulos. Carbon 32:771–776, 1994.

420. T Alvarez, AB Fuertes, JJ Pis, JB Parra, J Pajares, R Menéndez. Fuel 73:1358–1364, 1994.

421. H Teng, JA Ho, YF Hsu. Carbon 35:275–283, 1997.

422. MSA Baksh, RT Yang, DDL Chung. Carbon 27:931–934, 1989.
423. Y Kawabuchi, M Kishino, S Kawano, DD Whitehurst, I Mochida. Langmuir 12: 4281–4285, 1996.
424. Y Kawabuchi, C Sotowa, M Kishino, S Kawano, DD Whitehurst, I Mochida. Langmuir 13:2314–2317, 1997.
425. Y Kawabuchi, C Sotowa, M Kishino, S Kawano, DD Whitehurst, I Mochida. Carbon 36:377–382, 1998.
426. RC Bansal, JB Donnet, F Stoeckli. Active Carbon. New York: Marcel Dekker, 1988, Chapter 5.
427. CL Mangun, KR Benak, MA Daley, J Economy. Chem Mater 11:3476–3483, 1999.
428. BK Pradhan, NK Sandle. Carbon 37:1323–1332, 1999.
429. AB García, A Martínez-Alonso, CA Leon y Leon, JMD Tascón. Fuel 77:613–624, 1997.
430. DH Parker, K Chatterjee, P Wurz, KR Lykke, MJ Pellin, LM Stock. Carbon 30: 1167–1182, 1992.
431. LD Lamb, DR Huffman. J Phys Chem Solids 54:1635–1643, 1993.
432. H Singh, M Srivastava. Energy Sources 17:615–640, 1995.
433. JB Howard, AL Lafleur, Y Makarovsky, S Mitra, CJ Pope, TK Yadav. Carbon 30: 1183–1201, 1992.
434. EA Rohlfing, DM Cox, A Kaldor. J Chem Phys 81:3322–3330, 1984.
435. RE Haufler, Y Chai, LPF Chibante, J Conceicao, C Jin, LS Wang, S Maruyama, RE Smalley. Mater Res Soc Symp Proc 206:627–637, 1991.
436. RE Haufler, J Conceicao, LPF Chibante, Y Chai, NE Byrne, S Flanagan, MM Haley, SC O'Brien, C Pan, Z Xiao, WE Billups, MA Ciufolini, RH Hauge, JL Margrave, LJ Wilson, RF Curl, RE Smalley. J Phys Chem 94:8634–8636, 1990.
437. W Krätschmer, K Fostiropoulos, DR Huffman. Chem Phys Lett 170:167–170, 1990.
438. GN Churilov, LA Solovyov, YN Churilov, OV Chupina, SS Malcieva. Carbon 37: 427–431, 1999.
439. T Sugai, H Omote, S Bandow, N Tanaka, H Shinohara. J Chem Phys 112:6000–6005, 2000.
440. L Fulcheri, Y Schwob, F Fabry, G Flamant, LFP Chibante, D Laplaze. Carbon 38: 797–803, 2000.
441. G Peters, M Jansen. Angew Chem Int Ed Engl 31:223–224, 1992.
442. H Feld, R Zurmühlen, A Leute, A Benninghoven. J Phys Chem 94:4595–4599, 1990.
443. G Brinkmalm, P Demirev, D Fenyö, P Håkansson, J Kopniczky, BUR Sundqvist. Phys Rev B 47:7560–7567, 1993.
444. EG Gamaly, LT Chadderton. Proc R Soc London A 449:381–409, 1995.
445. RF Bunshah, S Jou, S Prakash, HJ Doerr, L Isaacs, A Wehrsig, C Yeretzian, H Cynn, F Diederich. J Phys Chem 96:6866–6869, 1992.
446. CL Fields, JR Pitts, MJ Hale, C Bingham, A Lewandowski, DE King. J Phys Chem 97:8701–8702, 1993.
447. D Laplaze, P Bernier, G Flamant, M Lebrun, A Brunelle, S Della-Negra. J Phys B 29:4943–4954, 1996.

448. L Alvarz, T Guillard, G Olalde, B Rivoire, JF Robert, P Bernier, G Flamant, D Laplaze. Synth Metals 103:2476–2477, 1999.
449. JR Heath. ACS Symp Series 481:1–22, 1991.
450. HW Kroto, DRM Walton. In: E Osawa, O Yonemitsu, eds. Carbocyclic Cage Compounds: Chemistry and Applications. New York: VCH, 1992, Chapter 3.
451. JM Hunter, JL Fye, EJ Roskamp, MF Jarrold. J Phys Chem 98:1810–1818, 1994.
452. RF Curl, PC Haddon. Phil Trans R Soc Lond A 343:19–32, 1993.
453. NS Goroff. Acc Chem Res 29:77–83, 1996.
454. X Gao, J Gao. J Phys Chem 98:5618–5621, 1994.
455. RK Mishra, YT Lin, SL Lee. J Chem Phys 112:6355–6364, 2000.
456. SY Xie, RB Huang, J Ding, LJ Yu, YH Wang, LS Zheng. J Phys Chem A 104: 7161–7164, 2000.
457. G von Helden, MT Hsu, N Gotts, MT Bowers. J Phys Chem 97:8182–8192, 1993.
458. AA Shvartsburg, RR Hudgins, P Dugourd, R Gutierrez, T Frauenheim, MF Jarrold. Phys Rev Lett 84:2421–2424, 2000.
459. T Belz, H Werner, F Zemlin, U Klengler, M Wesemann, B Tesche, E Zeitler, A Reller, R Schlögl. Angew Chem Int Ed Engl 33:1866–1869, 1994.
460. WR Creasy. J Chem Phys 92:7223–7233, 1990.
461. IS Bitensky, P Demirev, BUR Sundqvist. Nucl Instr Meth B 82:356–361, 1993.
462. A Weston, M Murthy. Carbon 34:1267–1274, 1996.
463. L Láska, J Krása, L Juha, V Hamplová, L Soukup. Carbon 34:363–368, 1996.
464. P Gerhardt, S Löffler, KH Homann. Chem Phys Lett 137:306–310, 1987.
465. JB Howard, JT McKinnon, Y Makarovsky, AL Lafleur ME Johnson. Nature 352: 139–141, 1991.
466. M Ozawa, E Osawa. Carbon 37:707–709, 1999.
467. R Taylor, GJ Langley, HW Kroto, DRM Walton. Nature 366:728–731, 1993.
468. C Crowley, R Taylor, HW Kroto, DRM Walton, PC Cheng, LT Scott. Synth Metals 77:17–22, 1996.
469. T Minakata. Polym Adv Technol 6:586–590, 1995.
470. GM Jenkins, LR Holland, H Maleki, J Fisher. Carbon 36:1725–1727, 1998.
471. M Ehbrecht, M Faerber, F Rohmund, VV Smirnov, O Stelmach, F Huisken. Chem Phys Lett 214:34–38, 1993.
472. X Armand, N Herlin, I Voicu, M Cauchetier. J Phys Chem Solids 58:1853–1859, 1997.
473. S Petcu, M Cauchetier, X Armand, I Voicu, R Alexandrescu. Combust Flame 122: 500–507, 2000.
474. T Oyama, S Osawa, K Takeuchi. Carbon 36:1236–1238, 1998.
475. K Yoshie, S Kasuya, K Eguchi, T Yoshida. Appl Phys Lett 61:2783, 1992.
476. H Koinuma, T Horiuchi, K Inomata, HK Ha, K Nakajima, KA Chaudhary. Pure Appl Chem 68:1151–1154, 1996.
477. T Alexakis, PG Tsantrizos, YS Tsantrizos, JL Meunier. Appl Phys Lett 70:2102–2104, 1997.
478. T Baum, S Löffler, P Löffler, P Weilmünster, KH Homann. Int J Mass Spectrom Ion Processes 138:133–148, 1994.
479. CJ Pope, JA Marr, JB Howard. J Phys Chem 97:11001–11013, 1993.

480. C Yeretzian, JB Wiley, K Holczer, T Su, S Nguyyen, RB Kaner, RL Whetten. J Phys Chem 97:10097–10101, 1993.
481. PH Fang. Fullerene Sci Technol 6:167–173, 1998.
482. J Théobald, M Perrut, JV Weber, E Millon, JF Muller. Separ Sci Technol 30:2783–2819, 1995.
483. K Jinno, Y Saito. Adv Chromatogr 36:65–125, 1996.
484. N Coustel, P Bernier, R Aznar, A Zahab, JM Lambert, P Lyard. J Chem Soc Chem Commun 1402–1403, 1992.
485. XH Zhou, ZN Gu, YQ Wu, YL Sun, ZX Jin, Y Xiong, BY Sun, Y Wu, H Fu, JZ Wang. Carbon 32:935–937, 1994.
486. RJ Doome, A Fonseca, H Richter, JB Nagy, PA Thiry, AA Lucas. J Phys Chem Solids 58:1839–1843, 1997.
487. JL Atwood, GA Koutsantonis, CL Raston. Nature 368:229–231, 1994.
488. RD Averitt, JM Alford, NJ Halas. Appl Phys Lett 65:374–376, 1994.
489. K Lynch, C Tanke, F Menzel, W Brockner, P Scharff, E Stumpp. J Phys Chem 99:7985–7992, 1995.
490. AR McGhie, JE Fischer, PA Heiney, PW Stephens, RL Cappelletti, DA Neumann, WH Mueller, H Mohn, HU ter Meer. Phys Rev B 49:12614–12618, 1994.
491. C Möschel, M Jansen. Chem Ber–Recl 130:1761–1764, 1997.
492. C Möschel, M Jansen. Z Anorg Allg Chem 625:175–177, 1999.
493. SW McElvany. J Phys Chem 96:4935–4937, 1992.
494. T Hirata, R Hatakeyama, T Mieno, N Sato. J Vac Sci Technol A 14:615–618, 1996.
495. EEB Campbell, R Tellgmann, N Krawez, IV Hertel. J Phys Chem Solids 58:1763–1769, 1997.
496. T Almeida Murphy, T Pawlik, A Weidinger, M Höhne, R Alcala, JM Spaeth. Phys Rev Lett 77:1075–1078, 1996.
497. T Braun, H Rausch. Chem Phys Lett 237:443–447, 1995.
498. GE Gadd, PJ Evans, DJ Hurwood, PL Morgan, S Moricca, N Webb, J Holmes, G McOrist, T Wall, M Blackford, D Cassidy, M Elcombe, JT Noorman, P Johnson, P Prasad. Chem Phys Lett 270:108–114, 1997.
499. Y Rubin. Chem Eur J 3:1009–1016, 1997.
500. FD Weiss, JL Elkind, SC O'Brien, RF Curl, RE Smalley. J Am Chem Soc 110:4464–4465, 1988.
501. S Saito, S Sawada. Chem Phys Lett 198:466–471, 1992.
502. DE Clemmer, KB Shelimov, MF Jarrold. Nature 367:718–720, 1994.
503. KB Shelimov, MF Jarrold. J Am Chem Soc 118:1139–1147, 1996.
504. FZ Cui, DX Liao, HD Li. Phys Lett A 195:156–162, 1994.
505. EEB Campbell, R Ehlich, G Heusler, O Knospe, H Sprang. Chem Phys 239:299–308, 1998.
506. S Patchkovskii, W Thiel. J Am Chem Soc 120:556–563, 1998.
507. Y Rubin. Topics in Current Chemistry 199:67–91, 1999.
508. C Ray, M Pellarin, J Lermé, JL Vialle, M Broyer, X Blase, P Mélinon, P Kéghélian, A Perez. Phys Rev Lett 80:5365–5368, 1998.
509. CNR Rao, T Pradeep, R Seshadri, A Govindaraj. Indian J Chem A 31:F27–F31, 1992.

510. HJ Muhr, R Nesper, B Schnyder, R Kötz. Chem Phys Lett 249:399–405, 1996.
511. ZC Ying, JG Zhu, RN Compton, LF Allard, RL Hettich, RE Haufler. ACS Symp Ser 679:169–182, 1997.
512. KB Shelimov, DE Clemmer, MF Jarrold. J Chem Soc Dalton Trans 567–574, 1996.
513. J Mattay, G Torres-Garcia, J Averdung, C Wolff, I Schlachter, H Luftmann, C Siedschlag, P Luger, M Ramm. J Phys Chem Solids 58:1929–1937, 1997.
514. P Weis, RD Beck, G Bräuchle, MM Kappes. J Chem Phys 100:5684–5695, 1994.
515. M Pellarin, C Ray, J Lermé, JL Vialle, M Broyer, P Mélinon. J Chem Phys 112: 8436–8445, 2000.
516. A Hirsch. Synthesis (Stuttgart) 895–913, 1995.
517. W Sliwa. Transition Met Chem 21:583–592, 1996.
518. J Chlistunoff, D Cliffel, AJ Bard. Thin Solid Films 257:166–184, 1995.
519. CA Mirkin, WB Caldwell. Tetrahedron 52:5113–5130, 1996.
520. C Jehoulet, YS Obeng, YT Kim, F Zhou, AJ Bard. J Am Chem Soc 114:4237–4247, 1992.
521. W Koh, D Dubois, W Kutner, MT Jones, KM Kadish. J Phys Chem 96:4163–4165, 1992.
522. FX Cheng, NQ Li, WJ He, ZN Gu, XH Zhou, YL Sun, YQ Wu. J Electroanal Chem 408:101–105, 1996.
523. AF Hebard, TTM Palstra, RC Haddon, RM Fleming. Phys Rev B 48:9945–9948, 1993.
524. A Fartash. Appl Phys Lett 64:1877–1879, 1994.
525. D Stifter, H Sitter. Appl Phys Lett 66:679–681, 1995.
526. EA Katz, D Faiman, S Shtutina, A Isakina. Thin Solid Films 368:49–54, 2000.
527. RL Meng, D Ramirez, X Jiang, PC Chow, C Diaz, K Matsuishi, SC Moss, PH Hor, CW Chu. Appl Phys Lett 59:3402–3403, 1991.
528. JZ Liu, JW Dykes, MD Lan, P Klavins, RN Shelton, MM Olmstead. Appl Phys Lett 62:531–532, 1993.
529. M Manfredini, C Castoldi, G Casamassima. P Milani, Mater Sci Eng B39:62–65, 1996.
530. S Henke, KH Thürer, JKN Lindner, B Rauschenbach, B Stritzker. J Appl Phys 76: 3337–3340, 1994.
531. ZM Ren, ZF Ying, XX Xiong, MQ He, YF Li, FM Li, YC Du. J Phys D Appl Phys 27:1499–1503, 1994.
532. K Tanigaki, S Kuroshima, TW Ebbesen. Thin Solid Films 257:154–165, 1995.
533. J Kong, AM Cassell, H Dai. Chem Phys Lett 292:567–574, 1998.
534. P Nikolaev, MJ Bronikowski, RK Bradley, F Rohmund, DT Colbert, KA Smith, RE Smalley. Chem Phys Lett 313:91–97, 1999.
535. ML Terranova, S Piccirillo, V Sessa, P Sbornicchia, M Rossi, S Botti, D Manno. Chem Phys Lett 327:284–290, 2000.
536. R Saito, G Dresselhaus, MS Dresselhaus. Physical Properties of Carbon Nanotubes. London: Imperial College Press, 1998, Chapter 5.
537. K Tanaka, M Endo, K Takeuchi, WK Hsu, HW Kroto, M Terrones, DRM Walton. In: K Tanaka, T Yamabe, K Fukui, eds. The Science and Technology of Carbon Nanotubes. Amsterdam: Elsevier, 1999, Chapter 12.
538. TW Ebbesen, PM Ajayan. Nature 358:220–222, 1992.

539. X Zhao, M Ohkohchi, M Wang, S Iijima, T Ichihashi, Y Ando. Carbon 35:775–781, 1997.
540. AG Rinzler, J Liu, H Dai, P Nikolaev, CB Huffman, FJ Rodriguez-Macias, PJ Boul, AH Lu, D Heymann, DT Colbert, RS Lee, JE Fischer, AM Rao, PC Eklund, RE Smalley. Appl Phys A 67:29–37, 1998.
541. GS Duesberg, J Munster, HJ Byrne, S Roth, M Burghard. Appl Phys A 69:269–274, 1999.
542. XF Zhang, XB Zhang, G Van Tendeloo, S Amelinckx, M Op de Beeck, J Van Landuyt. J Crystal Growth 130:368–382 1993.
543. J Bernholc, C Brabec, M Buongiorno Nardelli, A Maiti, C Roland, BI Yakobson. Appl Phys A 67:39–46, 1998.
544. OA Louchev, Y Sato. Appl Phys Lett 74:194–196, 1999.
545. S Amelinckx, XB Zhang, D Bernaerts, XF Zhang, V Ivanov, JB Nagy. Science 265:635–639, 1994.
546. A Fonseca, K Hernadi, JB Nagy, Ph Lambin, AA Lucas. Carbon 33:1759–1775, 1995.
547. P Nikolaev, MJ Bronikowski, RK Bradley, F Rohmund, DT Colbert, KA Smith, RE Smalley. Phys Chem Lett 313:91–97, 1999.
548. T Guo, P Nikolaev, A Thess, DT Colbert, RE Smalley. Phys Chem Lett 243:49–54, 1995.
549. YH Lee, SG Kim, D Tománek. Phys Rev Lett 78:2393–2396, 1997.
550. CH Kiang, WA Goddard. Phys Rev Lett 76:2515–2518, 1996.
551. CH Kiang. J Chem Phys 113:4763–4766, 2000.
552. F Banhart, T Füller, Ph Redlich, PM Ajayan. Chem Phys Lett 269:349–355, 1997.
553. T Burger, J Fricke. Ber Bunsenges Phys Chem 102:1523–1528, 1998.
554. CJ Brinker, GW Scherer. Sol-Gel Science. San Diego: Academic Press, 1990.
555. RW Pekala, FM Kong. J Phys Paris Colloq C 4:33, 1989.
556. RW Pekala. J Mater Sci 24:3221–3227, 1989.
557. AWP Fung, ZH Wang, K Lu, MS Dresselhaus, RW Pekala. J Mater Res 8:1875–1885, 1993.
558. GAM Reynolds, AWP Fung, ZH Wang, MS Dresselhaus, RW Pekala. J Non-Crystalline Solids 188:27–33, 1995.
559. H Tamon, H Ishizaka, M Mikami, M Okazaki. Carbon 35:791–796, 1997.
560. R Saliger, U Fischer, C Herta, J Fricke. J Non-Crystalline Solids 225:81–85, 1998.
561. V Bock, A Emmerling, J Fricke. J Non-Crystalline Solids 225:69–73, 1998.
562. Y Hanzawa, K Kaneko, N Yoshizawa, RW Pekala, MS Dreselhaus. Adsorption 4:187–195, 1998.
563. SQ Zhang, J Wang, J Shen, ZS Deng, ZQ Lai, B Zhou, SM Attia, LY Chen. Nanostruct Mater 11:375–381, 1999.
564. E Hoinkis. In PA Thrower, ed. Chemistry and Physics of Carbon, Vol. 25. New York: Marcel Dekker, 1997, pp. 71–241
565. YG Andreev, T Lundström. J Appl Cryst 28:534–539, 1995.
566. T Suzuki, K Kaneko. Carbon 26:742–746, 1998.
567. K Kaneko, T Suzuki, K Kakei. Langmuir 5:879–881, 1989.
568. C Ishii, T Suzuki, N Shindo, K Kaneko. J Porous Solids 4:181–186, 1997.

569. K Kaneko, T Suzuki, H Kuwabara, K Kakei. In: AB Mersman, SE Scholl, eds. Fundamentals of Adsorption. New York: Engineering Foundation, 1991, pp. 343–353.
570. K. Kaneko, T Suzuki, Y Fujiwara, K Nishikawa. In: F. Rodríguez-Reinoso et al., eds. Characterization of Porous Solids III. Amsterdam: Elsevier, 1991, pp. 389–398.
571. C Ishii, K Kaneko. Prog Org Coating 31:147–152, 1997.
572. K Kaneko, C Ishii, Y Hanzawa, N Setoyama, T Suzuki. Adv Colloid Interface Sci 77:295–320, 1998.
573. T Iiyama, M Ruike, K Kaneko. Chem Phys Lett (in press).
574. T Iiyama, K Nishikawa, T Suzuki, T Otowa, M Hijiriyama, Y Nojima, K Kaneko. J Phys Chem B 101:3037–3042, 1997.
575. T Iiyama, K Nishikawa, T Otowa, K Kaneko. J Phys Chem 99:10075–10076, 1995.
576. S Ergun. Carbon 14:139–150, 1976.
577. A Manivannan, M Chirila, NC Giles, MS Seehra. Carbon 37:1741–1747, 1999.
578. BE Warren NS Gingrich. Phys Rev 46:368–372, 1934.
579. JS Kasper, K Lonsdale. International Tables for X-ray Crystallography, Vol. IV. Dordrecht: Kluwer Academic, 1985.
580. D Shindo, K Hiraga. High-Resolution Electron Microscopy for Materials Science. Tokyo: Springer-Verlag, 1998.
581. M von Heimendahl. Electron Microscopy of Materials. New York: Academic Press, 1980.
582. N Uyeda, T Kobayashi, E Suito, Y Harada, M Watanabe. J Appl Phys 43:5181–5189, 1972.
583. S Iijima. J Appl Phys 42:5891–5893, 1971.
584. S Horiuchi. Fundamentals of High-Resolution Transmission Electron Microscopy. New York: North-Holland, 1994.
585. T Baird, JR Fryer, B Grant. Nature 233:329–330, 1971.
586. T Baird, JR Fryer, AR Arbuthno, B McEnaney, EV Riddell, D Walker. Carbon 12:381–390, 1974.
587. A Oberlin, G Terriere, JL Boulmier. Tanso 29, 1975.
588. GR Millward, JM Thomas. Carbon 17:1–5, 1979.
589. D Auguie, M Oberlin, A Oberlin, P Hyvernat. Carbon 18:337–346, 1980.
590. GR Millward, DA Jefferson. In: PL Walker Jr, PA Thrower, eds. Chemistry and Physics of Carbon, Vol. 14. New York: Marcel Dekker, 1978, pp. 1–82.
591. A Oberlin, G Terriere, JL Boulmier. Tanso 153, 1975.
592. A Oberlin. Carbon 17:7–20, 1979.
593. JR Fryer. Carbon 19:431–439, 1981.
594. H Marsh, D Crawford, TM O'Grady, A Wennerberg. Carbon 20:419–426, 1982.
595. RW Innes, JR Fryer, HF Stoeckli. Carbon 27:71–76, 1989.
596. K Kaneko. Carbon 38:287–303, 2000.
597. M Huttepain, A Oberlin. Carbon 28:103–111, 1990.
598. K Oshida, K Kogiso, K Matsubayashi, K Takeuchi, S Kobayashi, M Endo, MS Dresselhaus, G Dresselhaus. J Mater Res 10:2507–2517, 1995.
599. K Oshida, M Kobayashi, T Furuta, M Endo, A Oberlin. Electron Comm Jpn 81:64–70, 1998.

600. M Endo, T Furuta, F Minoura, C Kim, K Oshida, G Dresselhaus, MS Dresselhaus. Supramol Sci 5:261–266, 1998.
601. M Sato, T Sukegawa, T Suzuki, S Hagiwara, K Kaneko. Chem Phys Lett 181:526–530, 1991.
602. M Sato, T Sukegawa, T Suzuki, K Kaneko. J Phys Chem 101:1845–1850, 1997.
603. N Yoshizawa, Y Yamada, M Shiraishi. J Mater Sci 33:199–206, 1998.
604. M Shiraishi, G Terriere, A Oberlin. J Mater Sci 13:702–710, 1978.
605. KSW Sing, DH Everett, RAW Haul, L Moscou, RA Pierotti, J Rouquerol, T Siemieniewska. Pure Appl Chem 57:603–619, 1985.
606. SJ Gregg, KSW Sing. Adsorption, Surface Area and Porosity. London: Academic Press, 1982, p. 154.
607. M Ruike, T Kasu, N Setoyama, K Kaneko. J Phys Chem 98:9594–9600, 1994.
608. K Murata, K Kaneko. Chem Phys Lett 321:342–348, 2000.
609. S Suzuki, R Kobori, K Kaneko. Carbon 38:630–633, 2000.
610. Y Hanzawa, K Kaneko. Langmuir 13: 5802–5804, 1997.
611. EA Müller, LF Rull, LF Vega, KE Gubbins. J Phys Chem 100:1189–1196, 1996.
612. CL McCallum, TJ Bandosz, SC McGrother, EA Müller, KE Gubbins. J Langmuir 15:533–544, 1999.
613. T Iiyama, K Nishikawa, T Otowa, K Kaneko. J Phys Chem 99:10075–10076, 1995.
614. T Iiyama, K Nishikawa, T Suzuki, K Kaneko. Chem Phys Lett 274:152–158, 1997.
615. T Kimura, Y Hattori, T Suzuki, K Kaneko. J Phys Chem (in prep).
616. MM Dubinin, ED Zaverina, VV Serpinsky. J Chem Soc 2:1760–1766, 1955.
617. G Li, K Kaneko, S Ozeki, F Okino, H Touhara. Langmuir 11:716–717, 1995.
618. CM Yang, K Kaneko. Langmuir (in press).
619. H Kuwabara, T Suzuki, K Kaneko. J Chem Soc Faraday Trans 87:1915–1916, 1991.
620. N Setoyama, K Kaneko, F Rodríguez-Reinoso. J Phys Chem 100:10331–10336, 1996.
621. D Avnir. In: D. Avnir, ed. The Fractal Approach to Heterogeneous Chemistry. Chichester: Wiley, 1989, Chapter 4.
622. M Sato, T Sukegawa, T Suzuki, S Hagiwara, K Kaneko. Chem Phys Lett 186:526–530, 1991.
623. M Sato, S Sukegawa, T Suzuki, K Kaneko. J Phys Chem 101:1845–1850, 1997.
624. PJM Carrot, KSW Sing. In: KK Unger et al., ed. Characterization of Porous Solids. Amsterdam: Elsevier, 1988, p. 77.
625. F Rouquerol, J Rouquerol, KSW Sing. Adsorption by Powders and Porous Solids. San Diego: Academic Press, 1999, p. 19.
626. N Setoyama, T Suzuki, K Kaneko. Carbon 36:1459–1467, 1998.
627. J Miyawaki, T Kanda, T Suzuki, T Okui, Y Maeda, K Kaneko. J Phys Chem 102:2187–2192, 1998.
628. D Atkinson, AI McLeod, KSW Sing. J Chim Phys PCB 81:791–794, 1984.
629. T Ohba, T Suzuki, K Kaneko. Chem Phys Lett (in press).
630. Y Hanzawa, T Suzuki, K Kaneko. Langmuir 10:2857–2859, 1994.
631. ZM Wang, K Kaneko. J Phys Chem 102:2863–2868, 1998.
632. K Kaneko, Y Hanzawa, T Iiyama, T Kanda, T Suzuki. Adsorption 5:7–13, 1999.
633. NA Seaton, JPRB Walton, N Quirke. Carbon 27:853–861, 1997.

634. D Nicholson. J Chem Soc Faraday Trans 92:1–9, 1996.
635. MJ Bojan, WA Steele. Carbon 36:1417–1423, 1998.
636. PB Balbuena, KE Gubbins. Fluid Phase Equilibria 76:21–35, 1992.
637. C Lastoskie, KE Gubbins, N Quirke. J Phys Chem 97: 4785–4796, 1993.
638. C Lastoskie, KE Gubbins, N Quirke. Langmuir 9:2693–2702, 1993.
639. AV Neimark, PI Ravikovitch. Langmuir 13:5148–5160, 1997.
640. D Nicholson, N Quirke. In: KK Unger, G Kreysa, JP Gaselt, eds. Characterization of Porous Solids V. Amsterdam: Elsevier, 1999, p. 11.
641. JP Olivier. J Porous Mater 2:9–17, 1995.
642. S Brunauer, PH Emmett, E Teller. J Am Chem Soc 60:309–321, 1938.
643. MC Mittelmeijer-Hazeleger, JM Martin-Martinez. Carbon 30:695–709, 1992.
644. MT Gonzalez, M Molina-Sabio, F Rodriguez-Reinoso. Carbon 32:1407–1413, 1994.
645. J Kloubek. Carbon 31:445–450, 1993.
646. TW Zerda, X Yuan, SM Moore, CA Leon y Leon. Carbon 37:1999–2009, 1999.
647. ZM Wang, K Kaneko. J Phys Chem 99:16714–16721, 1995.
648. S Inagaki, Y Fukushima, K Kuroda. J Chem Soc Chem Commun 8:680–682, 1993.
649. CT Kresge, ME Leonowicz, WJ Roth, JC Vartuli, JS Beck. Nature 359:710–712, 1992.
650. PJ Branton, PG Hall, KSW Sing. J Chem Soc Chem Comm 16:1257–1258, 1993.
651. PL Llewellyn, Y Grillet, F Schüthe, H Reichert, KK Unger. Microporous Mater 3:345–349, 1994.
652. R Schmidt, M Stöcker, E Hansen, D Akporiaye, OH Ellestad. Microporous Mater 3:443–448, 1995.
653. PI Ravikovitch, SC ODomhnaill, AV Neimark, F Schüth, KK Unger. Langmuir 11:4765–4772, 1995.
654. PJ Branton, K Kaneko, KSW Sing. Chem Commun 7:575–576, 1999.
655. K Morishige, H Fujii, M Uga, D Kinukawa. Langmuir 13: 3494–3498, 1997.
656. M Kruk, M Jaroniec, A Sayari. J Phys Chem 101:583–589, 1997.
657. CG Sonwane, SK Bhatia, N Calos. Ind Eng Chem Res 37: 2271–2283, 1998.
658. S Inoue, Y Hanzawa, K Kaneko. Langmuir 14:3079–3081, 1998.
659. H Tanaka, T Iiyama, N Uekawa, T Suzuki, A Mastsumoto, M Grün, KK Unger, K Kaneko. Chem Phys Lett 292:541–546, 1998.
660. H Tanaka, S Inagaki, Y Fukushima, K Kaneko. In: A Sayari, M Jaroniec, TJ Pinnavaia, eds. Nanoporous Materials II. Amsterdam: Elsevier, 2000, p. 623.
661. K Liu, MG Brown, C Carter, RJ Saykally, JK Gregory, DC Clary. Nature 381: 501–503, 1996.
662. T Ohkubo, T Iiyama, K Nishikawa, T Suzuki, K Kaneko. J Phys Chem 103:1859–1863, 1999.
663. T Ohkubo, T Iiyama, K Kaneko. Chem Phys Lett 312:191–195, 1999.
664. LD Gelb, KE Gubbins, R Radhakrishnan, M Sliwinska-Bartkowiak. Rep Prog Phys 62:1573–1659, 1999.
665. MV Lopez-Ramon, J Jagiello, TJ Bandosz, NA Seaton. In: B McEnaney, TJ Mays, J Rouquerol, F Rouquerol, F Rodriguez-Reinoso, KSW Sing, KK Unger, eds. Characterization of Porous Solids IV. London: Royal Society of Chemistry, 1997, p. 73.

666. M Aoshima, K Fukasawa, K Kaneko, J Colloid Interface Sci 222:179–183, 2000.
667. BR Puri. In: PL Walker Jr, ed. Chemistry and Physics of Carbon, Vol. 6. New York: Marcel Dekker, 1970, pp. 191–282.
668. HP Boehm. In: DD Eley, H Pines, PB Weisz, eds. Advances in Catalysis, Vol. 16. New York: Academic Press, 1966, pp. 179–274.
669. CA Leon y Leon, LR Radovic. In: PA Thrower, ed. Chemistry and Physics of Carbon, Vol. 24. New York: Marcel Dekker, 1992, pp. 213–310.
670. G Kortum, W Vogel, K Andrussow. Dissociation Constants of Organic Acids in Aqueous Solutions. London: Butterworth, 1961.
671. EM Perdue, JH Reuter, RS Parrish. Geochim Consmochim Acta 48:1257–1263, 1984.
672. CA Leon y Leon, JM Solar, V Calemma, LR Radovic. Carbon 30:797–811, 1992.
673. JS Mattson, HB Mark Jr. Activated Carbon: Surface Chemistry and Adsorption from Solution. New York: Marcel Dekker, 1971.
674. K Tanabe. Solid Acids and Bases: Their Catalytic Properties. New York: Academic Press, 1970.
675. S Neffe. Carbon 25:441–443, 1987.
676. Y Matsumura, S Hagiwara, H Takahashi. Carbon 14:163–173, 1976.
677. E Papirer, E Guyon. Carbon 16:127–131, 1978.
678. TJ Bandosz, J Jagiello, JA Schwarz. Anal Chem 64:891–895, 1992.
679. J Jagiello, TJ Bandosz, JA Schwarz. Carbon 30:63–69, 1992.
680. TJ Bandosz, J Jagiello, JA Schwarz, A Krzyzanowski. Langmuir 12:6480–6486, 1996.
681. JW Hassler. Activated Carbons. New York: Chemical Publishing Company, 1963.
682. VA Garten, DE Weiss. Rev Pure Appl Chem 7:69–122, 1957.
683. F Adib, A Bagreev, TJ Bandosz. J Colloid Interface Sci 214:407–415, 1999.
684. F Adib, A Bagreev, TJ Bandosz. Langmuir 16:1980–1986, 2000.
685. DS Villars. J Am Chem Soc 69:214–217, 1947.
686. ML Studebaker, Proceedings of 5th Conference on Carbon, Vol. 2. New York: Pergamon Press, 1963, p. 189.
687. MO Corapcioglu, CP Huang. Carbon 25:569–578, 1987.
688. AS Arico, V Antonucci, M Minutoli, N Giordano. Carbon 27:337–347, 1989.
689. JS Noh, JA Schwarz. Carbon 28:675–682, 1990.
690. A Contescu, C Contescu, K Putyera, JA Schwarz. Carbon 35:83–94, 1997.
691. C Contescu, J Jagiello, JA Schwarz. Langmuir 9:1754–1765, 1993.
692. TJ Bandosz, J Jagiello, C Contescu, JA Schwarz. Carbon 31:1193–1202, 1993.
693. W Rudzinski, J Jagiello, Y Grillet. J Colloid Interface Sci 87:478–491, 1982.
694. J Jagiello. Langmuir 10:2778–2785, 1994.
695. J Jagiello, TJ Bandosz, JA Schwarz. Carbon 32:1026–1028, 1994.
696. J Jagiello, TJ Bandosz, K Putyera, JA Schwarz. J Colloid Interface Sci 172:341–346, 1995.
697. NR Laine, FJ Vasola, PL Walker Jr. J Phys Chem 67:2030–2034, 1963.
698. E Papirer, J Dentzer, S Li, JB Donnet. Carbon 29:69–72, 1991.
699. Y Otake, RG Jenkins. Carbon 31:109–121, 1993.
700. A Dandekar, RTK Baker, MA Vannice. Carbon 36:1821–1832, 1998.
701. F Adib, A Bagreev, TJ Bandosz. J Colloid Interface Sci 216:360–369, 1999.

702. JM Thomas, EL Evans, M Barber, P Swift. Trans Farad Soc 67:1875–1882, 1971.
703. EL Evans, JM Thomas, HP Boehm, H Marsh. Proceeding International Carbon Conference, Baden, 1972.
704. E Papirer, E Guyon, N Perol. Carbon 16:133–140, 1978.
705. A Proctor, PMA Sherwood. Carbon 21:53–59, 1983.
706. C Kozlowski, PMA Sherwood. Carbon 24:357–363, 1986.
707. JB Donnet, G Guilpain. Carbon 27:749–757, 1989.
708. A Yoshida, I Tanahashi, A Nishino. Carbon 28:611–615, 1990.
709. Y Nakayama, F Soeda, A Ishitani. Carbon 28:21–26, 1990.
710. I Hamerton, JN Hay, BJ Howlin, JR Jones, S-Y Lu, GA Webb. Chem Mater 9: 1972–1977, 1997.
711. S Biniak, G Szymanski, J Siedlewski, A Swiatkowski. Carbon 35:1799–1810, 1997.
712. T Tahahagi, A Ishitani. Carbon 26:389–396, 1988.
713. CA Baillie, JF Watts, JE Castle. J Mater Chem 2:939–944, 1992.
714. J Zawadzki. Carbon 18:281–285, 1980.
715. J Zawadzki. Carbon 19:19–25, 1981.
716. J Zawadzki. In: PA Thrower, ed. Chemistry and Physics of Carbon, Vol. 21. New York: Marcel Dekker, 1989, pp. 180–380.
717. CH Chang. Carbon 19:175–186, 1981.
718. PE Fanning, MA Vannice. Carbon 31:721–730, 1993.
719. PH Given, LW Hill. Carbon 7:649–658, 1968.
720. MM Dubinin. Carbon 18:355–364, 1980.
721. HF Stoeckli, F Kraehenbuehl, D Morel. Carbon 21:589–591, 1983.
722. F Rodríguez-Reinoso, M Molina-Sabio, MT Gonzalez. Langmuir 13:2354–2358, 1997.
723. MT Gonzalez, A Sepulveda-Escribano, M Molina-Sabio, F Rodríguez–Reinoso. Langmuir 11:2151–2155, 1995.
724. AJ Groszek. Proc Roy Soc London A314:473–498, 1970.
725. AJ Groszek. Carbon 25:717–722, 1987.
726. JR Conder, CL Young. Physicochemical Measurement by Gas Chromatography. New York: Wiley, 1979.
727. M Sidqi, G Ligner, J Jagiello, H Balard, E Papirer. Chromatographia 28:588–592, 1989.
728. J Jagiello, TJ Bandosz, JA Schwarz. J Colloid and Interface Science 151:433–445, 1992.
729. J Jagiello, TJ Bandosz, JA Schwarz. Carbon 30:63–69, 1992.
730. LD Gelb, KE Gubbins. Langmuir 14:2097–2111, 1998.
731. LD Gelb, KE Gubbins. Langmuir 15:305–308, 1999.
732. LD Gelb, KE Gubbins (in preparation).
733. TW Ebbesen, ed. Carbon Nanotubes: Preparation and Properties. Boca Raton: CRC Press, 1997.
734. N Hamada, S Sawada, A Oshiyama. Phys Rev Lett 68:1579–1581, 1992.
735. KG Ayappa. Langmuir 14:880–890, 1998.
736. GJ Tjatjopoulos, DL Feke, JA Mann Jr. J Phys Chem 92:4006–4007, 1988.
737. WA Steele, MJ Bojan. Adv Colloid Interface Sci 76–77:153–178, 1998.

738. PA Gordon, RB Saeger. Ind Eng Chem Res 38:4647–4655, 1999.
739. M Maddox, D Ulberg, KE Gubbins. Fluid Phase Equil 104:145–158, 1995.
740. M Maddox, KE Gubbins. Langmuir 11:3988–3996, 1995.
741. M Maddox, SL Sowers, KE Gubbins. Adsorption 2:23–32, 1996.
742. BK Peterson, JPRB Walton, KE Gubbins. J Chem Soc Farad Trans 2 82:1789–1800, 1986.
743. KG Ayappa. Chem Phys Lett 282:59–63, 1998.
744. IA Khan, KG Ayappa. J Chem Phys 109:4576–4586, 1998.
745. F Darkrim, D Levesque. J Chem Phys 109:4981–4984, 1998.
746. F Darkrim, D Levesque. J Phys Chem B 104:6773–6776, 2000.
747. Q Wang, JK Johnson. J Chem Phys 110:577–586, 1999.
748. Q Wang, JK Johnson. J Phys Chem B 103:4809–4813, 1999.
749. VV Simonyan, P Diep, JK Johnson. J Chem Phys 111:9778–9783, 1999.
750. YF Yin, T Mays, B McEnaney. Langmuir 15:8714–8718, 1999.
751. KA Williams, PC Eklund. Chem Phys Lett 320:352–358, 2000.
752. Z Mao, SB Sinnott. J Phys Chem B 104:4618–4624, 2000.
753. DW Brenner. Phys Rev B 42:9458–9471, 1990.
754. WG Mixter. Am J Sci (Series 4) 19:434–444, 1905.
755. WA Roth-Greifswald, H Wallasch, Z f Elektroch 21:1–5, 1915.
756. WH Bragg, WL Bragg. X-rays and Crystal Stucture. London: 1915.
757. P Debye, P Scherrer. Physik Zeitschr 18:291–301, 1917.
758. AW Hull. Phys Rev 10:661–696, 1917.
759. The less common rhombohedral form of graphite was subsequently discovered by Lipson and Stokes (Proc Roy Soc A 181:101–105, 1942) and latter confirmed by a number of groups as reviewed in Ruland (In: PL Walker, ed. Chemistry and Physics of Carbon, Vol. 4. New York: Marcel Dekker, 1969, pp. 1–84).
760. G Asahara. Jpn J Chem 1:35–41, 1922.
761. H Arnfeld. Arkiv F Mater Astron Fysik 23B:1, 1932.
762. Warren and coworkers were the first to use Fourier analysis to determine the C-C pair distribution function from the powder X-ray diffraction patterns: a) BE Warren, N Gingrich. Phys Rev 46:368, 1934); b) BE Warren. J Chem Phys 2:551–555, 1934. This was key to his confirming the turbostatic nature of the basic building blocks of noncrystalline carbons.
763. a) U Hofmann, D Wilm. Z Physik Chemie 18B:401, 1932); b) U Hofmann, W Lemcke. Gasmaske 5:129–134, 1933; c) D Wilm, U Hofmann. Kolloid-Z 70:21–24, 1935; d) U Hofmann, D Wilm. Z Elektrochem 42:504–522, 1936); e) U Hofmann, F Sinkel. Z Anorg Allgem Chemie 245:85–102, 1940.
764. It is worthwhile to note that many of the dimensions determined in these early years were subject to error in many cases due to the use of incorrect formulas for determining lattice spacings in noncrystalline materials (Warren. Phys Rev 59:693–698, 1941) and limited accuracy (Ekstein, Siegel. Acta Cryst 2:99, 1949).
765. WA Roth. Ber 56:530–536, 1927.
766. O Ruff, G Schmidt and O Werner. Z Anorg Allgem Chemie 148:313–331, 1925.
767. GL Clark. Appl X-Rays 177, 1927.
768. O Ruff, P Mautner, E Fritz. Z Anorg Allgem Chem 167:185–189, 1927.

769. M Oswald. Chim Ind 24:280–292, 1930.
770. WS Wesselowski, KW Wassiliew. J Phys Chem USSR 5:982–995, 1934.
771. J Gibson, M Holohan, HL Riley. J Chem Soc 456, 1946.
772. HL Riley. Q Rev 1:59–72, 1947.
773. WS Wesselowski, KW Wassiliew. Z-Krist 89:156–174, 1934.
774. P Krishnamurti. Indian J Phys 5:473–488, 1930.
775. WF Wolff. J Phys Chem 62:829–833, 1958.
776. WA Steele. The Interaction of Gases with Solid Surfaces. Glasgow, UK: Pergamon Press, 1974.
777. MJ Bojan, WA Steele. Surf Sci 199: L395–L402, 1988.
778. MJ Bojan, WA Steele. Langmuir 5:625–633, 1989.
779. MJ Bojan, WA Steele. Langmuir 9:2569–2575, 1993.
780. MJ Bojan, WA Steele. Carbon 36:1417–1423, 1998.
781. MJ Bojan, WA Steele. Mol Phys 95:431–437, 1998.
782. GM Davies, NA Seaton. Carbon 36:1473–1490, 1998.
783. MJ Bojan, R van Slooten, WA Steele. Sep Sci Technol 27:1837–1856, 1992.
784. MW Maddox, N Quirke, KE Gubbins. Mol Simul 19:267–283, 1997.
785. MW Maddox, N Quirke, KE Gubbins. In: MD LeVan, ed. Proceedings of the Fifth International Conference on Fundamentals of Adsorption. Boston: Kluwer Academic, 1996, pp. 571–578.
786. AR Turner, N Quirke. Carbon 36:1439–1446, 1998.
787. VA Bakaev. J Chem Phys 102:1398–1404, 1995.
788. K Kaneko, C Ishii, M Ruike, H Kuwabara. Carbon 30:1075–1088, 1992.
789. DA Wickens. Carbon 28:97–101, 1990.
790. JK Floess, Y Van Lishout. Carbon 30:967–973, 1992.
791. SG Chen, RT Yang. Langmuir 9:3259–3263, 1993.
792. M Sahimi, GR Gavalas, TT Tsotsis. Chem Eng Sci 45:1443–1502, 1990.
793. A Papadopoulou, A, F van Swol, U Marini Bettolo Marconi. J Chem Phys 97: 6942–6952, 1992.
794. E Kozak, G Chmiel, A Patrykiejew, S Sokolowski. Phys Lett A 189:94–98, 1994.
795. W Gac, A Patrykiejew, S Sokolowski. Thins Solid Films 298:22–32, 1997.
796. A Huerta, O Pizio, P Bryk, S Sokolowski. Mol Phys 98:1859–1869, 2000.
797. NA Seaton, SP Friedman, JMD MacElroy, BJ Murphy, Langmuir 13:1199–1204, 1997.
798. JMD MacElroy, SP Friedman, NA Seaton. Chem Eng Sci 54:1015–1027, 1999.
799. EI Segarra, ED Glandt. Chem Eng Sci 49:2953–2965, 1994.
800. R Eppenga and D Frenkel. Mol Phys 52:1303–1334, 1984.
801. WL Jorgensen, J Chandrasekhar, JD Madura, RW Impey, M L Klein, J Chem Phys 79:926–935, 1983.
802. JR Dahn, W Xing, T Gao. Carbon 35:825–830, 1997.
803. M Biggs, P Agarwal. Phys Rev A 46:3312–3318, 1992.
804. M Biggs, P Agarwal. Phys Rev E 49:531–537, 1994.
805. MJ Biggs. The Numerical Study of Transport and Reaction Within and Around a Porous Carbonaceous Particle in a Fluidized Bed, PhD thesis, University of Adelaide, Australia, 1996.

806. The hydrogen atoms are rarely included in classical molecular simulations but are instead treated via the united atom approach; see Dunfield et al., for example (LG Dunfield, AW Burgess, HA Scheraga. J Phys Chem 82:2609, 1978).

807. E Clar. Polycyclic Hydrocarbons, Vols. 1 and 2. New York: Academic Press, 1964.

808. A Buts, DC Williamson, MJ Biggs (in preparation).

809. RC Glen, AWR Payne. J Comput Aid Mol Design 9:181–202, 1995.

810. A completely independent cubic lattice is used for the calculation of the short-range fluid-solid interactions during any molecular dynamics or Monte-Carlo simulation of a fluid within the VPC.

811. CV Howard. Unbiased Stereology: Three Dimensional Measurement in Microscopy. Oxford, 1998.

812. EE Underwood. Quantitative Stereology. London: Addison-Wesley, 1970.

813. DH Everett. KK Unger, et al., eds. In: Characterization of Porous Solids. Amsterdam: Elsevier, 1988, pp. 1–21.

814. MR Landry, BK Coltrain, CJT Landry, JM O'Reilly. J Polym Sci B 33:637–655, 1995.

815. DF Watson. The Computer J 24:167–172, 1981.

816. NM Jackson, R Jafferali, DJ Bell, GA Davies. J Membrane Sci 162:23–43, 1999.

817. M Acharya, MS Strano, JP Mathews, SJL Billinge, V Petkov, S Subramoney, HC Foley. Phil Mag B 79:1499–1518, 1999.

818. JL Faulon, GA Carlson, PG Hatcher. Energy Fuels 7:1062–1072, 1993.

819. MS Kane, JF Goellner, HC Foley, R Di Francesco, SJL Billinge, LF Allard. Chem Mater 8:2159–2171, 1996.

820. V Petkov, RG DiFrancesco, SJL Billinge, M Acharya, HC Foley. Phil Mag B 79: 1519–1530, 1999.

821. RL McGreevy, L Pusztai. Mol Simul 1:359–367, 1988.

822. RL McGreevy, MA Howe, DA Keen, K Clausen. IOP Conf Ser 107:165–184, 1990.

823. FLB da Silva, B Svensson, T Akesson, B Jonsson. J Chem Phys 109:2624–2629, 1998.

824. JK Walters, JS Rigden, RJ Newport. J Phys Scir T57:137–141, 1995.

825. JS Rigden, RJ Newport. J Electrochem Soc 143:292–296, 1996.

826. RL McGreevy, MA Howe, VM Nield, JD Wicks, DA Keen. Physica B 180/181: 801–804, 1992.

827. FLB da Silva, W Olivares-Rivas, L Degrève, T Akesson. J Chem Phys 114:907–914, 2001.

828. G Tóth, A Baranyai. J Chem Phys 114:2027–2035, 2001.

829. B O'Malley, I Snook, D McCulloch. Phys Rev B 57:14148–14157, 1998.

830. K T Thomson, K E Gubbins. Langmuir 16:5761–5773, 2000.

831. J Pikunic, RJ-M Pellenq, KT Thomson, J-N Rouzaud, P Levitz, KE Gubbins. Stud Surf Sci Catal 132:647–652, 2001.

832. MP Allen, DJ Tildesley. Computer Simulation of Liquids. Oxford: Clarendon Press, 1987.

833. D Frenkel, B Smit. Understanding Molecular Simulation: From Algorithm to Applications. San Diego: Academic Press, 1996.

834. JDF Ramsay and BO Booth. In: AB Mersmann, SE Scholl, eds. Fundamentals of

Adsorption. Proceedings of the Third International Conference. New York: Engineering Foundation, 1991, pp. 701–714.

835. JDF Ramsay, PJ Russell, SW Swanton. In: F Rodríguez-Reinoso, J Rouquerol, KSW Sing, KK Unger, eds. Characterization of Porous Solids II. Amsterdam: Elsevier, 1991, pp. 257–265.

836. I-A Park, JMD MacElroy. Molec Simul 2:105–145, 1989.

837. JMD MacElroy, K Raghavan. J Chem Phys 93:2068–2079, 1990.

838. JMD MacElroy. Langmuir 9:2682–2692, 1993.

839. BP Feuston, SH Garofalini. J Chem Phys 89:5818–5824, 1988.

840. S Gavalda, K Kaneko, KT Thomson, KE Gubbins. Molecular modeling of carbon aerogels. Colloids Surf A (in press).

841. S Gavalda, KE Gubbins, K Kaneko, Y Hanzawa, KT Thomson. Langmuir (submitted).

842. G Mason. Disc Faraday Soc Lond A319:479, 1970.

843. H Hoshen, R Kopelman. Phys Rev B 14:3438–3445, 1976.

844. Y Hanzawa, K Kaneko, N Yoshizawa, RW Pekala, MS Dresselhaus. Adsorption 4:187–195, 1998.

845. C Chmiel, K Karykowski, A Patrykiejew, W Rzysko, S Sokolowski. Mol Phys 81: 691–703, 1994.

846. P Röcken, P Tarazona. J Chem Phys 105:2034–2043, 1996.

847. M Schoen, DJ Diestler. Chem Phys Lett 270:339–344, 1997.

848. P Röcken, A Somoza, P Tarazona, G Findenegg. J Chem Phys 108:8689–8697, 1998.

849. M Schoen, DJ Diestler Phys Rev E 56:4427–4440, 1997.

850. H Bock, M Schoen. Phys Rev E 59:4122–4136, 1999.

851. A Vishnyakov, EM Piotrovskaya, EN Brodskaya. Adsorption 4:207–224, 1998.

852. M Schoen, H Bock. J Phys Condens Mater 12:A333–A338, 2000.

853. H Bock, M Schoen. J Phys Condens Mater 12:1545–1568, 2000.

854. H Bock, DJ Diestler, M Schoen. J Phys Condens Mat (in press).

855. YK Tovbin, KV Votyakov. Langmuir 9:2652–2660, 1993.

856. EA Müller, LF Vega, KE Gubbins. Int J Thermophys 16:705–713, 1995.

857. LF Vega, EA Müller, LF Rull, KE Gubbins. Mol Simul 15:141–154, 1995.

858. MW Maddox, D Ulberg, KE Gubbins. Fluid Phase Equil 104:145–158, 1995.

859. LF Vega, EA Müller, LF Rull, KE Gubbins. Adsorption 2: 59–68, 1996.

860. EA Müller, KE Gubbins. Carbon 36:1433–1438, 1998.

861. C McCallum, TJ Bandosz, SC McGrother, EA Müller, KE Gubbins. Langmuir 15: 533–544, 1999.

862. EA Müller, FR Hung, KE Gubbins. Langmuir 16:5418–5424, 2000.

863. WL Jorgensen, CJ Swenson. J Am Chem Soc 107:1489–1496, 1985.

864. DE Ulberg, KE Gubbins. Mol Phys 84:1139–1153, 1995.

865. J Brennan, TJ Bandosz, KT Thomson, KE Gubbins. Colloids Surfaces A (in press).

866. J Brennan, KT Thomson, KE Gubbins. To be published.

867. J Sauer. Chem Rev 89:199–204, 1989.

868. PJ Pappano, JP Mathews, HH Schobert. In: 24th Biennial Conference on Carbon Extended Abstracts. American Carbon Society, 1999, pp. 202–203.

869. SG Chen, RT Yang. Energy Fuels 11:421–427, 1997.

870. SG Chen, RT Yang. J Phys Chem A 102:6348–6356, 1998.
871. X Ma, Q Wang, LQ Chen, W Cermignani, HH Schobert, CG Pantano. Carbon 35: 1517–1525, 1997.
872. T Kyotani, A Tomita. J Phys Chem B 103:3434–3441, 1999.
873. CM Goringe, DR Bowler, E Hernandez. Rep Prog Phys 60:1447–1512, 1997.
874. JC Charlier, JP Michenaud, PH Lambin. Phys Rev B 46:4540–4543, 1992.
875. C H Lee, WRL Lambrecht, B Segall, PC Kelires, T Frauenheim, U Stephan. Phys Rev B 49:11448–11451, 1994.
876. U Stephan, TH Frauenheim, P Blaudeck, G Jungnickel. Phys Rev B 49:1489–1501, 1994.
877. CZ Wang, SY Qui, KM Ho. Comp Mater Sci 7:315–323, 1996.
878. V Rosato, JC Lascovich, A Santoni, L Colombo. Int J Mod Phys C 9:917–926, 1998.
879. SE Stein, RL Brown. J Am Chem Soc 109:3721–3729, 1987.
880. T Kyotani, K Ito, A Tomita, LR Radovic. AIChE J 42:2303–2307, 1996.
881. PR Solomon, TH Fletcher. 25th Symposium (International) on Combustion, Pittsburgh: The Combustion Institute, 1994, pp. 463–474.
882. JM Jones, M Pourkashanian, CD Rena, A Williams. Fuel 78:1737–1744, 1999.
883. JP Mathews, PG Hatcher, AW Scaroni. Am Soc Preprints Fuel Chem Div 43:136, 1998.
884. LD Gelb, KE Gubbins. Mol Phys 96:1795–1804, 1999.
885. F Siperstein, KE Gubbins. Mol Sim (in press).

3

Adsorption of Water Vapor on Activated Carbon: A Brief Overview

D. Mowla* and D. D. Do

University of Queensland, Brisbane, Queensland, Australia

Katsumi Kaneko

Chiba University, Chiba, Japan

* *Current affiliation*: Shiraz University, Shiraz, Iran.

I. INTRODUCTION

Activated carbon (AC) has been used to remove and recover volatile organic compounds (VOCs) from contaminated air. AC is particularly attractive as an adsorbent due to its high surface area and good uptake capacity for VOCs. Also, it has a variety of morphology, such as fibers, films, and monoliths, in addition to traditional granules. Thereby, activated carbon has a high potential for further development in environmental engineering. Water vapor usually coexists with VOCs in the air. Due to the relatively hydrophobic nature of the AC, the adsorption equilibrium is not so much hindered by the coexistence of water vapor if the relative humidity is low. Unfortunately, the relative humidity of the coexisting water vapor is usually high, and this causes a reduction in the adsorption capacity of VOCs [33,107]. This points to the importance of the problem of water adsorption in chemical and environmental engineering. The adsorption of water vapor by microporous carbon is highly specific, being dependent on surface chemistry as well as on pore size and shape. Due to the low energy of dispersion between water molecules and aromatic surfaces, the carbonaceous surfaces are essentially hydrophobic. Here the hydrophobic term does not mean that the carbonaceous surfaces repel water molecules as there is always some dispersion attraction, but water molecules simply prefer to remain in the vapor phase rather than to be in the confined spaces of the carbon micropores (intermolecular interaction is greater than molecular-surface interaction). However, the water molecules can be adsorbed specifically on oxygen-containing sites, which typically cover only a small fraction of the total surface area of carbonaceous materials. So not only the pore size and structural characteristics of the carbon but also its surface chemistry could influence significantly the adsorptive properties. These properties are related both to AC crystalline constitution, which depends in turn on the raw material, and to the method and conditions used in its preparation.

Because of its technological importance, the adsorption of water vapor has been studied by many investigators [1–129]. It has been known for some time that water adsorption is especially sensitive to changes in the adsorbent structure and that the amount adsorbed is dependent on both surface chemistry and porosity. However, these combined effects are not easy to separate, and much remains to be done to improve our understanding of the mechanisms involved. Although several theories have been proposed to describe the adsorption isotherm of water on AC, none could satisfactorily be used for all types of AC and in different ranges of relative pressure. Very often only the experimental data on one particular AC were analyzed. No comprehensive comparison of data on various carbon samples has been made. A detailed review of the existing experimental data and theories is necessary due to the rapidly increasing awareness of our deteriorating air quality. The purpose of this study is to review the work done on this subject and give some guidelines as to the proper use and understanding of different mechanisms of water adsorption on AC. This should also be helpful in understanding a general concept of the hydrophobic interaction.

II. STATES OF WATER ADSORBED IN MICROPORES

The structural and calorimetric studies of adsorbed water are important in helping us to understand the adsorption mechanism of water in micropores. Recently, Iiyama et al. [66] developed an in situ X-ray diffraction technique to investigate the nature of water adsorbed in micropores of pitch-based activated carbon fibers (ACFs). The adsorption isotherms showed an explicit hydrophobicity character. The electron radial distribution function (ERDF) from the X-ray diffraction pattern was used to compare adsorbed water with bulk liquid water. The authors found that the ERDF of adsorbed water is considerably different from that of bulk liquid, indicating that adsorbed water at 303 K is highly ordered in comparison with the liquid. The temperature dependence of the ERDF was examined down to 143 K. In case of water adsorbed in micropores of pore width <0.8 nm, almost no change in the ERDF was observed. That is, water adsorbed in micropores <0.8 nm at 303 K should have a structure similar to that of a solid. The amplitude of the ERDF of water adsorbed in wider micropores became higher as the temperature decreased. Hence, water adsorbed in wider micropores has a less ordered structure than that in narrower micropores, but still water molecules are less mobile than those of liquid water. Such a study was done by another group [16] using X-ray and neutron diffraction, and they showed also that water adsorbed in activated carbon has a solid-like structure.

Kimura et al. [79] examined the thermodynamic state of adsorbed water with the aid of differential scanning calorimetry (DSC) over the temperature range 130–300 K. They observed an exothermic peak below 220 K due to freezing of water adsorbed in wider micropores, whereas there was no peak in the DSC image of water adsorbed in narrow micropores. This phase behavior agrees with the X-ray diffraction result. The enthalpy of freezing of adsorbed water was found to be much smaller than that of liquid water [92].

Hence, both quasi-liquid and quasi-solid states of water adsorbed in wider micropores should not be ordinary ones. The quasi-liquid is more ordered, whereas the quasi-solid is less perfect compared with the bulk states. Thus water adsorbed in micropores is neither liquid nor solid, and its structure depends on pore width. The measurement of the differential heats of adsorption often gave high values near the plateau of the adsorption isotherm, suggesting that reconstruction of the adsorbed water structure occurs near its saturation.

Although today we have better information on the state of water adsorbed in micropores, a complete understanding is still lacking. Further research in this area together with additional adsorption data should provide better insight into the adsorption mechanism of water in AC.

III. POROUS STRUCTURE OF ACTIVATED CARBON

As one of the most important adsorbents in industry, AC is prepared by carbonization and activation of a large number of raw materials of biological origin, such as coconut shells, wood, coal, and peat. During carbonization most of the noncarbon elements are removed by pyrolytic decomposition of the raw material, and the carbonized product is then formed, having a structure of more or less disordered elementary graphite-like crystallites with a poorly developed porous structure. During the activation process the spaces between the elementary crystallites become cleared of less organized carbonaceous compounds and, at the same time, carbon is also removed in part from the elementary crystallites. As a consequence, the porous structure of AC is a function of raw material used in its preparation, the type of activation process being employed, and the extent of activation. A more detailed and up-to-date discussion of these issues is provided in Chapter 2 of this volume.

Recently, ACs having special morphologies, such as fibers, beads, films, tubes, and monoliths, in addition to more conventional powders and granules, have been obtained. Activated carbon of specific morphology has been requested for further development of technologies.

As far as the porous structure of AC is concerned, several models have been proposed (see Chapter 1). The main feature common in all models is the existence of carbon layer planes with different degrees of order in two and three dimensions. The spaces between these layers constitute the porosity of the carbon. For most ordered layer graphite-like carbons, the interlayer spaces have apertures that are smaller than 2 nm. In the IUPAC classification, these micropores are further classified as ultramicropores (<0.7 nm) and supermicropores (from 0.7 to 2 nm). More disordered layers contain larger mesopores (from 2 nm to 50 nm) and macropores (>50 nm). It is worthwhile to note that the specific surface of the macropores, not exceeding 0.5 m^2 g^{-1}, is negligibly small when compared with 20–100 m^2 g^{-1} for mesopores and 200–450 m^2 g^{-1} for micropores [89]. At least 90–95% of the total surface area of a typical AC is contributed by micropores [120].

IV. CHEMICAL NATURE OF THE ACTIVATED CARBON SURFACE

Activated carbon usually contains elements other than carbon. These components are attached to the carbon surface in various arrangements, compositions, and quantities, affecting the chemical nature of the carbon surface [69]. Two basic types of noncarbon impurities in AC are mineral components or ash and heteroatoms. Ash does not chemically combine with the carbon surface; rather it occupies some of the open pores. The amount of ash present in any AC depends primarily on the type of raw material used. Heteroatom impurities are mainly oxygen, hydrogen, sulfur, and halogens. For relatively pure carbons, these are mainly oxygen atoms located at the edges of the carbon layer [37]; in less pure carbons, other heteroatoms, like nitrogen, could also contribute to the polar sites. These sites, especially those containing oxygen, are responsible for the adsorption of polar adsorbates. Hydrogen could be present as a residue of incomplete carbonization and could be distributed throughout the volume. The concentration of heteroatoms, especially oxygen, combined with the carbon could be changed considerably by different treatments. For example, the burn-off of carbon at temperatures above 1000°C could decrease, and oxidation of carbon could increase the amount of oxygen attached to the surface. The heteroatoms which combine with peripheral carbon atoms at the corners and edges of crystallite basal planes, form some functional groups that are very reactive to many reagents. These functional groups usually consist of more than one type of heteroatom, e.g., oxygen and hydrogen together as −OH or −COOH (see Chapter 2). The nature of these functional groups depends to a large extent on the method of activation as well as on the raw material from which the carbon is produced. The heteroatoms could also combine with the carbon atoms in the space between the basal planes and even in the highly defective zones of these planes. These heteroatoms are not reactive to other reagents due to both their inaccessibility and their mode of combination with the host carbon atoms [69,120]. Although the surface sites associated with functional groups constitute only a small portion of the total surface area, they may affect considerably the adsorption capacity [69,120].

Greater quantities of oxygen could be introduced into the surface of AC by subjecting it to chemical modification such as oxidation. The oxidizing agent used for this purpose could be a gaseous oxidizing agent, such as oxygen, ozone, air, water vapor, carbon dioxide, and nitrogen oxides, or a liquid oxidizing agent, such as nitric acid, a mixture of nitric and sulfuric acids, hydrogen peroxide, acidic potassium permanganate, chlorinated water, sodium hypochlorite, and ammonium persulfate [69]. The polar surface sites or surface functional groups are associated with various types of surface complex [3,21,36]. Properties such as acid strength and thermal stability, as well as the strength of the carbon-water interaction, vary from one complex to another [3,81]. In addition, the number and geometrical distribution of each will depend on the particular carbon studied and on the conditions of pretreatment [3,21,39]. In view of these factors, it is not surprising to find that water isotherms determined even on nonporous carbons exhibit a wide variety, with significant differences being found both in the level of uptake and in the shape of the isotherms.

The surface compounds containing oxygen are usually divided into two main groups: functional groups of acidic nature and functional groups of basic character. The first group may be of carboxyl, phenolic, quinonoid, normal lactone, fluorescein-type lactone, and anhydride originating from neighboring carboxyl groups [69,89]. These functional groups are shown in Fig. 1. In comparison with this group, the second group is not well characterized. Typical groups are chromene [58] or pyrone-like moieties [20], as shown in Fig. 2.

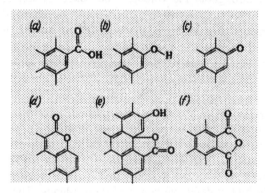

FIG. 1 Different types of acidic functional groups: (a) carboxyl, (b) phenolic, (c) quinonoid, (d) normal lactone, (e) fluorescein-type lactone, (f) anhydride originating from neighboring carboxyl groups [69].

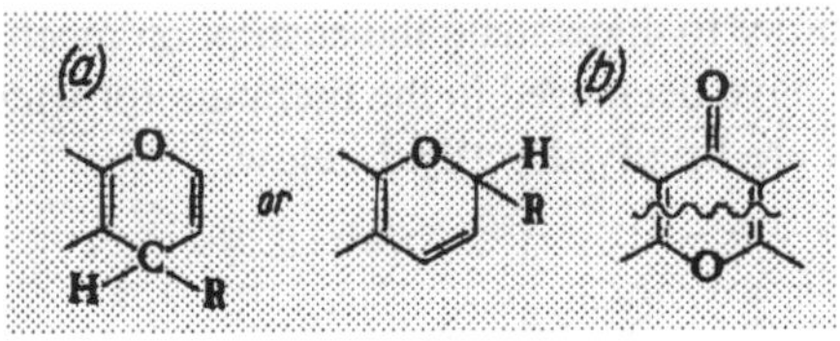

FIG. 2 Functional groups of basic character: (a) chromene [58], (b) pyrone-like [69].

ACs prepared at temperatures below 400–500°C typically contain acidic functional groups, while those prepared at room temperatures about 800–1000°C possess basic functional groups [89]. Even natural carbonaceous compounds contain different functional groups depending on their sources [87].

Recently a fluorinated AC was developed [85], and it was found that water cannot be adsorbed on such carbons [78]. This dramatically illustrates the fact that surface chemistry can control the adsorptivity of AC toward water. Hence, a more precise determination of the surface chemistry and a systematic modification of chemical structure of AC are strongly needed. An effort to summarize the current state of the art was published recently in this series (Leon y Leon and Radovic, Vol 24, 1994). For an authoritative brief review, see Ref. 21.

V. MECHANISM OF WATER ADSORPTION

As a water molecule has a large dipole moment, it is believed that its adsorption on AC near ambient temperature belongs to the general class of adsorption of polar molecules under subcritical conditions. However, water adsorption on AC has inherently unique characteristics and thereby we must define a separate category for it. For example, adsorption of other polar molecules, such as C_2H_5OH or SO_2, can be described by use of the Lennard-Jones type of interaction, while water adsorption does not conform to this description [128]. Hence, we must summarize many factors that could affect this process. Some of these are:

1. The quantity and the nature of the functional groups
2. The role of adsorbed molecules acting as sites for further adsorption
3. The structural characteristic of the adsorbent.

Although all of the above factors are important and must be accounted for in the mechanism of adsorption, the concentration of surface functional groups should be the critical factor. The mechanism of water adsorption on an activated carbon having a large quantity of surface functional groups is different from that on a highly hydrophobic AC. Thus, we must distinguish ACs according to the concentration of surface functional groups in order to understand the adsorption mechanism of water. The various models that have been used in the literature for water vapor adsorption on carbon adsorbents are briefly described below.

A. BET Model

Brunauer et al. [26] presented the famous BET model for vapor adsorption. According to this model, once a molecule is adsorbed on the surface it provides sites for more molecules to be adsorbed on it in a vertical direction, resulting in a tower-like structure. There is no horizontal interaction between adsorbed molecules (which is now known as a severe restriction). The number of original sites on the surface are fixed according to the well-known Langmuir [84] assumption and uniformly distributed throughout the surface. The major shortcoming of the BET model is the assumption that the molecules wet the surface. It means that the molecules interact more strongly with the surface than they do with each other. This model should not be used for the description of water adsorption on activated carbon.

B. Dubinin-Astakhov and Dubinin-Serpinsky Models

A well-known model for adsorption of vapors on solid surfaces is the Polanyi potential field model [103]. According to this model, the adsorbent surface induces a potential field in the adjacent region. The vapor molecules are captured in this potential field. So different layers with uniform concentration of vapor molecules are formed on the surface. The population density decreases with distance from the solid surface. Using the concept of the potential field, Dubinin and Astakhov [40] derived a semiempirical equation that could be used for the description of water adsorption on AC.

Based on a kinetic argument similar to the Langmuir model [84], Dubinin and his collaborators [37,39,42,44,50,53,105] presented another model for describing water adsorption on carbonaceous materials. In the Dubinin-Serpinsky model it was assumed that (1) water is initially adsorbed onto specific sites [3,36,60,81, 126] and (2) adsorbed water molecules behave as secondary sites for further adsorption to form hydrogen-bonded clusters. The phenomenon of cluster formation was also mentioned in other work [3,36,60,81,102]. In much of the literature for simplicity it is assumed that the clusters are three dimensional. However, in the work of Pierce [102] and Dubinin [43] it was concluded that the clusters spread out in two dimensions along the surface. A third stage after cluster formation is the merging of the clusters within the micropores, forming a continuous adsorbed phase. Although this model is not consistent thermodynamically at the limit of zero relative pressure [117], it has been used for water adsorption on AC.

C. Talu and Meunier Association Model

More recently, Talu and Meunier [118] suggested a new theory, called theory of association in micropores. They adapted the Dubinin-Serpinsky approach in which the water molecules are primarily adsorbed on the active sites, then took into account the cooperative molecule-to-molecule lateral interaction proposed by Dubinin [42] in a finite micropore volume. This last concept results in the existence of a plateau region of the adsorption isotherm at high relative pressure. In the derivation of their model these authors used the concepts of chemical equilibrium and equation of state for describing the behavior of the surface phase and phase equilibrium for linking the unmeasurable surface phase properties to the measurable properties of the bulk phase.

D. Other Models

Another mechanism was proposed by Kaneko and coworkers [67,74] and was modeled by Do [35]. In this model, carbon materials are assumed to be composed of graphitic carbon and amorphous carbon. Micropores are interstices between the graphitic units composed of a few graphene layers, and the functional groups are located at the edges of the basal planes (graphene layers) of the graphite units. These functional groups are accommodated within the amorphous region, but not inside the micropores. Due to the stronger chemisorption interaction of a water molecule with the functional group that the dispersive force with the graphene layers, the initial water adsorption occurs at the functional groups, and further water adsorption occurs on top of the chemisorbed water molecules via hydrogen bonding. This process results in the growth of a water cluster around the chemisorption site. When this cluster grows beyond a critical size, a cluster of five water molecules [67,74] attains a sufficient dispersive energy with the carbon atoms to enter the micropores. This micropore adsorption proceeds with increasing pressure until the micropore volume is filled with water. Since the water inside the micropore exists as a cluster of five water molecules, the packing is not as effective as that of the water molecules in bulk liquid water. This is why the micropore volume calculated from water adsorption is always smaller than the volume calculated from nitrogen adsorption [67]. Another model using computer simulation [91] is also useful in understanding the adsorption of water.

E. Discussion

As it is evident, the basic views in the different mechanisms presented for the adsorption of water vapor could be divided into two main concepts: the concept of micropore filling [70,90,101] and the concept of specific interaction of water with primary active centers (PACs) [43,102]. Even in recent work done on water adsorption, these mechanisms have not been completely clarified. This is because of the studies using physical methods such as X-ray diffraction or small angle X-ray scattering. Some investigators used the equations that are based on pore filling, whereas others draw attention to the fact that the presence of acidic oxygen–containing surface complexes increases the adsorption capacity and facilitates micropore filling [11,69,70]. Vartapetyan and Voloshchuk [124] attempted to clarify the effects of the surface active sites and the microporous structure of the AC on the adsorption capacity, the nature of PACs, the interaction phenomena of the adsorbed molecules with active sites, and the mechanism of pore volume filling. By analysis of different viewpoints given in the literature, it can be concluded that at low relative pressure the water molecules react with the functional groups that are situated at the edges and corners of the basal planes and form some type of nuclei for further adsorption, creating clusters around the functional group sites. For nonporous carbons, depending on the number of PACs and their separation, either separated clusters or a continuous film of water could be formed on the surface. The first case occurs when the distance between the PACs is larger than a limiting cluster size. In this case the limiting water uptake is proportional to the concentration of the PACs and does not depend on the specific surface area of the adsorbent. The limiting cluster size is calculated from the slope of a straight line of the variation of the total amount of water adsorbed at saturation versus the number of PACs, as presented in Fig. 3 [124]. This plot gives a value of 150 water molecules per active site at saturation although there is no theoretical basis for the linearity. Assuming a hemispheric shape for this cluster results in a value of about 3 nm for the limiting cluster size. The second case occurs when the

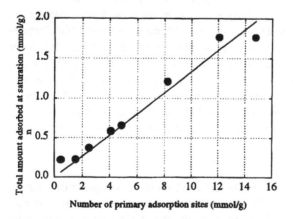

FIG. 3   Variation of total amount of water adsorbed at saturation with the number of PACs on graphitized carbon [124].

distance between the PACs is less than the limiting cluster size. In this case, a statistical number of layers, 1.7 ± 0.3, is formed on the surface and the limiting water uptake should be proportional to the specific surface area of the adsorbent. For porous carbons, in addition to the concentration of the PACs and their distance, the width of the pores could also affect the adsorption behavior. In this case, if the distance between PACs is smaller than both the width of a pore and the cluster limiting size, the clusters will join each other and form a continuous film of water on the surface. If the distance between the PACs is larger than the cluster limiting size and the pore width is small, the adsorbed molecules on the opposite walls of the pore could interact and form some types of bridge bond [124]. At higher relative pressure the clusters grow and finally merge into the micropores and cause micropore filling. The theory of volume filling of micropores (TVFM) was discussed in detail by many investigators [1,17,18,38,41, 46,47,49,51,52,80]. Barton and Koresh [6,8,9,11,82] suggested that the adsorption of water molecules by AC thus depends both on the amount of existing surface oxides and the micropore size. The former plays a major role at relative pressures below 0.3 while the latter manifests itself in the relative pressure of 0.3–0.5.

VI. ISOTHERM EQUATIONS FOR WATER ADSORPTION

Based on the different mechanisms, several isotherm equations have been proposed for the description of adsorption of water vapor on AC. Some of them are presented below.

A. D'Arcy-Watt Equation

Starting from the free energy of adsorption and using statistical thermodynamics, D'Arcy and Watt [34] proposed the following equation for the adsorption of water on heterogeneous adsorbents:

$$W = \sum_{i=0}^{l} \frac{K_i' K_i(p/p_0)}{1 + K_i(p/p_0)} + C(p/p_0) + \frac{k'k(p/p_0)}{1 - k(p/p_0)} \tag{1}$$

Here W represents the weight of sorbate taken up by 1 g sorbent; $K_i' = mn_i/N$ is the number of primary sites of type i (n_i), multiplied by the molecular weight of sorbate (m), and divided by Avogadro's number N; $k_i = \alpha\beta_i/p_0$ is a constant which is a measure of the attraction of the sites for the sorbate. K_0' and K_0 are both equal zero. $l = n - j$ is the number of different types of sorption sites for primary adsorption defined by a Langmuir isotherm; $C = m K/(Np_0)$ is a constant for the linear approximation to Langmuir adsorption on specific sites; $k' = mD/$

N and $k = \alpha\beta_m p_0$. In these definitions of α and β are temperature-dependent constants, D is the number of sites for secondary adsorption, and j is the number of Langmuir expressions that have been approximated by the single straight line. Although this equation was used by some investigators [12], its applications are not widespread due to the large number of parameters.

B. Dubinin-Astakhov Equation

In an attempt to extend the generality of the Dubinin-Radushkevich (D-R) equation [44], Dubinin and Astakhov [40] have put forward the following equation:

$$N = N_0 \exp[-(A/E)^n] \tag{2}$$

where N represents the amount of water adsorbed (mmol g^{-1}) at temperature T and relative pressure p/p_0, N_0 is the limiting amount adsorbed in the micropores, $A = RT \ln p/p_0$, and n and E are the characteristics of the system. Although this equation was used by some investigators [24,112,113] due to its simplicity, it does not reflect the physical mechanism of water adsorption. Furthermore, the S shape of the DA equation (by choosing small enough value of E) is not sufficient to explain many adsorption data [112,113] because most data show a favorable behavior at very low pressure, followed by a hyperbolic behavior and a Langmurian behavior at intermediate pressure.

C. Dubinin-Serpinsky Equation

Based on a kinetic model, Dubinin and Serpinsky [53] postulated the following relation for the adsorption of water on ACs:

$$\frac{p}{p_0} = \frac{C_\mu}{c(1 - kC_\mu)(C_{\mu 0} - C_\mu)} \tag{3}$$

In this equation, C_μ represents the amount of water (mmol g^{-1}) adsorbed at relative pressure p/p_0, $C_{\mu 0}$ is the concentration of primary sites (mmol g^{-1}) of solid, and c is the ratio of the rate constants of adsorption and desorption. The term $(1 - kC_\mu)$ takes into account the decrease in the number of adsorption sites with increasing micropore filling, while the parameter k itself represents the loss of the secondary sites with the progress of adsorption and hence affects the maximal capacity of adsorption. It does not have any influence on the initial adsorption rate. The D-S equation has been used to fit numerous data in the literature in an attempt to relate the parameters $C_{\mu 0}$, c, and k to the properties of the adsorbent. These parameters are listed in Table 1 for different types of AC. As seen in the table, there is no particular trend for the variation of these parameters for different ACs. Although extensive efforts have been made to correlate these parameters to other physically measurable properties of carbon, such as the density of oxygen surface sites [5], extent of burn-off during activation [24,42], and heat of immersion of carbon [121], no meaningful results could be derived.

D. Talu-Meunier Equation

Based on chemical equilibrium, equation of state, and phase equilibrium, Talu and Meunier [118] proposed the following equation for the adsorption isotherm of water on activated carbons:

$$P = \frac{H\psi}{(1 - K\psi)} \exp\left(\frac{\psi}{N_m}\right) \tag{4}$$

$$\psi = \frac{-1 + \sqrt{1 + 4K\xi}}{2K} \tag{5}$$

$$\xi = \frac{N_m \cdot N}{N_m - N} \tag{6}$$

In these equations, P is the equilibrium pressure, N is the amount of water adsorbed, H is Henry's law constant, K is the reaction constant for cluster formation in micropores, and N_m is the saturation capacity. Like the Dubinin-Serpinsky parameters, the different values of N_m, H, and K, the Talu-Meunier parameters, are also obtained by fitting the equation to the experimental data available in the literature. These parameters are given in Table 1 for different types of AC. As in the D-S equation, there is no particular trend for the variation of these parameters with the properties of AC.

VII. ENTHALPY OF IMMERSION

It has been shown by Stoeckli et al. [114] that the number of primary surface sites $C_{\mu 0}$ is directly related to the enthalpy of immersion of pure AC in liquid water. They found that the adsorption sites left on the surface after outgassing near 400°C containing a uniform type of hydrophilic sites, probably of the carbonyl type, contribute to the enthalpy of immersion by -25 kJ/mol, compared with -0.6 kJ/mol for the bulk of the water filling in the micropores at 307 K. They proposed the following relation:

$$\Delta h_i = -25C_{\mu 0} - 0.6(C_{\mu s} - C_{\mu 0}) \tag{7}$$

The quantity Δh_i represents the change (in J g^{-1}) of the enthalpy associated with the transfer of $C_{\mu s}$ mmol of water at room temperature, from the liquid state into 1 g of AC. Kraehenbuehl et al. [83] have modified the above equation and expressed the enthalpy of immersion of AC into water in terms of two parameters,

TABLE 1 Dubinin-Serpinsky and Talu-Meunier Parameters for Different Activated Carbons

Type of AC	Characteristics of AC				Dubinin—Serpinsky parameters			Talu–Meunier parameters			Remarks	Ref.
	BET area (m²/g)	Total pore volume (cm³/g)	Micropore volume (cm³/g)	Pore width (nm)	$C_{\mu o}$ (mmol/g)	c	k (g/mmol)	N_m (mmol/g)	H (g/mmol)	K		
BPL1					0.871	3.132	0.026	27.820	0.028	0.415	BPL oxidized in HNO_3 for 0.5 h	12
BPL2					2.723	2.738	0.031	25.558	0.117	0.214	BPL oxidized in HNO_3 for 1.0 h	12
BPL3					1.878	2.956	0.028	27.348	0.147	0.287	BPL oxidized in HNO_3 for 2.0 h	12
BPL4					4.263	2.401	0.030	27.852	0.089	0.146	BPL oxidized in HNO_3 for 4.0 h	12
BPL5					0.893	3.636	0.042	18.822	0.029	0.468	BPL oxidized in HNO_3 for 7.0 h	12
BPL					0.133	1.604	0.066	6.986	5.155	6.493	At 293 K	25
BPL					0.256	2.061	0.021	0.463	57.664	86.13	BPL	54
BPL1					2.402	2.835	0.027	0.522	6.520	12.17	BPL oxidized for 2 h by HNO_3	54
BPL2					1.118	3.423	0.040	0.341	11.403	25.97	BPL oxidized for 7 h by HNO_3	54
BPL	90–1000	0.94	0.44	1.40	0.233	2.277	0.021	28.944	0.694	59.86	BPL	10
Red BPL					0.072	1.858	0.016	28.389	2.176	168.6	BLP reduced in H_2-N_2 for 5 h at 500°C	10
CAL					0.292	1.977	0.019	30.035	0.938	1.278	CAL	121
CAL–methanol			0.425	1.83	0.352	1.943	0.019	28.658	1.308	1.813	CAL preadsorbed with 0.076 mmol/g methanol	121
CAL–methanol					0.338	1.948	0.019	28.259	1.386	1.945	CAL preadsorbed with 0.128 mmol/g methanol	121
CAL–benzene		0.416		1.90	0.187	1.958	0.019	30.495	1.011	1.350	CAL preadsorbed with 0.019 mmol/g benzene	121
CAL–benzene		0.425		1.92	0.311	1.874	0.019	29.542	0.844	1.108	CAL preadsorbed with 0.299 mmol/g benzene	121
Carbon Cloth					0.160	3.367	0.091	8.149	0.912	1.870	TCM-128 (Carbone Lorraine, Paris)	11
CC-OX-2					1.369	6.243	0.131	7.940	0.059	0.186	TCM-128 oxid. in HNO_3, 2 h	11
CC-OX-3					2.068	6.784	0.077	12.432	0.053	0.216	TCM-128 oxid. in HNO_3, 3 h	11
CC-OX-4					2.551	6.789	0.070	15.390	0.020	0.063	TCM-128 oxid. in HNO_3, 6 h	11
CC-OX-6					3.303	4.956	0.058	17.730	0.035	0.089	TCM-128 oxid. in HNO_3, 6 h	11
CC					0.154	3.440	0.092	8.480	0.739	1.465	TCM128 (Carbone Lorraine)	11
CC-OX-400-4					0.526	4.480	0.080	10.470	0.249	0.656	TCM-128 air oxid. at 400° for 4 h	6
CC-OX-400-8					0.664	4.683	0.075	11.250	0.204	0.574	TCM-128 air oxid. at 400° for 8 h	6
CC-OX-400-12					0.394	4.807	0.069	18.185	0.222	0.613	TCM-128 air oxid. at 400° for 12 h	6

Sample	A	B	C							Description	Ref.
CC-OX-400-18				1.133	4.439	0.048	16.940	0.145	0.412	TCM-128 air oxid. at 400° for 18 h	6
CC-OX-400-24				1.392	3.890	0.036	22.462	0.123	0.295	TCM-128 air oxid. at 400° for 24 h	6
CC-OX-400-72				3.027	2.426	0.0212	37.299	0.114	1.180	TCM-128 air oxid. at 400° for 72 h	6
CC-OX-HNO$_3$-2				1.280	6.362	0.133	8.075	0.048	0.137	TCM oxid. in HNO$_3$ for 2 h after evacuation at 40°C	6
CC-OX-HNO$_3$-3				2.843	6.508	0.078	12.964	0.032	0.121	TCM oxid. in HNO$_3$ for 3 h after evacuation at 40°C	6
CC-OX-HNO$_3$-4				2.365	6.860	0.072	15.160	0.023	0.068	TCM oxid. in HNO$_3$ for 4 h after evacuation at 40°C	6
CC-OX-HNO$_3$-6				3.764	4.513	0.058	17.940	0.037	0.087	TCM oxid. in HNO$_3$ for 6 h after evacuation at 40°C	6
CC-OX-HNO$_3$-2				0.178	3.173	0.073	10.120	0.648	1.529	TCM oxid. in HNO$_3$ for 2 h after evacuation at 40°C	6
CC-OX-HNO$_3$-3				0.601	3.704	0.045	16.200	0.384	0.886	TCM oxid. in HNO$_3$ for 3 h after evacuation at 40°C	6
CC-OX-HNO$_3$-4				0.435	4.029	0.048	15.400	0.355	0.932	TCM oxid. in HNO$_3$ for 4 h after evacuation at 400°C	6
CC-OX-HNO$_3$-6				0.883	2.833	0.031	21.451	0.317	0.672	TCM oxid. in HNO$_3$ for 6 h after evacuation at 400°C	6
CC-OX-HNO$_3$-2				1.266	6.406	0.132	7.562	0.072	0.252	TCM oxid. in HNO$_3$ for 2 h after evacuation at 400°C	6
CC-OX-HNO$_3$-3				2.732	5.718	0.076	12.238	0.061	0.236	TCM oxid. in HNO$_3$ for 3 h after evacuation at 40°C	6
CC-OX-HNO$_3$-4				3.436	6.116	0.069	15.219	0.025	0.079	TCM oxid. in HNO$_3$ for 4 h after evacuation at 40°C	6
Char LOP$_3$	754	0.18	0.51	0.384	2.522	0.053	15.109	0.488	0.641		2
Charcoal cloth				0.702	2.561	0.027	26.224	0.238	0.397	VK-50 after polymer deposition followed by outgassing at 523 K	28
Charcoal cloth	200	1.11		0.709	2.111	0.011	60.644	0.255	0.343	VK-92 unmodified sample	28
Charcoal cloth	20	0.35		0.349	4.946	0.058	13.556	0.231	40.89	VK-20	29
Charcoal cloth	30	0.47		0.384	3.780	0.038	18.556	0.310	44.69	VK-31 at 298 K	29
Charcoal cloth	30	0.67		0.469	2.588	0.020	27.889	0.506	54.61	VK-50	29
Charcoal cloth	200	1.11		0.431	2.175	0.012	43.778	0.448	38.79	VK-92	29
Charcoal cloth	50	0.85		0.306	2.559	0.014	39.722	0.418	43.08	VK-70	29
Ungraphitized charcoal				0.190	1.817	0.013	30.778	4.875	369.2		60
Charcoal C				0.002	1.661	0.009	35.558	11.718	0.903		60
Charcoal A				1.406	2.472	0.022	33.667	0.234	0.021		60
Charcoal B				0.951	1.675	0.010	55.447	0.468	0.032		60
Carbon cloth				0.249	3.314	0.090	7.731	0.923	2.069	As received carbon cloth	6

TABLE 1 Continued

Type of AC	Characteristics of AC				Dubinin—Serpinsky parameters			Talu-Meunier parameters			Remarks	Ref.
	BET area (m^2/g)	Total pore volume (cm^3/g)	Micropore volume (cm^3/g)	Pore width (nm)	$C_{\mu o}$ (mmol/g)	c	k (g/mmol)	N_m (mmol/g)	H (g/mmol)	K		
Carbon cloth ox					0.185	3.613	0.729	9.793	0.816	1.991	ox-HNO$_3$-2 h after evacuation at 400°C	6
Carbon cloth ox					1.201	6.396	0.130	7.369	0.095	0.362	ox-HNO$_3$-2 h after evacuation at 40°C	6
PIT-ACF-2	1230		0.50	1.10	0.523	3.180	0.027	24.261	0.522	0.065	PITACF oxidized at 573 K	76
PIT-ACF-3	1246		0.512	1.12	1.335	3.389	0.022	30.481	1.422	0.024	PITACF oxidized at 773 K	76
PIT-ACF-4	1422		0.618	1.2	0.166	2.500	0.018	28.308	0.846	0.148	PITACF oxidized at 1173 K	76
PIT-10-373	1230		0.50	0.81	0.121	2.787	0.025	23.433	0.54	0.089	PIT10 oxidized at 373 K	74
PIT-10-573	1230		0.50	0.81	0.412	3.382	0.029	24.329	0.162	0.065	PIT-10 oxidized at 573 K	74
PIT-10-773	1250		0.51	0.82	1.162	3.439	0.023	30.730	0.882	0.021	PIT10 oxidized at 773 K	74
PIT-10-973	1280		0.56	0.88	0.394	2.687	0.022	32.292	0.504	0.104	PIT10 oxidized at 973 K	74
PIT-10-1173	1422		0.69	0.87	0.377	2.436	0.018	33.810	0.594	0.048	PIT10 oxidized at 1173 K	74
PIT-5	900		0.34	0.75	0.115	3.027	0.036	16.996	0.036	0.070		74
Coal tar pitch	1100			0.5–0.6	0.377	2.436	0.018	33.810	0.504	0.048	Coal tar pitch kilned at 1200°C	60
PVDC		0.4	0.4		0.031	2.845	0.029	20.254	4.444	2.448	Polyvinylidene chloride carbon	7
PVDC-1123K		0.29	0.24		0.273	1.530	0.041	10.119	0.580	5.683	PVDC treated in argon at 1673 K	25
PVDC (5850)		0.41	0.38		0.415	2.801	0.041	16.192	0.277	1.150	Untreated PVDC char	65
PVDC (5850)		0.54	0.48		0.942	2.328	0.023	28.816	1.121	0.449	65% burn-off	65
PVDC-600					0.210	2.452	0.026	20.625	0.094	1.149	Degassed at 600°C	3
PVDC-600					2.908	2.909	0.036	24.277	0.072	0.094	Degassed at 300°C	3
PVDC-600					5.603	2.497	0.034	24.104	0.085	0.079		3
PVDC-600					3.865	2.763	0.035	24.338	0.115	0.089	K$_2$-S$_2$O$_8$ oxidized	3
PVDC-600					4.802	3.685	0.033	23.020	10.967	0.212		3
PVDC-600					0.001	1.904	0.0211	19.911	0.083	9.201	Degassed at 1000°C	3
PVDC-600					5.181	2.588	0.0360	22.393	0.397	1.100		3
VK50	30	0.67			0.078	2.678	0.0207	28.393	0.504	0.763		11
SC-11 carbon					0.705	1.237	0.3153	15.395	1.591	1.597	Coconut shell based at 293 K	24
Saran 600					0.291	2.821	0.0379	18.391	0.583	0.586		3
Saran 600					0.077	2.369	0.020	27.476	2.572	2.344	Steam activated at 850°C	3
Vegetable-based AC					0.359	2.082	0.013	40.066	0.474	0.712		114

Sample											Description	Ref.
Anthracite-based AC					0.195	1.939	0.018	28.856	1.352	1.887		114
CEL	1470		0.6	0.8 ± 0.1	0.549	3.040	0.019	34.023	0.306	0.034	Cellulose-based ACF	73
GAC	2500	3.50			0.010	2.937	0.028	23.467	0.648	0.068	Coconut shell granular AC	75
AX-21	724		1.30	0.95	0.069	1.792	0.007	48.190	2.088	0.166		108
HGI	91		0.85		3.616	0.957	0.003	29.778	1.910	126.841		101
Nonporous carbon black					0.035	2.684	0.279	3.337	2.925	3.877	Powder-like nonporous carbon black	37
Carbon B4					0.012	2.117	0.025	16.006	6.181	10.53	19.4% burn-off	37
Carbon B2					0.056	2.518	0.052	11.116	3.413	6.129	5.7% burn-off	37
Takeda 5Å					0.030	3.397	0.656	1.063	5.550	13.00		31
Kevlar char					16.421	0.001	0.013	41.333	0.022	0.437		55
NOM1	470		0.21		0.961	7.53	0.061	17.833	0.023	3.698	Steam-activated Norox char 10% burn-off	55
NOM8	1043	0.74	0.44		0.261	1.921	0.006	70.722	0.602	49.03	Steam-activated Norox char 80% burn-off	55
NOM1-CO_2	470	0.21	0.21		1.189	9.340	0.072	12.778	0.045	14.19	CO_2-activated Norox char 10% burn-off	55
PAN	920		0.40	0.9 ± 0.1	2.522	4.015	0.044	18.471	0.108	0.017	Polyacrylonitrile-based ACF	73
AG2	460		0.322	0.699	0.221	1.962	0.027	20.278	1.898	149.4		37
VPI-5			0.361		1.461	0.003	0.049	30.500	0.003	0.015	Aluminophosphate microporous carbon	31
D5-A	1249		0.65	1.18	1.401	2.297	0.016	39.383	0.204	0.334		2
Carbon black E_2					0.262	0.985	0.448	1.046	5.904	6.299	Carbon black Elftex 120 degassed at 450°C then cooled to 25°C in oxygen	30
Carbon black E_3	33.2				0.338	3.051	0.623	1.552	0.859	1.630	Elftex 120 degassed at 450°C, maintained 2 h at 450°C in oxygen, then cooled to 95°C	30
U-02		0.45	0.43	1.2	0.001	1.962	1.722	28.722	2.052	155.5	at 293 K microporous carbon	112
CMS		0.25	0.24	0.8	0.001	2.598	1.879	0.293	86.464	166.8	at 275 K microporous carbon	112
PCB-2			0.426	0.94	0.782	2.379	0.023	28.389	0.305	27.67		122
CEP19/91	736		0.21	0.57	0.069	2.506	0.045	15.798	0.406	1.364		111
UO-2					0.333	2.092	0.013	39.292	0.539	0.823	at 293 K	3
MSC5-1					0.983	3.008	0.025	30.249	0.200	0.357	at 293 K containing 26.5% primary sites	3
MSC5-2					0.165	1.942	0.018	28.219	2.009	2.829	at 293 K containing 4.6% primary sites	3
UO3-1					0.045	1.775	0.014	29.814	4.43	6.110	at 293 K	83

$C_{\mu 0}$ and c, both appearing in the D-S equation (3). They proposed the following equations:

$$c = c_0 \exp(\Delta H_i / RT) \tag{8}$$

$$\Delta H_i = \Delta h_i / C_{\mu s} \tag{9}$$

and suggested that the adsorption branch of the water adsorption isotherm can be obtained from the data of immersion enthalpy and the micropore volume.

VIII. CHARACTERISTICS OF ISOTHERM AND HYSTERESIS

The adsorption isotherm of water on AC has a sharp adsorption uptake accompanied by a clear adsorption hysteresis occurring over a medium or high relative pressure range. McBain et al. [90] associated this noticeable uptake with capillary condensation, while Dubinin [42] attributed it to the cluster growth of water on the hydrophilic sites and the coalescence of these clusters as the pressure increases. Both the geometry and the chemistry of micropores are the important parameters that govern the shape of the adsorption isotherm and the hysteresis loop.

A. Shape of Adsorption Isotherm

Most authors [2,3,25,42,60,61,86,106,108] found the shape of water adsorption isotherms on microporous carbons to be invariably of type V in the IUPAC classification. They attributed this character not only to the microporous structure but to the surface chemistry of carbonaceous adsorbents. However, some other work [106,110,124] suggests that the micropore size does affect the shape of the isotherm and that adsorption to a large extent is controlled by a cooperative pore-filling mechanism. This fact was shown by Vartapetyan et al. [124] by presenting different shapes of water adsorption isotherms for different types of AC. Their isotherms were limited between a convex shape for ultramicroporous and a concave one for mesoporous AC, as shown in Fig. 4. In order to show the relative influence of surface chemistry, pore size, and pore shape on the form of the isotherm, Carrott et al. [31] studied a wide range of microporous carbons, zeolites, and aluminophosphates. They found that the microporous carbons have much lower affinity for water vapor, compared to sodium and calcium zeolites. This is mainly due to the small number of specific adsorption sites on the carbon surface. They also concluded that the shape of the isotherm for different microporous carbons may be regarded as type V but the point of inflection varies greatly between carbons. Fitting their experimental data to the D-S equation led them to believe that adsorption of water vapor at low relative pressure is largely depen-

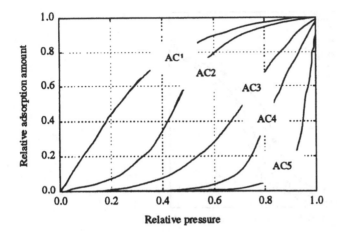

FIG. 4 The shape of water adsorption isotherm on different ACs [124]. AC1, ultramicroporous, based on polyacrilonitrile; AC2, peat based; AC3, coal based, activated by steam; AC4, furfural based; AC5, Mesoporous form molybdenum carbide.

dent on specific adsorbent-adsorbate interactions, while at higher relative pressures the micropore size and shape control the extent of adsorption. The same results were obtained by Vartapetyan et al. [124] for carbons with low number of PACs, and by Stoeckli and Kraehenbuehl [115] and by Bradley and Rand [25] for microporous carbon. The shape of the isotherm on carbon black was also studied by Kiselev and Kovaleva [81]. They showed that in this case the isotherm is of type III, whose initial part is mainly affected by the number of PACs, and it is shifted to the right when chemisorbed oxygen is removed, in agreement with the more detailed subsequent study of Walker and Janov [126]

1. Effect of Temperature on Isotherm Shape

The temperature dependence of the isotherm shape was studied by several investigators [44,64,94,95,97–99]. In an attempt to analyze the equilibrium of water on AC, Haggahalli and Fair [64] found that the saturation equilibrium uptake is independent of temperature, provided that the variation of condensate density with temperature is not considerable. They attributed this behavior to the fact that the adsorbent pores are filled with condensate and not with layers of adsorbed molecules. But they found that the onset of pore filling is dependent on temperature. The same results are shown in the work of Morimoto and Miura [94,95,97–99], who found that the dependence of adsorption on temperature is significant and the onset of pore condensation occurs at higher reduced pressure for higher temperature. This is summarized graphically in Fig. 5 for water adsorption on BPL-activated carbon at four different temperatures. The onset of pore condensa-

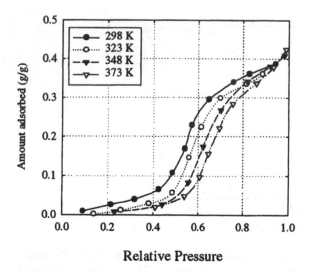

FIG. 5 Adsorption isotherm of water on AC (BPL) at different temperatures [107].

tion occurs at a reduced pressure of 0.4 for 298 K and at 0.6 for 373 K. In the case of typical pore filling of organic vapors in micropores, the temperature difference is removed by using the relative pressure. The marked difference of the isotherm with temperature change suggests a pore-filling mechanism that is different from the classical capillary condensation in mesopores. What really occurs in the case of water is the pore filling of micropores [85].

B. Hysteresis Loop

In water adsorption on AC once the saturation vapor pressure is reached, desorption usually follows a different path from adsorption and a hysteresis loop is therefore observed. The desorption branch is above the adsorption branch. The width and shape of the hysteresis loop are different for different ACs. The hysteresis loop may be limited to moderate to high relative pressure (high-pressure hysteresis, HPH), but in some cases it extends to the low-pressure region (low-pressure hysteresis, LPH) [120]. Different parameters are attributed to the appearance, shape, and width of hysteresis loop. For example, Dubinin [42] reported that the width of the hysteresis loop was related to the degree of activation of AC, and hence to the functional groups in the micropores. Although hysteresis is generally believed to stem from capillary condensation [109], the hysteresis of the water adsorption isotherm is not attributed to capillary condensation. The Kelvin equation, which describes well the capillary condensation phenomena, is not effective for water adsorption [62]. For nonporous carbon blacks, hysteresis is not observed [30], confirming that the micropores are the source of the hysteresis.

1.   Effects of Pore Structure on Hysteresis Loop

In order to study the effects of pore structure, Dubinin [37] examined the water adsorption isotherm for a microporous carbonaceous adsorbent with different degrees of burn-off. He found no hysteresis loop for samples with a burn-off of less than 8%, while at a burn-off of about 11% a small hysteresis loop appears in the range of high relative pressure, and at 20% burn-off or greater, a complete hysteresis loop is observed. These results are reproduced in Fig. 6.

In fact, increasing the percentage of burn-off usually means increasing the pore width of the sample. Kaneko et al. [74] found a similar result by comparing the adsorption isotherms for two different pitch-based ACs. They found that the width of the hysteresis loop depends on the pore width. Although no hysteresis loop is observed for small pore size samples (pore width < 0.8 nm), a wide hysteresis loop exists for samples having larger pore size. Indeed for small micropores, due to the space limitation, the water molecules exit the pore the same way they enter the pore, hence no hysteresis occurs. On the other hand, for larger micropores the water clusters merge into a stable solid-like structure; the adsorbed water structure on adsorption and that on desorption should thus be different. The water structure after the saturated adsorption should be more stabilized and the evaporation pressure is different from adsorption pressure, giving the hysteresis. Iiyama et al. [65] investigated this phenomenon using in situ small-angle X-ray scattering (SAXS). They found a distinct difference in adsorption processes in narrow and wide micropores. In case of the pore width < 0.8 nm, the SAXS intensity versus p/p_0 has no hysteresis, coinciding with the adsorption isotherm. On the other hand, the SAXS intensity versus p/p_0 has a marked hysteresis similar to that of the isotherm. They interpreted the hysteresis mechanism using the density fluctuation derived from the Ornstein-Zernike theory for the SAXS data.

2.   Effects of Adsorption Cycling on Hysteresis Loop

Lin and Nazaroff [86] studied the isotherm and hysteresis loop of water adsorption on a granular AC and proposed a model for adsorption-desorption kinetics and equilibrium partitioning of water molecules between gas and adsorbent surface. They found a type V isotherm with a hysteresis loop. The width of the hysteresis loop was a function of the relative humidity at which the desorption is initiated. The higher was the relative pressure the wider was the hysteresis loop, but the lower closure point in any case occurred at $p/p_0 \cong 0.3$. Similar hysteresis behavior and lower closure points have previously been found by Tsunoda [123] for several ACs. Vartapetyan and Voloshchuk [124] attributed the existence of the hysteresis loop to irreversible chemisorption, which occurs si-

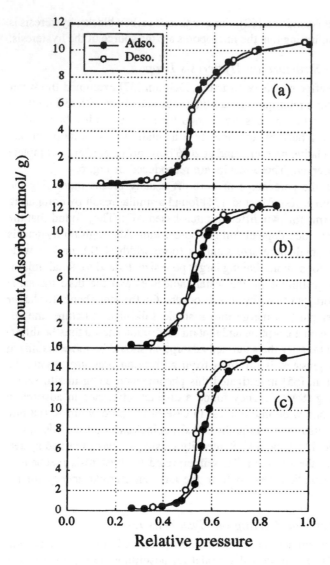

FIG. 6 Water adsorption isotherm for (a) 5.7% burn-off, (b) 11.3% burn-off, and (c) 19.5% burn-off carbons.

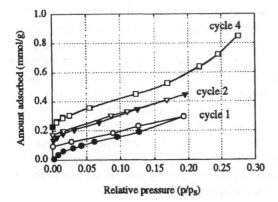

FIG. 7 Effect of adsorption-desorption cycle numbers on the adsorption isotherm of water on AC [124]. Closed symbols, adsorption; open symbols, desorption.

multaneously with reversible physisorption of water molecules. They showed that the width of the hysteresis loop is decreased if the sample is subjected to successive adsorption-desorption processes and it disappeared completely after several cycles. They concluded that the second adsorption cycle did not result in any increase in the amount of firmly bound water, which causes the open hysteresis loop. By examining their results, shown in Fig. 7, we conclude that successive adsorption-desorption cycles could increase the adsorption capacity due to the formation of new adsorption sites by chemisorption. The particular carbon used in their study is probably insufficiently activated and has narrow micropores. The water adsorption can irreversibly enlarge the micropores. Kaneko et al. [71] showed, using X-ray diffraction and SAXS [57,116], that this swelling is substantial in some activated carbons. In the case of hydrophobic ACF the closure point is close to $p/p_0 = 0.4$. The closure point should be associated with the stability of the adsorbed water. Hence, the in situ X-ray examination and heat of adsorption measurements near the closure point are desirable.

IX. CARBON PRETREATMENT

The process of activation causes changes in carbon structure or carbon porosity and is expected to increase the adsorptive capacity of a given carbon. Activation can also increase the noncarbon content of the structure, like that of chemisorbed oxygen or other residual heteroatoms from impregnation treatments. These may be the active sites which are capable of specific interactions with polar components such as water. Activation may be done physically or chemically. In physical activation of carbon with either steam or carbon dioxide, some carbon is oxidized or is removed as oxides. Chemical activation is achieved by mixing a suitable activating agent, such as zinc chloride or phosphoric acid, with the original material before carbonization is carried out. Treatment of carbon with oxidizing gases such as oxygen, ozone, nitrous oxide and nitric oxide, or oxidizing solutions like nitric acid, alkaline permanganate, hydrogen peroxide, acidic permanganate, percholoric acid, and acidic dichromate, may also lead to the formation of surface oxygen complexes. The degree of oxidation, the type of the resulting oxygen complex and the adsorption properties of the treated carbon depend on several factors, including the type of chemical agent used as oxidant, temperature of oxidation, duration of treatment, and type of carbon that is oxidized [6]. The mechanism of oxidation reactions was investigated by several authors [4,68,74, 104,123] (see also Chapter 1 in this volume).

A. Effects of Heat Treatment

Heat treatment of carbons in an inert environment could modify the structure such that pores could be narrowed or even collapsed. This also removes some of the functional groups, thus decreasing the polarity of the carbon surface. As an illustration, Fig. 8 shows the adsorption isotherm of water on peat-based ACs that are pretreated at different temperatures [112]. Morimoto and Miura [94–99] have studied the effects of high-temperature treatment, autoclave treatment, oxidation treatment, hydrogen treatment, and ozone treatment of graphite on the shape of the water adsorption isotherm and water adsorption capacity. The following results were obtained:

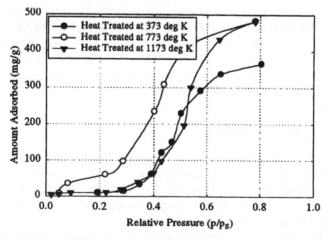

FIG. 8 Adsorption isotherm of water on oxidized PIT-10 at 30°C [112].

1.　The high-temperature heat treatment changes the shape of the isotherm from type II to type III. The adsorption capacity decreases remarkably, which could be attributed to the destruction of micropores.
2.　The autoclave treatment (under 8.6 MPa and 300°C) causes a considerable decrease in the amount of surface functional groups and, as a result, in the adsorbability for water.
3.　Oxidation treatment or partial burning causes some reduction in the specific surface area. Interestingly, it also reduces the amount of surface functional groups and the water adsorption capacity.
4.　Hydrogen treatment at 1000°C causes a drastic decrease in the amount of surface oxides and hence in the adsorption capacity.
5.　Although ozone treatment of hydrogenated graphite increases the amount of surface oxides and the hydrophilicity of the samples, further heat treatment in vacuo reduces the amount of surface oxides and the water adsorption capacity.

In order to compare the water adsorption capacity of ACs regenerated in different manners, Hassan et al. [63] found that, although the isotherm shape on BPL carbon is of type V, the relative pressure at which the isotherm rises increases with heat treatment temperature. A strong hydrophobic effect is observed as a result of heat treatment, as expected. In a similar study, Mitropoulos et al. [93] studied the effects of high-temperature treatment (in air) on two different high-volatility bituminous coals, by water adsorption and SAXS. In addition to the expected significant evolution of volatile material, resulting in a partial collapse of the pore structure as well as a reduction in pore volume, surface area, and adsorption capacity, they found a strong hydrophobic effect that extends up to a relative pressure of ≈ 0.1. This treatment affects not only the concentration of the surface functional groups but also the acid-base properties, as evidenced by adsorption measurements of NH_3 and SO_2 in addition to water [76].

B.　Effects of CO_2 Activation

The effects of CO_2 activation and heat treatment on three different carbons— coal-based BPL, nutshell based SC11, and a polymer-derived (polyvinylidene chloride, PVDC) carbon—were studied by Bradley and Rand [23]. They found that the surface polarity decreases with increasing heat treatment temperature due to the progressive desorption of chemisorbed oxygen and increases by oxidation of the carbon structure with burn-off (more intensive CO_2 activation) as more basal-plane edge area is exposed. It should be mentioned that at typical activation temperatures (>800°C), the degree of surface oxygen incorporation is lower than during treatment in air (<400°C).

Freeman et al. [55] investigated the behavior of PVDC char in water adsorption. The samples were either activated by CO_2 to different degrees of burn-off or heat-treated in argon at different temperatures. The adsorbed water volume increased with burn-off, compared with the untreated char, while higher heat treatment had the opposite effect, as expected. The adsorbed water uptakes were also lower than the nitrogen saturation volumes. Although this phenomenon could be attributed to a lower density of water in the adsorbed phase compared with that of the normal bulk liquid at the same temperature, Gregg and Sing [60] have suggested that this may be due to the hydrogen bonding nature of water.

C. Effects of Oxidizing Acid Concentration

Tamon and Okazaki [119] have studied the adsorption of a variety of compounds on AC (CAL, Calgon Carbon Corp.) oxidized by nitric acid. They found that the adsorption capacity of oxidized AC for polar compounds such as water and ammonia increases upon mild oxidation, while a higher concentration of oxidizing solution could produce adverse results. These phenomena are shown in Fig. 9. The effects of oxidation on the adsorptive properties of BPL carbon (Calgon Carbon Corp.) have also been studied [54]. For example, oxidation with nitric acid was found to change the adsorption behavior of water at both low and high relative pressures. Although similar to untreated BPL, the mildly oxidized sample yielded a type V isotherm, and the steep rise in uptake occurred at lower relative pressure. For more severely oxidized samples, the isotherms departed from type V shape to become type IV. The deviations at low relative pressure can be attributed to the introduction of new oxygen-containing surface complexes, while at high relative pressure the changes are attributed to the removal of carbon by oxidation which modifies the pore structure and hence the sorptive capacity. Thus, even though wet oxidation increases the concentration of hydrophilic sur-

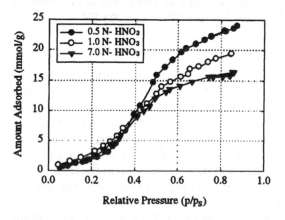

FIG. 9 Adsorption isotherm of water on CAL carbon at 25°C [119].

face oxides, the surface area and micropore volume could be destroyed by the severe oxidation, resulting in larger pore widths and reduced water uptakes.

Carbons derived from different sources can have wide variation in their oxidation susceptibility [104,129]. For example, Choma and Jaroniec [32] oxidized five different carbonaceous adsorbents using 68% nitric acid at its boiling point. For microporous ACs, the amount of water adsorbed was greater than for untreated carbons at relative pressures below 0.5. The opposite effect was observed in the high relative pressure range, where adsorption occurs by the pore-filling mechanism.

A comparative study of both water and N_2 adsorption at 77 K is needed to determine the adsorbed density of water in micropores of AC. However, we will try to explain water adsorptivity changes with the degree of carbon oxidation in a qualitative way here. We propose a simple model in which AC is composed of parallel layer planes formed by carbon atoms ordered in regular hexagons and held together by weak van der Waals forces. The spaces between these graphene layers form micropores and the various functional groups are located on their edges. This simple model is shown in Fig. 10. If l and w represent the length and width of one unit cell, α is the concentration of functional groups (number per unit length), and N is the number of unit cells per unit mass of AC, then the total micropore volume V_μ and total functional groups number S are calculated as follows:

$$V_\mu = Nl^2w \tag{10a}$$

$$S = 8Nl\alpha \tag{10b}$$

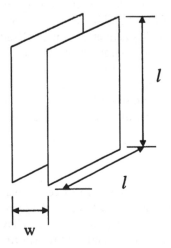

FIG. 10 A simple schematic model of a slit-shaped pore for water adsorption on AC.

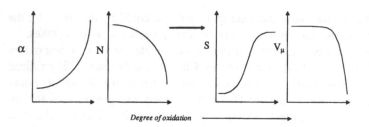

FIG. 11   Schematic representation of variation of S and V_μ with the degree of oxidation.

The treatment of AC by oxidation could to some extent increase the concentration of functional groups. The pore length and width may not be affected by mild oxidation, but upon severe oxidation they might decrease because of carbon consumption. Further oxidation could even reduce the number of cells per unit mass because of the complete removal of some unit cells, resulting in adsorption capacity decrease. These phenomena are represented shematically in Fig. 11 as a function of the degree of oxidation. The increase in the number of functional group gives rise to a sharper isotherm at very low pressure (due to greater adsorption of water molecules around the functional groups) while the decrease in the pore volume results in a lower adsorption capacity at higher pressures.

D.   Effects of Aging

Aging is another problem that could change significantly the properties of ACs, especially for water adsorption [25,27]. In order to elucidate the effects of aging, Carrasco-Marin et al. [27] analyzed the amount of acidic and basic surface complexes in fresh and aged samples of a heat-treated CO_2-activated demineralized Spanish lignite. They found that the acidic groups to a large extent, and basic groups to a lesser extent, are stabilized by aging. They also observed that the aged samples tend to restore the adsorption behavior of the original oxidized activated carbons. This may be attributed to chemisorption of oxygen at room temperature, which results in the formation of different oxygen-containing complexes of varying thermal stabilities [3,19].

E.   Effects of Mechanoactivation

Activation could also be achieved by mechanical processes such as centrifugation. Vartapetyan and Voloshchuk [125] studied the mechanoactivation of a granular AC prepared from furfural and its effects on the micropore structure. They observed that, although with increasing mechanoactivation treatment time (in air) the pore volume increased, the average pore width and the specific surface area decreased. This was accompanied by a sharp increase in the number of active sites for water adsorption. These phenomena are explained by the fracture of crystallites and mechanically induced structural changes, which causes a collapse of micropores and supermicropores and significant oxygen chemisorption.

F. Effects of Preadsorption

Tsunoda [121] studied the adsorption of water on a CAL carbon (Calgon Carbon Corp.) preadsorbed with various amounts of methanol and benzene. He found that, although at low relative pressure the uptake on methanol-preadsorbed AC was much lower than on untreated AC, the total water adsorption capacity at very high humidity remained the same. On the contrary, preadsorption of benzene had little effect. Using the Dubinin-Serpinsky equation, he concluded that methanol molecules are adsorbed on the surface complexes by hydrogen bonds between the hydroxyl groups of methanol and the carbon-oxygen surface complexes, thus occupying some of the primary sites for water adsorption. As humidity increases, the adsorbed methanol molecules act as secondary sites for water adsorption in such a way that the total water adsorption capacity is not affected. On the other hand, benzene molecules are adsorbed on the basal plane and so do not occupy the primary sites for water adsorption. Fujie et al. [56] examined the water adsorption and desorption behavior of NO-preadsorbed ACF and reported the possibility of formation of a NO/H_2O hydrate that is otherwise unstable at subatmospheric pressure. Hence, micropores of AC can be used for stabilization of unstable hydrates.

X. FORM OF ADSORBED WATER MOLECULES

Both the McBain capillary condensation mechanism and the Dubinin hydrophilic site association mechanism have assumed a liquid state of water adsorbed in the nanopores (micropores and smaller mesospores), without sufficient structural study. The X-ray diffraction and SAXS studies by Suzuki et al. [116], Kaneko et al. [75], and Fujiwara et al. [57] suggest that the micrographitic structure of cellulose-based ACFs changes as water accumulates in the micropores. As mentioned in Section II, by studying the ERDF obtained for water molecules confined in micropores and bulk liquid, Kaneko et al. [74] found that water adsorbed in micropores is in liquid form and even has a more ordered and rigid structure as the pore size is decreased. The compressed water molecules in the hydrophobic pores are dimerized more than the water molecules in the bulk liquid phase. The dimers grow in the micropores to form a larger cluster such as a pentamer, $(H_2O)_5$. The size of the pentamer is about 0.5 nm. The neutral pentamers may have a Lennard-Jones type of interaction with the carbon pores and be adsorbed. Sing [108] emphasized the dependence of the extent of molecular packing in very narrow pores on both pore size (in relation to molecular diameter) and pore shape. He stated that the mesopores are completely filled by the adsorbate in its normal liquid state, but this could not be exactly the same for micropores. Therefore, he introduced the concept of effective micropore volume for each particular adsorbate in a given type of adsorbent.

The nature of water associated with coal may also be relevant in this context. It was examined, for example, by Allardice and Evans [1]. They raised the temperature of coal after water adsorption from ambient to slightly above the boiling point of water. Their results showed that weakly associated water molecules within the coal matrix are lost at room temperature, while the strongly associated water molecules are removed as the temperature is increased to 110°C.

XI. CONCLUSION

The most acceptable mechanism that can explain the adsorption of water on AC seems to be the filling of the micropore volume as postulated by Dubinin several decades ago. According to this mechanism, water molecules are first adsorbed on the primary active sites on the edges of the basal planes of carbon. The adsorbed water molecules then act as secondary sites for further adsorption of water to form a cluster. As the relative pressure increases, those clusters grow and finally merge to fill the micropores. So although at low relative pressure the controlling factor in water adsorption is the surface chemistry of carbon, at high relative pressure the structural characteristics do play an important role.

Although several equations have been proposed to describe the water adsorption isotherm, only those of Dubinin-Serpinsky and Talu-Meunier reflect this mechanism well, and they satisfactorily describe some experimental data with three parameters.

The shape of the water adsorption isotherm on AC is of type V, but the position of the inflection point depends on the type of AC. The isotherm usually has a hysteresis loop which occurs at a medium to high relative pressure. The width of the hysteresis loop depends on the porous structure of carbon and for very small micropores the hysteresis loop does not exist. For a given type of AC the hysteresis loop would disappear if subjected to successive adsorption-desorption cycles. The occurrence of a hysteresis loop is attributed to the ease with which water molecules can enter and exit the pores and also to the irreversible phenomenon of chemisorption that occurs in conjunction with reversible physisorption.

The details of the activation of carbons have profound effects on their water adsorption behavior. Although the activation process often increases the concentration of primary active sites and may clear the pores from some inorganic impurities, and thus improves the adsorption characteristics, more severe conditions of activation, such as high-temperature treatment or oxidation with concentrated acids, could narrow or even destroy the pores and so may have adverse effects on adsorption properties.

Although the exact state of the adsorbed water molecules is not very clear, and different investigators have different views of this issue, the most probable form is a cluster structure that is more ordered and more rigid than the normal liquid state of water. The rigidity of such clusters depends primarily on the pore size of the carbon.

ACKNOWLEDGMENT

Support from Australian Research Council is gratefully acknowledged.

REFERENCES

1. DJ Allardire, DG Evans. Fuel 50:236 (1977).
2. D Atkinson, AI McLeod, KSW Sing, A Capon. Carbon 20:339 (1982).
3. RC Bansal, JB Donnet, F Stoeckli. Active Carbon. New York: Marcel Dekker, 1988.
4. TJ Bandosz, J Jagiello, JA Schwarz. Langmuir 9:2528 (1993).
5. SS Barton, MJB Evans, J Holland, JE Koresh. Carbon 22, 3:265 (1984).
6. SS Barton. J Chem Soc, Faraday Trans 79:1147 (1983)
7. SS Barton, MJB Evans, BH Harrison. Colloid Interf Sci 45, 3:542 (1973).
8. SS Barton. J Chem Soc, Faraday Trans 1, 79:1157 (1983).
9. SS Barton. J Chem Soc, Faraday Trans 1, 79:1147 (1983).
10. SS Barton. J Chem Soc, Faraday Trans 1, 79:1165 (1983).
11. SS Barton, MJB Evans. Carbon 29, 8:1099 (1991).
12. SS Barton, JE Koresh. J Chem Soc, Faraday Trans 1, 79:1173 (1983).
13. SS Barton, MJB Evans, JAF MacDonald. Langmuir 10, 11:4250 (1994).
14. SS Barton, JR Dacey, BH Harrison, JR Sellors. Colloid Interf Sci 71:367 (1979).
15. SS Barton, GL Boulton, JR Dacey, MJB Evans, BHJ Harrison. Colloid Interf Sci 44:50 (1973).
16. MC Bellissent-Fonel, R Sridi-Porbez, L Bosio. J Chem Phys 104:10023 (1996).
17. BP Bering, MM Dubinin, VV Serpinsky. Colloid Interf Sci 38:185 (1972).
18. BP Bering, MM Dubinin, VV Serpinsky. Colloid Interf Sci 21:378 (1966).
19. HP Boehm. Adv Catal 16:179 (1966).
20. HP Boehm. M Voll. Carbon 8:227 (1970).
21. HP Boehm. Carbon 32:759 (1994).
22. RH Bradley, B Rand. Carbon 31, 2:269 (1993).
23. RH Bradley, B Rand. J Colloid Interf Sci 169:167 (1995).
24. RH Bradley, B Rand. Carbon 29:1165 (1991).
25. LB Adams, CR Hall, RJ Holmes, RA Newton. Carbon 26(4):452–459 (1988).
26. S Brunauer, PH Emmett, E Teller. J Am Chem Soc 60:309 (1938).

27. F Carrasco-Marin, J Rivera-Utrilla, JP Joly, C Moreno-Castilla. J Chem Soc, Faraday Trans 92, 15:2779 (1996).
28. PJM Carrott. Carbon 29, 4/5:499 (1991).
29. PJM Carrott. Carbon 29, 4/5:507 (1991).
30. PJM Carrott. Carbon 30, 2:201 (1992).
31. PJM Carrott, MB Kenny, RA Roberts, KSW Sing, CR Theocharis. Characterization of Porous Solids II. Amsterdam: Elsevier, 1991, pp 685–692.
32. J Choma, M Jaroniec. Ads Sci Technol 16, 4:295 (1998).
33. JC Crittenden, RD Cortright, B Rick, SR Tang, D Perrram. J AWWA 80:73 (1988).
34. RL D'Arcy, IC Watt. Trans Faraday Soc 66:1236 (1970).
35. DD Do, H Do. Carbon 38:767 (2000).
36. JB Donnet, A Voet. Carbon Black, New York: Marcel Dekker (1976)
37. MM Dubinin. Carbon 18:355 (1980).
38. MM Dubinin. In Chemistry and Physics of Carbon, Vol 2 (PL Walker, ed.). New York: Marcel Dekker 1966, pp 51–120.
39. MM Dubinin, ED Zaverina. Zh Fiz Khim 21:1373 (1947).
40. MM Dubinin, VA Astakhov. Adv Chem Ser 102:69 (1971).
41. MM Dubinin. Zh Fiz Khim 39:1305 (1965).
42. MM Dubinin. Chem Rev 60:235 (1960).
43. MM Dubinin, ED Zaverina, VV Serpinsky. J Chem Soc 1760 (1955).
44. MM Dubinin, ED Zaverina, LV Radushkevich. Zh Fiz Khim 21:1351 (1947).
45. MM Dubinin, GM Plavnik. Carbon 6:183 (1968).
46. MM Dubinin. Izv AN SSSR Ser Khim 18 (1980).
47. MM Dubinin. Izv AN SSSR Ser Khim 1691 (1979).
48. MM Dubinin, AA Isirikian. Izv AN SSSR Ser Khim 13 (1980).
49. MM Dubinin, GM Plavnik. Carbon 2:261 (1964).
50. MM Dubinin, ED Zaverina. Zh Fiz Khim 23:57 (1949).
51. MM Dubinin, HF Stoeckli. J Colloid Interf Sci 75:34 (1980).
52. MM Dubinin, VV Serpinsky. Dokl AN SSSR 99:1033 (1954).
53. MM Dubinin, VV Serpinsky. Carbon 19:402 (1981).
54. MJB Evans. Carbon 25:81 (1987).
55. JJ Freeman, JB Tomlinson, KSW Sing, CR Theocharis. Carbon 33, 6:795 (1995).
56. K Fujie, S Minagawa, T Suzuki, K Kaneko. Chem Phys Lett 236:427 (1995).
57. Y Fujiwara, K Nishikawa, T Iijima, K Kaneko. J Chem Soc Faraday Trans 87: 2763 (1991).
58. VA Garten, DE Weiss. Austral J Chem 10:309 (1957).
59. A Gil, G De la Puente, P Grange. Microporous Mater 12:51 (1997).
60. SJ Gregg, KSW Sing. Adsorption, Surface Area and Porosity, 2nd ed London: Academic Press, 1982.
61. PG Hall, RT Williams. J Colloid Interf Sci 113:301 (1986).
62. Y Hanzawa, K Kaneko. Langmuir 13:5802 (1997).
63. NM Hassan, TK Ghosh, AL Hines, SK Loyalka. Carbon 29:681 (1991).
64. M Huggahalli, JR Fair. Ind Eng Chem Res 35:2071 (1996).
65. T Iiyama, M Ruike, K Kaneko. Chem Phys Lett 331:359 (2000).
66. T Iiyama, K Nishikawa, T Suzuki, K Kaneko. Chem Phys Lett 274:152 (1997).
67. T Iiyama, K Nishikawa, K Otowa, K Kaneko. J Phys Chem 99:10075 (1995).

68. J Jagiello, TJ Bandosz, JA Schwarz. Carbon 30:63 (1992).
69. H Jankowska, A Swiatkowski, J Choma. Active Carbon. Chichester: Ellis Horwood, pp 75–85.
70. AJ Juhola. In Extended abstracts, 15th Biennial Conference on Carbon, Pennsylvania State University, 1981, p 216.
71. K Kaneko, T Suzuki, Y Fujiwara, K Nishikawa. In Characterization of Porous Solids II (F Rodriguez-Reinoso et al., eds). Amsterdam: Elsevier 1991, p 389.
72. K Kaneko, Y Hanzawa, T Iiyama, T Kanda, T Suzuki. Pacific Basin Workshop on Adsorption Science and Technology, Kisarazu, Japan, May 7–10, 1997.
73. K Kaneko, Y Fujiwara, K Nishikawa. J Colloid Interf Sci 127, 1:298 (1989).
74. K Kaneko, Y Hanzawa, T Iiyama, T Handa, T Suzuki. Presented at the Pacific Basin Workshop on Adsorption Science and Technology, May 7–10, 1997, Kisarazu, Japan.
75. K Kaneko, T Suzuki, Y Fujiwara, K Nishikawa. In Characterization of Porous Solids II, F Rodriguez-Reinoso et al., eds. Amsterdam: Elsevier, 1991, p 389.
76. K Kaneko, T Katori, K Shimizu. J Chem Soc Faraday Trans 88(9):1305 (1992).
77. K Kaneko, T Katori, K Shimizu, N Shindo, T Maeda. J Chem Soc, Faraday Trans 88, 9:1305 (1992).
78. K Kaneko, C Yang, T Ohkubo, T Kimura, T Iiyama, H Touhara. In The Second Pacific Basin Conference on Adsorption Science and Technology, Brisbane, Australia, May 14–18, 2000.
79. T Kimura, J Miyawaki, M Merraoui, T Iiyama, T Suzuki, K Kaneko. Int Symp Carbon (Tokyo, 1998) p 612.
80. AV Kiselev. Second International Congress on Surface Activity. London: Butterworths, 1957, p 219.
81. AV Kiselev, NV Kovalera. Izvest Akad Nauk SSSR, Otd Khim Nauk 955 (transl) 1959.
82. JE Koresh, TH Kim, DRB Walker, WJ Koros. J Chem Soc, Faraday Trans 1 85, 12:4311 (1989).
83. F Kraehenbuehl, C Quellet, B Schmitter, HF Stoeckli. J Chem Soc, Faraday Trans 1, 82:3439 (1986).
84. I Langmuir. J Am Chem Soc 40:1361 (1918).
85. G Li, K Kaneko, F Okino, H Touhara, R Ishikawa, M Kanda. J Colloid Interf Sci 172:539 (1995).
86. T Lin, WW Nazaroff. J Environ Eng 122:176 (1996).
87. OP Mahajan, PL Walker. Fuel 50:308 (1971).
88. KJ Master, B McEnaney. Carbon 22, 6:595 (1984).
89. JS Mattson, HB Mark Jr. Activated Carbon, New York: Marcel Dekker (1971).
90. JW McBain, JLK Porter, RF Sessions. J Amer Chem Soc 55:2294 (1993).
91. C McCallum, T Bandosz, SC McGrother, E Muller, K Gubbins. Langmuir 15:533 (1999).
92. A Watanabe, T Iijama, K Kaneko. In International Symposium on Carbon. Tanso, AP11-11 (1998).
93. AC Mitropoulos, JM Haynes, RM Richardson, TA Steriotis, AK Stubos, NK Kanellopoulos. Carbon 34, 6:775 (1996).
94. K Miura, T Morimoto. Langmuir 7:374 (1991).

95.   K Miura, T Morimoto. Langmuir 10:807 (1994).
96.   K Miura, T Morimoto. Langmuir 4:1283 (1988).
97.   T Morimoto, K Miura. Langmuir 1:658 (1985).
98.   T Morimoto, K Miura. Langmuir 2:43 (1986).
99.   T Morimoto, K Miura. Langmuir 2:824 (1986).
100. A Oberlin, M Viley, A Combaz. Carbon 18:347 (1980).
101. M Okazaki. Proceeding of the IVth Int Conf on Fundamentals of Adsorption, May 17–22, Kyoto, Japan (1992), pp 13–26.
102. C Pierce, RN Smith. J Phys Chem 57:64 (1953).
103. M Polanyi. Trans Faraday Soc 28:316 (1932).
104. BR Puri. In Chemistry and Physics of Carbon (PL Walker Jr, ed). New York: Marcel Dekker, 1970, Vol 6, p 191.
105. LV Radushkevich. Zh Fiz Khim 23:1410 (1949).
106. RA Roberts. PhD thesis, Brunel University, 1988.
107. EN Rudisill, JJ Hacskaylo, MD LeVan. Ind Eng Chem Res 31, 4:1122 (1992).
108. KSW Sing. Third Int Conf Fund Adsorption, pp 60–83, Sonthofen, Germany, May 5–9, 1989.
109. KSW Sing, DH Everett, RAW Haul, L Moscou, RA Pierotti, J Rouquerol, T Siemieniewska. Pure Appl Chem 57(4):603 (1985).
110. KSW Sing. Characterization of Porous Solids II. Amsterdam: Elsevier, 1991, pp 1–9.
111. F Stoeckli, D Huguenin. J Chem Soc Faraday Trans 88:737 (1992).
112. F Stoeckli, TS Jakubov, AJ Lavanchy. J Chem Soc, Faraday Trans 90:783 (1994).
113. F Stoeckli, L Currit, A Laederach, TA Centeno. J Chem Soc, Faraday Trans 90: 3680 (1994).
114. HF Stoeckli, F Kraehenbuehl, D Morel. Carbon 21:589 (1983).
115. HF Stoeckli, F Kraehenbuehl. Carbon 19:353 (1981).
116. T Suzuki, K Kaneko. Chem Phys Lett 191:569 (1992).
117. O Talu, AL Myers. AIChE J 34:1887 (1988).
118. O Talu, F Meunier. AIChE J 42:809 (1996).
119. H Tamon, M Okazaki. Carbon 34:741 (1996).
120. F Rodriguez-Reinoso, A Linares-Solano. Chemistry and Physics of Carbon, Vol 21 (PA Thrower, ed). New York: Marcel Dekker, 1989, pp 1–146.
121. R Tsunoda. J Colloid Interf Sci 188:224 (1997).
122. R Tsunoda, J Ando. J Colloid Interf Sci 146, 1:291 (1991).
123. R Tsunoda. J Colloid Interf Sci 137, 2:563 (1990).
124. R Sh Vartapetyan, AM Voloshchuk. Russian Chemical Reviews 64, 11:985 (1995).
125. R Sh Vartapetyan, AM Voloshchuk, GM Plavnik, YP Toporov, GS Khrustaleva. Russ J Phys Chem 69, 10:1667 (1995).
126. PL Walker, J Janov. J Colloid Interf Sci 28:449 (1968).
127. A Watanabe, T Iiyama, K Kaneko. Int Symp Carbon, Tokyo, Japan, 1998, p 624.
128. Z Wang, K Kaneko. J Phys Chem 102:2863 (1998).
129. AM Youssef, TM Ghazy, Th El-Nabarawy. Carbon 20:113 (1982).

4

Coal-Tar Pitch: Composition and Pyrolysis Behavior

Marcos Granda, Ricardo Santamaría, and Rosa Menéndez

Instituto Nacional del Carbón (CSIC),
Oviedo, Spain

I. INTRODUCTION

The industry of commercial coal-tar pitches is closely associated with that of metallurgical coke and aluminium industries. Changes in the latter industries have a profound impact on pitch production and/or pitch consumption. At the present time, coal-tar pitches have a prosperous market in the aluminum industry (carbon anodes) where coal tar pitch totally monopolizes the binder market. However, in recent times the situation has changed, mainly because of the dramatic reduction in coal-tar supplies and the introduction of rigorous environmental laws.

In the late 1980s and early 1990s the closing down of numerous coke ovens in the United States due to economic and environmental pressures produced an acute decline in the coal tar supply in this country [1,2]. This situation is expected to continue in the future (Fig. 1). The panorama in Europe is less worrying as the supply of coal-tar is guaranteed, at least in the short-term.

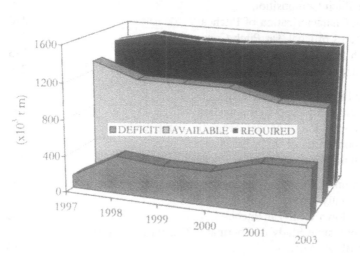

FIG. 1 North American (USA, Canada, and Mexico) coal tar supply trend. (Based on Ref. 2.)

However, a problem common both to the United States and Europe is the severe regulations concerning the emission and exposure to carcinogens at work [3–5]. This problem in particular affects coal-tar pitches. The transformation of a coal-tar pitch into a carbon material produces the emission of considerable amounts of volatiles, many of which are considered to be carcinogens [6]. For this reason, the use of coal-tar pitches in traditional fields, such as aluminium industries, which operate with Söderberg technology, is now in doubt.

With this dark horizon, it might be thought that coal-tar pitch is destined to become a thing of the past. On the contrary, despite these negative circumstances coal-tar pitch still has a bright and promising future.

It cannot be denied that tar production has decreased and that coal-tar pitch contains a number of environmentally unfriendly compounds, but it is no less true that coal-tar pitch is a unique product as a carbon precursor, and if suitably processed, environmental hazards can be reduced or even avoided. From the scientific point of view, it might be thought that not much is left to be discovered about coal-tar pitch composition and its pyrolysis behavior, in view of the large number of publications that have come out over the last 20 years. However, despite the great advances in analytical techniques, coal-tar pitch is still not fully known or completely understood. This is due to the complexity of its composition: it is made up of thousands of compounds, some of which are nonsoluble, covering a range of size that is difficult to determine by any of the analytical techniques at present available. New forms of carbon are continually being found through the manipulation of pitch composition and processing conditions, and it is the possibility of such new discoveries that maintains the interest of researchers around the world in this branch of science.

Within the framework of the chemistry and physics of carbon, this is the first review specifically devoted to coal-tar pitches. It serves as a complement to that of Greinke [7], which focused on petroleum pitches. Over the years, several other reviews concerned with carbonization have been published in these volumes. In 1968, Brooks and Taylor [8] stated the bases of mesophase formation taking into consideration different types of graphitizing carbon precursors. Fitzer et al. [9] reviewed the chemistry of the pyrolytic conversion of many organic compounds to carbon residues below 1000°C. The chemical kinetics of mesophase formation, as a first stage in the formation of graphitizable carbons, was discussed by Marsh and Walker [10]. As a complement to the latter review, Forrest et al. [11] studied the mechanisms of anisotropic carbon formation. More recently, Oberlin et al. [12] stressed the role of the basic structural units during the carbonization and graphitization of various carbon precursors (e.g., kerogens, coal, pitches).

In the present review, we have not only tried to give an overview of the state of the art of pitch composition and carbonization behavior. We have also made an effort to clarify the concepts and contradictions of different theories and to transmit to the readers our persistent faith in the promising future of coal-tar pitch as a valuable precursor of carbon materials.

II. COAL-TAR PITCH PREPARATION

Pitches can be obtained from several sources. Coal and petroleum are the two most important raw materials. Other sources of pitches include some aromatic compounds, such as naphthalene [13], methylnaphthalene [14], anthracene [15], and polyvinyl chloride [16]. Pitches obtained by the polymerization of these compounds are expensive and are only applied in specific fields (carbon fibers [17], polygranular carbons [18], ion-lithium batteries [19]) where the amount of pitch required is small. In these fields, the value of the final product justifies the high cost of the precursors.

The preparation of pitches from coal tar is based on a distillation process. By means of distillation, coal-tar yields a series of liquid fractions, each one of which has several industrial applications, and a solid residue (at room temperature) called coal tar pitch.

There are many alternatives to distilling coal tar for obtaining pitch. However, these alternatives basically follow the same sequence of steps (Fig. 2). In the first stage, a blend of coal tars is loaded into a distillation column where BTX (benzene, toluene, and xylenes) and water are removed. This fraction constitutes <6 wt % of the tar. Afterwards, the dehydrated tar is made to pass through a second column where it is rectified again and naphtha, naphthalene oil, and washing

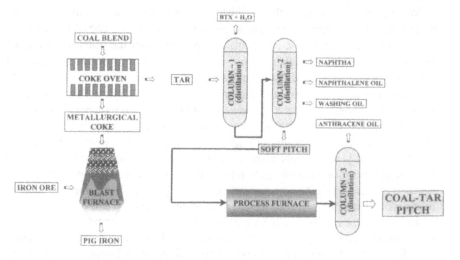

FIG. 2 Coal coking and coal tar refining to coal-tar pitch.

oil are recovered. These fractions represent ~20 wt % of the dehydrated tar. Naphthalene oil is the most relevant component of this fraction from an industrial point of view because it is the source of naphthalene. The fraction at the bottom of the second column is soft pitch, which in a subsequent step is fed into a process reactor where it is heated to ~400°C. Finally, the heated soft pitch is loaded into a third column (with or without vacuum) where the pitch parameters have been previously set (mainly the softening point) depending on the intended use of the pitch. If the column has capacity for further thermal treatment, toluene insolubles and beta resin (see Section III.C.1 for definition) can also be controlled. The third distillation yields a pitch with the required final parameters and a fraction called anthracene oil (~20 wt % of dehydrated tar). Anthracene oil is industrially used as a precursor for anthraquinone, colorants, and carbon black. The global yield of the process for obtaining pitch is ~50 wt %.

TABLE 1 Industrial Specifications for a Typical Binder and a Typical Impregnating Coal-Tar Pitch

Specification	Binder	Impregnating
Softening point (Mettler, °C)	108–115	85–90
Toluene insolubles (wt %)	25–35	<20–25
Quinoline insolubles (wt %)	10 ± 2	<5
Fixed carbon yield (Sers, wt %)	>52	>40
Ash (wt %)	<0.3	<0.2
Moisture (wt %)	—	<1
Na (ppm)	<200	—
Fe (ppm)	<300	—
Si (ppm)	<300	—
Ca (ppm)	<100	—
Zn (ppm)	<300	—
Sulfur (wt %)	<0.6	—
Mesophase (vol %)	Undetectable[a]	—
Volatile matter content (distillation, wt %):		
0–270°C	<1	—
0–360°C	<4	—
Aromaticity (C/H)	>1.75	—
Density at 20°C (g cm^{-3})	>1.310	1.260
Viscosity (cp):		
160°C	2000	
200°C	200	
225°C	—	10–15

[a] By optical microscopy.
Courtesy of Industrial Química del Nalón, S.A.

Two different types of pitches are obtained by the fractionated distillation of tar: binder pitches and impregnating pitches. Binder coal-tar pitches are primarily used as glue for joining petroleum coke grains in the manufacture of carbon anodes and graphite electrodes. Impregnating coal-tar pitches are used for densifying graphite electrodes. These two types of pitches account for ~87% of the consumption of coal-tar pitch in the USA, ~75% of which is consumed in the production of carbon anodes for the aluminum industry, and ~12% in the manufacture of graphite electrodes for electrical arc furnaces [20]. Table 1 summarizes the parameters required for a typical binder and a typical impregnating coal-tar pitch. When compared, binder pitches are denser, less fluid, and have a higher softening point and a greater quinoline-insoluble and fixed carbon content (carbon yield) than the latter.

These parameters provide valuable information to coal-tar pitch makers and consumers alike, and they are currently taken as user guidelines in the field of carbon anodes and graphite electrodes. However, they offer no more than an empirical approach to the suitability of a pitch for a specific application. Moreover, for conventional applications there is a lack of the necessary knowledge to explain and predict the behavior of a pitch during its technological processing. This is because the relationship between pitch parameters—such as solubility, softening point, and carbon yield—and the chemical composition of the pitch has not been fully clarified. Thus, pitches showing similar values for these parameters may behave in a very different way. As an example, Fig. 3 shows the pyroly-

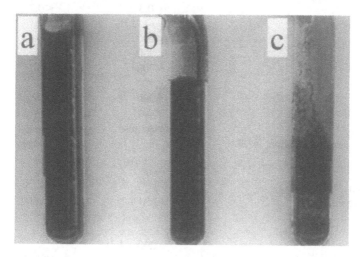

FIG. 3 Photographs of pyrolyzed products obtained at 450°C/30 min from (a) air-blown anthracene oil, (b) air-blown/thermally treated anthracene oil, and (c) a typical impregnating coal-tar pitch.

sis products obtained from three pitches with similar characteristics, at 450°C/ 30 min. Although their softening points and carbon yields are very close, swelling on carbonization is considerably higher in the case of the two anthracene oil–based pitches. In addition, the evolved gases are different as shown by the coloration of the tube walls.

The different behaviors of pitches with similar characteristics make the acquisition of a deeper physicochemical knowledge of pitch essential.

III. COAL-TAR PITCH COMPOSITION

Coal-tar pitches are extremely complex mixtures made up of thousands of compounds [21] that occur in different concentrations depending on the pitch. These compounds, which are polynuclear and aromatic, differ from each other in their molecular weight, functionality, and molecular structure. As an example, Fig. 4

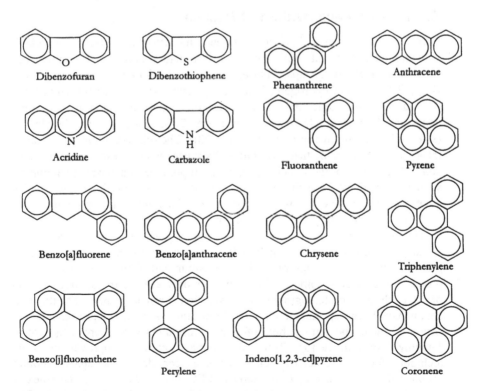

Dibenzofuran   Dibenzothiophene   Phenanthrene   Anthracene

Acridine   Carbazole   Fluoranthene   Pyrene

Benzo[a]fluorene   Benzo[a]anthracene   Chrysene   Triphenylene

Benzo[j]fluoranthene   Perylene   Indeno[1,2,3-cd]pyrene   Coronene

FIG. 4 Representative compounds in coal-tar pitch.

shows some of the most representative compounds identified in a commercial coal-tar pitch by gas chromatography [22].

In the last few decades many research programs have been carried out with the common aim of establishing relationships among the composition, structure, and thermal reactivity of pitch components. However, as stated in the introduction, pitch composition is still not fully known and some compounds, especially those of large size, still remain unidentified. Traditional procedures such as solubility tests and elemental analysis prove to be insufficient for predicting pitch thermal reactivity or coke microstructure, as the heteroatoms can be in different functionalities and the molecules having similar C/H ratios may adopt different structures.

It is possible to approach pitch characterization from the bulk material, which gives an average information, or from fractions obtained by means of previous separation into classes of compounds of similar characteristics. This procedure mitigates, to some extent, the complexity of the pitch but is more time consuming. The use of one or another procedure will depend on the type of information required.

A. Bulk Characterization of Pitches

Pitch characterization by spectroscopic and chromatographic techniques yields valuable qualitative and semiquantitative information about pitch components, e.g., main functional groups (Fourier transformed infrared spectroscopy, FTIR [23]); distribution of hydrogen (^{1}H nuclear magnetic resonance [24]) and carbon (^{13}C nuclear magnetic resonance [25]); identification of individual pitch components (gas chromatography/mass spectrometry [26]); molecular mass distribution (size exclusion chromatography, SEC [27,28]). The understanding and detailed information provided by these techniques have been widely described in the literature [29,30]. Use of these techniques makes it possible to determine parameters such as aromaticity and ortho substitution indices or condensation degree, which are especially relevant because they have a significant incidence on pitch pyrolysis behavior and carbon materials structure.

The individual use of each of these techniques gives only partial information about pitch composition. A combination of such techniques, therefore, would provide a better knowledge of pitch composition and pitch behavior during its processing. However, some of the techniques only analyze the soluble fraction of the pitch (gas chromatography, size exclusion chromatography). For this reason, in recent years special attention has been paid to achieving a greater solubilization of pitch. Thus, in the field of size exclusion chromatography, the use of 1-methyl-2-pyrrolidone [31,32] was an important step in the improvement of the technique for the characterization of complex coal-derived samples. Another problem that is still a subject of study in SEC, due to the inherent complexity of pitch composition, is related to the use of adequate calibration standards. The development of new characterization techniques, e.g., matrix-assisted laser desorption ionization mass spectrometry (MALDI-MS), has given a great boost to pitch characterization. Recent attempts to determine "safe" high-mass limits in MALDI mass spectra with reference to instrument noise have led to conservative upper mass limit estimates of between 40,000 and 60,000 amu for a coal-tar pitch [33]. The use of laser desorption mass spectrometry has considerably extended the range of molecular masses that can be detected in coal-tar pitches by other analytical methods, such as gas chromatography/mass spectrometry (GCMS) or probe mass spectrometry whose upper mass limits for coal-derived products are about 300 amu (for aromatics) [34] and 600 amu [35], respectively.

B. Features of Coal-Tar Pitch Composition

As previously mentioned, very heavy ion mass fragments have been estimated for a coal-tar pitch. However, most of the coal-tar pitch compounds range between 150 and 2500 amu [36], as determined by preparative SEC in conjunction with vapor pressure osmometry (Fig. 5). However, despite the complexity of

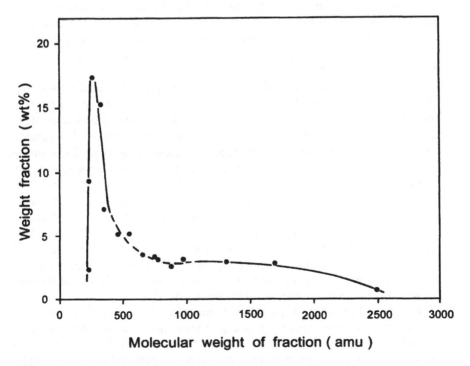

FIG. 5 Molecular weight distribution in a coal-tar pitch [36].

coal-tar pitches which is basically due to the large number of compounds, and the broad distribution of molecular weights, their constituents can be grouped into a relatively small number of compound classes [37]:

- Polycyclic aromatic hydrocarbons (PAHs)
- Alkylated PAHs
- PAHs with cyclopentene moieties (acenaphthylene types)
- Partially hydrogenated PAHs
- Oligoaryls and oligoaryl methanes
- Heterosubstituted PAHs: NH_2, OH
- Carbonyl derivatives of PAHs
- Polycyclic heteroaromatic compounds (benzologs of pyrrole, furan, thiophene, and pyridine)

PAHs are the predominant class of compounds in coal-tar pitches ($\sim$2/3). Polycyclic heteroaromatic systems, derived from furan, pyrrole, thiophene, and pyridine, are less concentrated than PAH. It has been observed that both sulfur and nitrogen are likely to be present in all molecules with a molecular mass greater than a few hundred, at levels of $\sim$0.5%, and all compounds with molecular masses greater than a few thousand are virtually certain to contain several atoms of sulfur and nitrogen [38]. Heteroaromatic compounds, despite being in lower amounts than PAHs, are of great importance because of their high thermal reactivity [39–41]. The rest of the coal-tar pitch compounds are found in lower concentrations than the classes just mentioned.

The molecular structure of PAHs derives from different building principles [37] as shown in Fig. 6. To establish these criteria the centers of the benzenic rings must be substituted by points. Adjacent points are connected by lines. When the resulting feature is an open-continuous system, the compound is catacondensed. On the other hand, when the feature is a closed or an open-discontinuous system, the compounds are pericondensed. Catacondensed compounds can be divided into two categories: branched and nonbranched (Fig. 6). Pericondensed compounds can be alternant or nonalternant (Fig. 6). The molecular structure of the compounds is very important from the viewpoint of thermal/chemical reactivity and solubility. Thus, branched systems are thermodynamically more stable and chemically less reactive than nonbranched systems of the same size [37]. These differences are patent even for compounds which belong to the same group. For example, the solubility of tetracene in dichloromethane is almost 50 times higher than that of its isomer, benzo[a]anthracene [41].

Recently, Martín et al. [42] developed a procedure for separating and quantifying coal-tar pitch components according to their molecular structure by means of high-performance liquid chromatography. They distinguished four classes of compounds (three types of catacondensed and one of pericondensed compounds), which were closely related to their thermal reactivity.

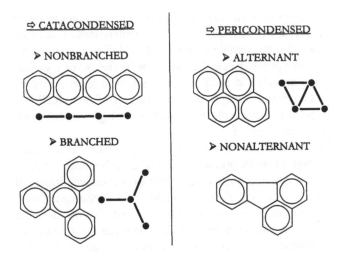

FIG. 6 Classification of the polycyclic aromatic hydrocarbons (PAHs) according to their molecular structure. (Based on Ref. 37.)

C. Pitch Fractionation

A first approach to simplification of pitch characterization could be its fractionation into classes of compounds according to the main functionalities of the constituents, molecular size or reactivity, and the characterization of the resultant fractions. Sequential solvent extraction, liquid chromatography, charge-transfer reactions, and extrography are currently the most frequent preparative fractionating methods used for this purpose.

1. Solvent Extraction

By means of solvent extraction, the pitch components are fractionated according to their solubility, which in turn depends on molecular size, topology, and degree of planarity [41]. Toluene and quinoline are widely used to determine pitch solubility parameters of special importance in industrial processes: α resin (quinoline-insoluble fraction), β resin (toluene-insoluble and quinoline-soluble fraction), and γ resin (toluene-soluble fraction). Toluene and 1-methyl-2-pyrrolidinone have been successfully used to monitor the degree of coal-tar pitch polymerization by thermal treatment and air blowing [43,44]. Solvent extraction offers the possibility of obtaining different fractions depending on the solvents and their sequential order. As an example, Fei et al. [45] propose a fractionation procedure for the effective separation of pyrroles and phenols in a coal-tar pitch by extraction with methanol and subsequent precipitation with *n*-hexane. Solvent extraction usually takes place at the boiling point of the solvent (reflux and Soxhlet extraction).

Consequently, these techniques have the drawback of needing high temperatures (pyridine b.p. 115°C, quinoline b.p. 237°C), with possible alterations in the pitch components due to the fact that the sample is subjected to high temperatures for prolonged periods [46]. These problems can be overcome by the use of a sonication procedure. Sonication operates at room temperature for a short period of time. However, the small amount of sample used and the lack of reproducibility when sonication is performed with solvents of low density or high viscosity [47] are the main limitations of this method.

Although solvent extraction has been used extensively because it is simple and because it is the first step in sample preparation for analysis by means of other techniques (i.e. liquid chromatography), its suitability as a chemical/structural characterization technique is very limited. The solubility of a substance not only depends on its C-H skeleton and chemical functionality, but also on the other pitch components, which may act as cosolvents. The cosolubilization effects often lead to the undesirable situation in which the same compounds can be found in different fractions.

2. Liquid Chromatography

The two main techniques based on liquid chromatography that are used for pitch characterization are liquid adsorption and SEC. The fractionation of pitches, after solvent extraction, by these techniques makes it possible to obtain simpler fractions (made up of a smaller number of compounds), which can therefore be more easily studied by other analytical techniques (e.g., mass spectrometry, infrared spectroscopy).

Liquid adsorption chromatography makes it possible to fractionate pitch according to chemical functionalities [48–50]. This separation is achieved by sequential elution with solvents of increasing polarity and/or solvent capacity [51], and by using different stationary phases (alumina, silica gel, etc.). Fractionation by liquid adsorption has a number of weak points: (1) it is an extremely slow procedure that requires long periods of time not only during fractionation but also prior to fractionation (activation of stationary phase, column filling, etc.), (2) the use of complex schemes of fractionation and small amounts of sample and (3) the need for large volumes of high-purity solvents.

Preparative SEC allows pitches to be fractionated mainly according to molecular size [31,36,52–54]. In an ideal SEC system, the mechanisms of separation involve the elution of compounds of increasing molecular size with time. However, in complex samples, i.e. coal-tar pitches, other factors (sample adsorption by the stationary phase and associations between sample, eluent and stationary phase) play an important role in the separation mechanisms, resulting in deviation from the ideal behavior [48]. Some of these problems have been surmounted by the use of powerful eluents at high temperature (e.g. 1-methyl-2-pyrrolidinone at 80°C [32]). Using pyridine at 60°C, Boenigk et al. [36] propose a SEC procedure for coal-tar pitch fractionation (Fig. 7) in which a linear correlation between

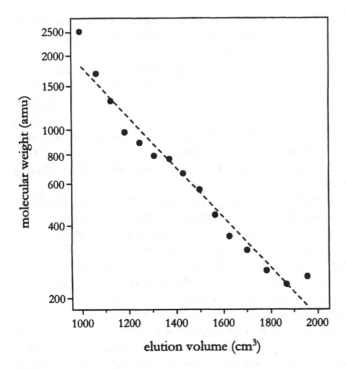

FIG. 7 Log_{10} molecular weight of coal-tar pitch fractions v. elution volume using the SEC Sephadex LH-20/pyridine system [36].

molecular weight and elution volumes reveals that the fractionation of a pitch is mainly due to size exclusion with any interference of polar adsorption in the stationary phase being disrupted.

3. Charge-Transfer Reactions

The π-electron donor aromatic compounds present in pitches (basic nitrogen-containing compounds) can form charge-transfer complexes with electron acceptor species such as picric acid and iodine [30,55,56]. These complexes are usually stable, and consequently, can be isolated by filtration. Charge-transfer reactions are extremely effective for the separation of compounds according to their chemical and thermal reactivity [41]. With this procedure, Zander [57] has successfully proved that the concentration of basic nitrogen compounds in coal-tar pitch fractions increases with the increasing molecular weight of the fractions.

4. Extrography

Extrography has been specifically developed for the fractionation of petroleum derivatives [58]. However, because of the good quality of the separations and the versatility of this technique, it was applied with success to the fractionation of other samples, such as coal-tar pitches [59–64], coal liqu. ls [65], coal-derived oils [66], and other complex mixtures [67].

Basically, extrography is a technique where the sample is fractionated by solvent extraction using a chromatographic column. The sample is suspended in a solvent, which acts as an impregnating agent, and an adsorbent (i.e. silica gel or alumina) is added to the suspension. After mixing and solvent removal, the silica gel impregnated with the sample is packed into a glass column, so that the sample is homogeneously spread throughout the stationary phase. The elution of a sequence of solvents of increasing polarity results in a series of fractions differing in chemical composition.

The mechanisms of separation by extrography have been stated by Bogdoll and Halász [68]. Extrography involves the preseparation of the sample components, in decreasing order of polarity, on the polar surface of the adsorbent particles which are then orderly separated, in increasing order of polarity, by the elution of the solvents. The desorption of the molecules is explained in terms of energies of interaction between molecule-adsorbent and molecule-solvent. When the former prevails over the latter, molecules are retained by the adsorbent, whereas when the latter prevails the molecules are eluted with the solvent.

Fig. 8 shows the extrographic procedure used for the fractionation of coal-tar pitches by Granda et al. [69]. This procedure shows certain differences in comparison with that initially used by Halász [58]. These include a different sequence of solvents (larger number) and the use of nonimpregnated silica gel at the bottom

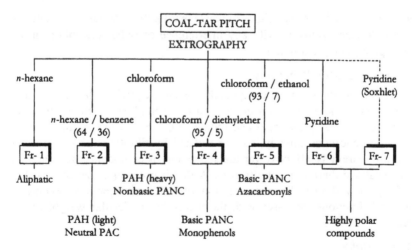

FIG. 8 Sequence of eluents and chemical composition of extrographic fractions.

of the column, which improves the efficiency of the fractionation. Seven fractions of different chemical composition are obtained: (1) Fr-1 consists of aliphatic hydrocarbons, and it is virtually nonexistent in the case of pitches derived from high-temperature coal-tar; (2) Fr-2 is mainly composed of PAHs and neutral sulfur and oxygen heterocyclic compounds (thiophene and furan derivatives); (3) Fr-3 contains PAHs of higher molecular mass than those of Fr-2 and nonbasic polycyclic aromatic nitrogen compounds (PANCs), benzologs of pyrrole; (4) Fr-4 is predominantly made up of basic PANCs (pyridine structure derivatives) and, to a lesser extent, of monophenols; (5) Fr-5 consists of a mixture of basic PANCs, which are more condensed than those of Fr-4 [70], and azacarbonyls; (6, 7) Fr-6 and Fr-7 are difficult to analyze due to their low solubility. Nevertheless, from FTIR spectra and elemental analysis it can be said that these fractions are made up of highly polar compounds [69]. The differences between the chemical compositions of the extrographic fractions are confirmed by the FTIR spectra of Fig. 9.

The mechanisms involved in the procedure used by Granda et al., as in the others, are mainly dependent on the polarity of the constituents. Nitrogen and

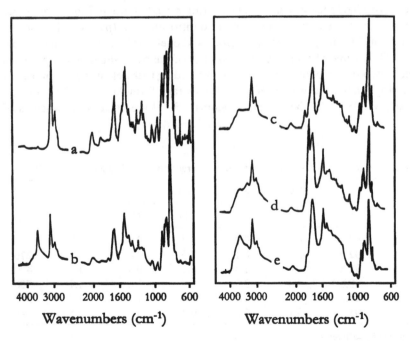

FIG. 9 FTIR spectra of (a) Fr-2, (b) Fr-3, (c) Fr-4, (d) Fr-5, and (e) Fr-6 extrographic fractions.

oxygen functionalities are, therefore, mostly in fractions that elute with highly polar solvents. However, polarity is due not only to the functionality but also to the molecular size and aromaticity of the components. That is why PAHs are in Fr-2 and Fr-3, and basic PANCs in Fr-4 and Fr-5. However, compounds in these fractions are not identical because they differ either in molecular mass (Fr-2/Fr-3) or in condensation degree (Fr-4/Fr-5). Consequently, a broad distribution of molecular weight is found in the extrographic fractions of pitches [71]. It has been observed that with the increase in the number of fractions, the average molecular weight and hydrogen content of the fractions decrease (Table 2).

Taking into account the elemental analysis, average molecular weight (Table 2), and results obtained by ^{1}H NMR and solid-state ^{13}C-NMR (Table 3) of the fractions, an attempt to describe the typical average structure of the compounds of the fractions has been made. The average Fr-2 molecule would appear to be a hydrocarbon containing four or five rings, with half of the molecules bearing methyl substituents and one molecule in four having a methylene bridge. The average Fr-3 molecule is probably an eight-ring aromatic system, with half of the molecules having a methylene bridge, a methyl and an ethyl substituent. Moreover, one-third of the molecules of this fraction may contain a nonbasic nitrogen atom (pyrrole derivative). Fr-4 is constituted by an average molecule made up of nine to ten aromatic rings bearing a methyl group. More than 90 wt % of the molecules probably bears a basic nitrogen atom (pyridine derivative); one-third of these may have an ethyl group and one-fourth may have a methylene bridge. The degree of aromatic condensation of the rings suggests the presence of one or two Ar-Ar' bonds per molecule. According to the elemental composition and average molecular weight of Fr-5, the average formula of this fraction may be $C_{43}H_{26}N_{1.8}OS_{0.25}$. Of the 43 carbon atoms, 23–24 are quaternary, whereas of

TABLE 2   Characteristics of the Extrographic Fractions of a Binder Coal-Tar Pitch

| Sample | Elemental Analysis (wt %) | | | | | C/H[b] | M_{av}[c] |
	C	H	N	S	O[a]		
Pitch	93.4	4.5	0.9	0.4	0.8	1.7	—
Fr-2	92.0	5.0	0.1	0.5	2.4	1.5	270
Fr-3	92.8	4.6	1.3	0.9	0.3	1.7	420
Fr-4	90.8	4.4	2.5	0.7	1.6	1.7	530
Fr-5	87.2	4.3	4.4	1.4	2.7	1.7	590
Fr-6	91.4	3.1	2.0	0.4	3.1	2.5	—

[a] Determined by difference.
[b] Carbon/hydrogen atomic ratio.
[c] Average molecular weight.

TABLE 3 ^{1}H NMR and ^{13}C NMR Data (%) for the Extrographic Fractions of a Binder Coal-Tar Pitch

	Factor	Fr-2	Fr-3	Fr-4	Fr-5	Fr-6
^{1}H NMR	H_{ar}	85.2	76.5	79.1	81.9	—
	H_{al}	14.8	23.5	20.9	18.1	—
	H_{ar}/H_{al}	5.8	3.2	3.8	4.5	—
	$H_{\alpha,2}$	3.7	4.0	1.7	3.8	—
	H_α	10.1	15.6	15.0	12.0	—
	H_β	1.0	3.8	3.7	1.7	—
	H_γ	0.0	0.1	0.5	0.6	—
^{13}C NMR	C_{al}	6.8	3.6	4.0	5.0	2.0
	C_{ar}	93.2	96.4	96.0	95.0	98.0
	C_{ar}/C_{al}	13.7	26.8	24.0	19.0	49.0
	C_q [a]	35.4	—	44.8	54.8	45.6

[a] Quaternary carbon.

the 26 hydrogen atoms, 21–22 are aromatic, (three H_α and one $H_{\alpha,2}$). The main characteristic of this fraction, therefore, is the high heteroatom content and a degree of ring condensation (quaternary carbon) higher than that of the other fractions. Finally, the analytical data available for Fr-6 indicate that this fraction is rich in aromatic carbon. However, the molecular structures of its components are not so highly condensed as might be expected from their low hydrogen content.

Extrography may be considered as a very selective technique. The distribution into fractions is a characteristic of each pitch, relating to its origin (high-temperature coal tar, low-temperature coal tar, petroleum, etc.), industrial application (impregnating or binder), and preparation conditions (vacuum distillation, thermal treatment, etc.). Figure 10 shows the distribution of the extrographic fractions of three different coal-tar pitches. The low-temperature coal-tar-derived pitch (LT) is characterized by the presence of aliphatic compounds, which elute in Fr-1. This fraction is completely absent in the case of coal-tar pitches derived from high-temperature tar. LT is also characterized by a considerable amount of compounds that elute in Fr-4 and Fr-5 (basic PANCs, azacarbonyls, and monophenols). Monophenols are scarce in impregnating (CTP-I) and binder coal-tar pitches (CTP-B) [64]. The effect of the coking temperature on monophenols, azacarbonyls, and basic PANCs is to cause the partial loss of heteroatoms, giving rise to PAHs and/or arylaryl ethers (in the case of the monophenols) via the condensation of two or more molecules. These types of compounds, which elute in Fr-2, are more concentrated in high-temperature coal-tar pitches (Fig. 10).

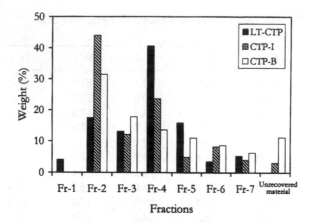

FIG. 10 Distribution of extrographic fractions of a low-temperature coal-tar–derived pitch (LT-CTP), an impregnating coal-tar pitch (CTP-I), and a binder coal-tar pitch (CTP-B).

Extrography also differentiates between pitches obtained from high-temperature coal-tar for use in different applications. Impregnating and binder coal-tar pitches differ mainly in the distribution of PAHs. These compounds are lighter in CTP-I than in CTP-B, as can be seen from their different Fr-2 and Fr-3 contents.

These results clearly highlight the valuable information about pitch composition provided by extrography. Extrography offers considerable advantages over other fractionation techniques: simplicity of equipment, shorter time involved and low volume of solvents required, good repeatability and reproducibility, the possibility of fractionating large amounts of sample, and good fractionation depending on the chemical composition of the compounds.

D. Primary Quinoline Insolubles

Coal-tar pitches contain dispersed carbonaceous particles produced by the thermal cracking of volatile products during the coal coking process. These particles are dragged into the coke oven by the gases that are incorporated first into the tar and then into the pitch. The particles are called primary quinoline insolubles (QIs) because they are generated in a primary process of coal conversion, and their percentage in pitch is determined by quinoline extraction. QI particles can assume different shapes [72], and can be conglomerated with each other to form chains and clusters [73]. However, when isolated from the rest of the pitch they are seen as discrete spherical particles under the scanning electron microscope (~1 μm diameter) (Fig. 11a).

 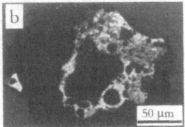

FIG. 11 Scanning electron and optical micrographs of (a) primary QI particles and (b) carryover particle.

Because of their insolubility little research has been carried out to establish the chemical composition of primary QIs. It is generally accepted that QIs have a C/H atomic ratio >3.5 [30,74,75]. Figueiras et al. give the primary QIs of an industrial binder coal-tar pitch (11 wt % of QI), an elemental composition of 93.4 wt % C, 1.8 wt % H, 0.9 wt % N, 0.8 wt % S, and 3.1 wt % O. However, the C/H atomic ratio and elemental composition of QIs change considerably from one primary QI to another [76,77].

The nature and amount of primary QIs in tars and pitches is determined by the characteristics of the coal and the coking conditions (coal charge, temperature, coking time, etc. [74]). Depending on these factors, primary QIs can be contaminated by larger particles (carryover) that come from coal, coke, and inorganic matter (Fig. 11b). These particles constitute a minority fraction of primary QIs. Nevertheless, they are undesirable because they are not only abrasive to equipment (pumps impellers and piping) [20] but they also have a negative effect on the mechanical and electrical properties [78] and oxidation resistance [20] of pitch-based materials.

The primary QI content is a parameter that is currently used by pitch makers and consumers alike because it is related to pitch quality [20]. Moreover, the rheological properties of pitches, mesophase development, and optical texture of the resultant cokes are greatly affected by the presence of QI particles, as will be discussed in Section IV.

IV. COAL-TAR PITCH PYROLYSIS BEHAVIOR

A. Principles of Pitch Carbonization

The transformation of isotropic pitch into anisotropic carbon requires controlled heating, in an inert atmosphere, at temperatures >600°C (carbonization). On carbonization, a series of physicochemical processes take place. These lead to the formation of a liquid crystal phase, called mesophase, which finally solidifies into an anisotropic coke. The microstructure and properties of the coke are dictated by the characteristics of the mesophase (shape, size, viscosity, fluidity), which primarily depend on the chemical composition of the pitch and the experimental conditions used.

The process of pitch carbonization is extremely complex, as might be expected, due to the complexity of pitch composition and the different reactivities exhibited by the pitch components. Since the most significant effects are molecular growth and the formation of graphite-like structures, the carbonization process can be schematized as illustrated in Fig. 12.

It is well known that when pitch is heated, a homogeneous, isotropic fluid phase is produced. The increase in temperature involves the distillation of those pitch components that are stable at their boiling points and the thermal polymerization (molecular size growth) of the reactive pitch components. This polymerization leads to the formation of planar macromolecules with a laminar and ordered structure (mesogens). Mesogens are grouped together in parallel stacking by means of van der Waals forces. Although these stackings are not connected, they show a high degree of compaction as they are able to segregate from the isotropic phase to produce small anisotropic microspheres (mesophase) [8]. The formation of mesophase is a spontaneous and homogeneous process within the isotropic pitch phase. With the increase in temperature or residence time, a greater amount of molecules acquire the necessary dimensions to be incorporated into the existing mesophase microspheres [8]. When the carbonization process takes place rapidly, new microspheres are formed [8]. On the other hand, when carbonization occurs slowly and viscosity conditions are adequate, the microspheres interact with each other, giving rise to microspheres of larger size (mesophase coalescence) [8]. The growth and coalescence of mesophase progresses until the viscosity of the system is excessively high, and consequently, mobility ceases. At this stage, mesophase solidifies into coke. The microstructure of the coke is that of the mesophase at the solidification point. The size, shape, and orientation

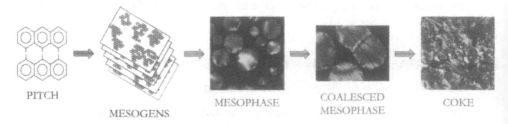

PITCH   MESOPHASE   COALESCED   COKE
          MESOGENS   MESOPHASE

FIG. 12   Schematic process of pitch carbonization.

of the microcrystalline structures observed through the optical microscope constitute the optical texture of the coke [10,79].

The onset temperature for mesophase initiation and mesophase solidification, and how the mesophase develops, depends on the experimental conditions used (heating rate, use of pressure, stirring, etc.) and the chemical composition of the pitch. The way in which mesophase develops is of special relevance because it controls not only the operational parameters for pitch processing [75] but also the optical texture of the coke [80] and coke properties, such as mechanical strength [81], electrical conductivity [82], and stability against oxidation [83]. Consequently, the right choice of pitch and processing conditions makes it possible to obtain materials whose microstructure and properties are in accordance with future applications.

B. Carbonization of Model Aromatic Compounds

Basically, the carbonization process involves a dehydrogenative polymerization process via hydrogen transfer. However, as mentioned above, the complexity of pitch composition makes pitch carbonization an exceedingly complex process. As an example, the carbonization of anthracene yields eleven dimeric structures and oligomers ranging from dimers to octamers [84]. This problem is more pronounced in the case of pitch because of the existence of thousands of compounds, covering a broad distribution of molecular weights, which are capable not only of generating free radicals but also of polymerizing and copolymerizing. Moreover, the rest of the components affect the reactivity of a single compound. In this respect, Zander et al. [85] observed that the conversion degree of perylene at 450°C in an inert matrix (2,2'-binaphthyl) was 13%. This increased to 74% when it was part of a coal-tar pitch. Hurt and Hu [86] formulated a quantitative multicomponent thermodynamic model in which they evidenced the interactions between components. An additional problem when carbonizing model compounds is that most of these compounds are stable at their boiling point and, consequently, distill before reacting. This limitation is usually overcome by enhancing polymerization at lower temperatures by means of catalysts [15,87] and/ or by using pressure to avoid distillation.

Nevertheless, most of the knowledge concerning the chemistry involved in pitch carbonization has been obtained from the study of model aromatic compounds. Considerable research has been carried out using model compounds with the aim of gaining an insight into the mechanisms of carbonization [9,21,84,85,88–95]. Research has been mainly focused on the chemical transformation of the compounds during the initial stages of polymerization. At these stages a series of processes that lead to a greater aromatization of the structures take place. These processes include [88]: (1) C-H and C-C bond cleavage to form reactive free radicals, (2) molecular rearrangement, (3) thermal polymerization,

(4) aromatic condensation, (5) elimination of side chains and small molecules (H_2), etc. Although from a mechanistic point of view these processes can be considered separately, they may in fact occur simultaneously on carbonization. A global idea of the complexity of the chemical reactions which take place during the carbonization of model aromatic compounds can be illustrated with the pyrolysis of acenaphthylene (Fig. 13) [9].

A study of polymerized products provides information on the possible structure of mesogens. Using gel permeation chromatography and field desorption mass spectrometry, Lewis [84,88] identified compounds with molecular mass ~1400 amu (octamers) along with smaller compounds (dimers, trimers, etc.). Most of these compounds have an oligoaryl structure. Mochida et al. [87] obtained mesophase by the thermally assisted polymerization of aromatic compounds (e.g., naphthalene, methylnaphthalene, anthracene, etc). They also sug-

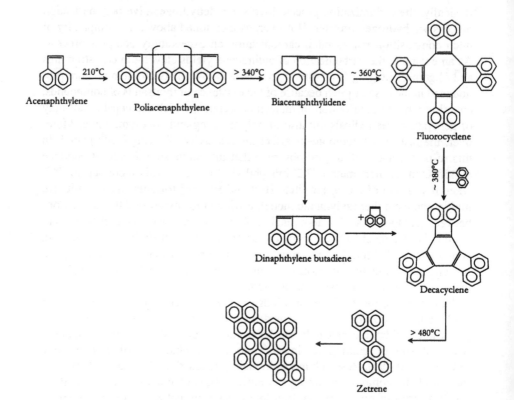

FIG. 13 Mechanism of pyrolysis of acenaphthylene [9].

gested the formation of oligoaryl structures in the mesogens derived from these compounds. Similar conclusions were reached by Hurt and Hu [86] who stated that fully condensed aromatics do not form discotic liquid crystals in their pure state, so that liquid crystalline mesophase can only be rationalized by considering partially condensed structures (e.g. oligoaryl structures) and/or multicomponent mixtures. This may seem surprising given the general belief that mesophase is made up of planar aromatic macromolecules. Zander [41] explains this apparent contradiction in terms of ring closure reactions, which require very low activation energy. Oligoaryl structures are predominant in the isotropic phase. However, when ring closure reactions take place, planar aromatic systems are formed, constituting the anisotropic phase.

Most of the carbonization studies of model compounds have been concerned with polycyclic aromatic hydrocarbons (PAHs), which are the most abundant compounds in pitches. Alkyl-substituted PAHs and heterocycles have received less attention. However, it might be worthwhile evaluating the effects of the substituents and heteroatoms on carbonization.

Although alkyl groups and partially hydrogenated structures are present in low proportions in coal-tar pitches, their role during carbonization cannot be underestimated. Alkyl groups accelerate the rate of coking [9,91,95,96], an effect that is the more pronounced the greater the number and the length of the alkyl groups [9]. The alkyl groups occupy the positions where the free radicals are formed. This can be attributed to the activating effect and the greater facility with which the bonds of the alkyl groups break. Greinke [97] found that the loss of side chains through thermal cracking during carbonization had only a minor influence on mesophase formation. However, the optical texture of the resultant carbons seems to be dependent on the length of the side chain. Thus, when the length of the alkyl group is too great, the size of the aromatic ring increases rapidly, resulting in a less fluid system. Consequently, the optical texture of the derived carbon is of a smaller size (mosaics) [95].

Heterocyclic aromatic compounds containing nitrogen, sulfur, and oxygen were carbonized in order to study the effect of heteroatoms on the reactivity of the PAHs, the optical texture of the coke, and the temperature of heteroatoms released. The latter is particularly important in the case of heterocycles containing nitrogen and sulfur because there is a relation between the release of heteroatoms on carbonization and the undesirable effect of puffing. Mochida et al. [98,99] carbonized several heterocycles, using aluminum trichloride as a catalyst. They observed that the amount of catalyst used exerted a significant influence on carbonization rate and, consequently, on the size of the optical texture of the cokes. The tendency to yield anisotropic cokes decreased in the order of sulfur, nitrogen, and oxygen. Marsh et al. [100] studied the pressurized cocarbonization of fluorene and carbazole in the presence of several nitrogen, oxygen, and sulfur heterocyclic compounds. The addition of selected oxygen containing molecules (phthalimide, phthalic anhydride, and pyromellitic dianhydride) to fluorene and carbazole caused the development of anisotropic carbon, whereas none was formed during the carbonization of the single compounds. On the contrary, phenol severely retarded mesophase growth [8]. Nitrogen- and sulfur-containing compounds with nitrogen and sulfur contents greater than 3 wt % hindered mesophase growth. This is possibly because a Lewis base, such as acridine, produces active ionic intermediates by removing protons from PAHs, thus increasing reactivity [21,39,101]. However, it must be taken into account that when nitrogen compounds are in a multicomponent system, such as a pitch or a pitch fraction, the behavior of these compounds may differ considerably on carbonization because of the interactions between two components.

It is a well-established fact that the formation of anisotropic coke occurs via mesophase. Although this is undoubtedly true for liquid phase carbonization, it has recently been demonstrated that PAH with a high melting point and an appropriate structure (Fig. 14) can be transformed into anisotropic coke by an exclusively solid-state carbonization process, without passing through an intermediate liquid state [102]. This is possible because at sufficiently high temperatures, in the absence of oxygen, the thermally induced removal of hydrogen and the formation of anisotropic coke may occur before melting. Solid-state carbonization for generating graphitizable carbons opens up a new way of producing high-purity graphite crystals for specific applications. Compounds in Fig. 14 have a molecular weight of 522 (a) and 596 amu (b). Compounds with a molecular weight greater than 600 amu can be expected in significant amounts in coal-tar pitch (see Fig. 5). Moreover, large polycyclic aromatic hydrocarbons with a similar molecular weight and structure to those of Fig. 14 have also been proposed, and in some cases identified, in coal-tar pitches [103]. It would not be surprising, therefore,

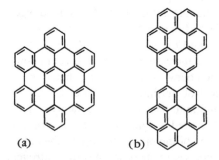

(a)   (b)

FIG. 14   Structures of high-melting PAHs susceptible to carbonization by solid-state: (a) hexabenzo[bs,ef,hi,kl,no,qr]coronene and (b) benzo[1,2,3,-bc:4,5,6-b'c']dicoronene [102].

if large PAHs present in coal-tar pitches similar to those of Fig. 14 behaved in a similar manner.

C. Carbonization of Pitch Fractions

The study of the carbonization behavior of pitch fractions may represent a further step in the study of pitch behavior on carbonization [36,104,105]. Pitch fractionation yields fractions with a simpler composition. Although pitch might best be understood as a bulk material, initially the global composition of a pitch can be approached via separate fractions. Extrographic fractionation may be a helpful technique for studying the relationships between the composition and pyrolysis behavior of the different classes of compounds, as a previous step for establishing these relationships in pitches.

Thermogravimetric analysis (TG), derivative thermogravimetry (DTG), differential thermal analysis (DTA), and differential scanning calorimetry (DSC) provide valuable information not only about the extent and rate of weight loss but also about exothermal and endothermal changes associated with the different processes which occur on carbonization [9,88,93,105–122]. Interpretation of the thermal analysis curves is not easy, however, especially in the case of the DTA and DSC. As mentioned above (Section IV.B), during pitch carbonization many physico-chemical processes take place (e.g., devolatilization, polymerization, condensation, cracking, isomerization, molecular rearrangement, hydrogen transfer, molecular association, ring closure, aromatization, cyclization). Moreover, most of these effects are concurrent and produce opposite thermal effects. It would be of great interest, therefore, to be able to assign the different exothermic and endothermic effects to different well-known reactions so that they could be extrapolated to pitch fractions and finally to bulk pitches.

Acenaphthylene is one of the most suitable model compounds for the study of the carbonization process it can produce graphitizable carbon through a liquid crystal phase at atmospheric pressure, and also because the mechanisms involved and intermediate products formed are well known (Fig. 13) [9,123,124]. Martínez-Alonso et al. [106] monitored acenaphthylene carbonization by TG/DTG/DTA and assigned the peaks to the different reactions involved. This study led to the conclusion that, of the various types of phenomena expected to occur during pitch carbonization, polymerization and condensation reactions would be exothermic whereas depolymerization, distillation, and cracking would be endothermic. This is in agreement with results obtained by others [9,119] using similar techniques.

1. Thermogravimetry of Pitch Fractions

To study the effects of the chemical composition on the physicochemical changes involved in carbonization, Bermejo et al. [105] used the seven fractions obtained

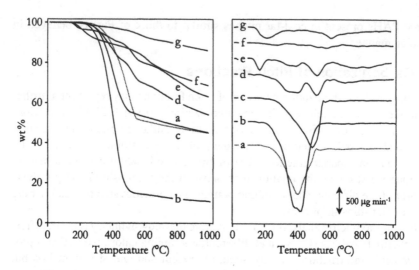

FIG. 15 TG (left) and DTG (right) curves of extrographic fractions obtained from a binder coal-tar pitch: (a) parent pitch, (b) Fr-2, (c) Fr-3, (d) Fr-4, (e) Fr-5, (f) Fr-6, and (g) Fr-7.

by extrography of a binder coal-tar pitch (the composition of these fractions is summarized in Fig. 8). Figure 15 shows the TG and DTG curves of these fractions. A massive loss of weight occurs between 200°C and 600°C. Losses below 200°C are insignificant, except for Fr-5 (basic polycyclic aromatic nitrogen compounds (PANC) and azacarbonyls) and Fr-7 (highly polar compounds). These fractions exhibit a peak of maximum weight loss rate at 169°C and 224°C, respectively (Fig. 15). This weight loss at low temperature could be due to incomplete solvent elimination and/or the presence of polar compounds of low molecular weight.

The weight loss of the fractions is not homogeneous, although it increases in decreasing order of elution (Fig. 15 and Table 4), except for Fr-7, which shows an intermediate behavior between Fr-5 and Fr-6 (highly polar compounds). Fr-3 (heavy PAH and nonbasic PANCs) is the fraction that retains the greatest similarity to the parent coal-tar pitch, especially in the value of the carbonaceous residue at 1000°C (45.2 and 45.0 wt % for Fr-3 and parent pitch, respectively). In the 200–600°C range weight loss in the parent coal-tar pitch is basically due to the contribution of Fr-2 (light PAHs and neutral polycyclic aromatic compounds containing furan and thiophene structures) and to a lesser extent Fr-3 and Fr-4 (basic PANCs and monophenols). Between 600°C and 1000°C Fr-2 loses 3.9 of its total weight, contributing 1.2% to the total weight loss of the parent pitch. On the other hand, Fr-4, Fr-5, Fr-6, and Fr-7 lose 34.2, 45.4, 50.0, and 45.7%

TABLE 4 Evolution of Weight Loss of a Binder Coal-Tar Pitch and Its Extrographic Fractions During Carbonization to 1000°C at 10°C min^{-1}

Sample	$-\Delta W_{200}$	$-\Delta W_{400}$	$-\Delta W_{600}$	$-\Delta W_{1000}$	$-\Delta W_{1000\text{-}600}{}^{a}$	$-\Delta W_r{}^{b}$
			Weight loss (wt %)			
Pitch	0.0	23.1	46.9	55.0	8.1	14.7
Fr-2	0.3 (0.1)	44.1 (15.4)	85.6 (30.0)	89.1 (31.2)	3.5 (1.2)	3.9 (1.4)
Fr-3	0.0 (0.0)	11.3 (2.0)	48.9 (8.5)	54.8 (9.5)	5.9 (1.0)	10.8 (1.9)
Fr-4	0.0 (0.0)	13.1 (2.5)	30.2 (5.7)	45.9 (4.6)	15.7 (3.0)	34.2 (6.4)
Fr-5	3.2 (0.2)	8.3 (0.6)	20.1 (1.4)	36.8 (2.5)	16.7 (1.2)	45.4 (3.1)
Fr-6	0.4 (0.0)	2.6 (0.2)	7.1 (0.6)	14.2 (1.2)	7.1 (0.6)	50.0 (4.4)
Fr-7	2.6 (0.1)	10.9 (0.5)	17.2 (0.7)	31.7 (1.4)	14.5 (0.6)	45.7 (2.0)
ΣFr	0.4	21.1	46.8	54.4	7.6	19.1

^{a} Weight loss between 600 and 1000°C (wt %).
^{b} $(-\Delta W_{1000\text{-}600}/-\Delta W_{1000}) \cdot 100$.
Note: Numbers in parenthesis refer to extrographic fractions weight loss corrected with the values of the extrographic fractions content.

of their total weight, respectively, contributing 5.4% to the total weight loss of the parent pitch. It seems evident that fractions containing heteroatoms contribute significantly to the losses of the parent pitch in the transition from semicoke (600°C) to coke (1000°C).

It should be noted that the weight loss undergone by the parent pitch is practically the same as that obtained from the addition of the individual extrographic fractions. This might suggest that no interactions occur between the components of the fractions or, if they do occur, they do not affect weight loss.

2. Differential Thermal Analysis of Pitch Fractions and Optical Microscopy of the Resulting Cokes

Figure 16 shows the DTA curves of a binder coal-tar pitch and its extrographic fractions. The thermogravimetric and differential thermal analysis of the acenaphthylene pyrolysis suggests that the considerable weight loss of the PAHs present in Fr-2 (Fig. 15) must be associated with the broad and deep endothermic peak in the DTA curve of this fraction between (~200°C and ~325°C. This is indicative of the considerable thermal stability of the light aromatic polycyclic compounds that constitute this fraction. Above 325°C, distillation proceeds but it overlaps the opposite effect of polymerization. Consequently, the DTA curve shows a significant exothermic effect highlighted by a peak at ~500°C. The magnitude of the exothermic effect between 325–500°C contrasts with the equivalent exothermic effects of polymerization observed in the other fractions in this temperature range (Fig. 16). This can be attributed to the occurrence of multiple

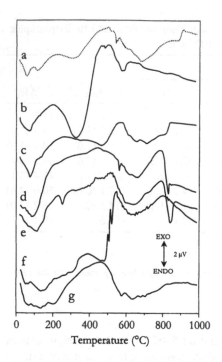

FIG. 16 DTA curves of extrographic fractions obtained from a binder coal-tar pitch: (a) parent pitch, (b) Fr-2, (c) Fr-3, (d) Fr-4, (e) Fr-5, (f) Fr-6, and (g) Fr-7.

polymerization steps. The components of this fraction (average molecular weight <300 amu [70]) require several polymerization steps to achieve the molecular weight, and hence the molecular structure, neeeded to generate mesogens and subsequent mesophase. The higher average molecular weight of the components of the other fractions reduces the number of steps involved in their polymerization. Consequently, the exothermic effect is less pronounced.

The coke obtained from this fraction (Fig. 17) clearly shows evidence of this multistep polymerization. At 500°C and 1 h of soaking time [40], Fr-2 generates a partially anisotropic material with an optical texture of flow domains. Small isolated mesophase spheres coexisting with larger distorted coalesced mesophase spheres embedded in the isotropic phase are observed. This indicates that the reactive components of Fr-2 (those that do not distill on carbonization) probably need more severe carbonization conditions to be converted to anisotropic coke. Interestingly, under the same experimental conditions the parent pitch yields an anisotropic coke, which demonstrates the occurrence of interactions between the different kinds of pitch components during carbonization.

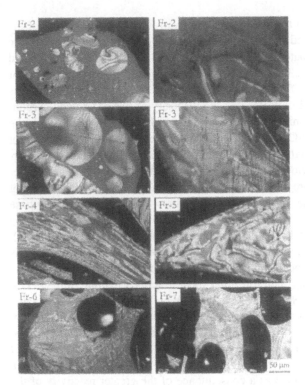

FIG. 17 Optical micrographs of pyrolysis products of extrographic fractions obtained from a binder coal-tar pitch.

In the other fractions distillation takes place at higher temperatures, thus overlapping with polymerization (Fig. 16). In Fr-3, which is composed of nonbasic PANCs and PAHs (>300 amu), a massive release of volatiles occurs between ~400°C and 550°C (Fig. 15). The predominance of the exothermic effect due to polymerization is detected at 450°C and extends to 550°C, which is the highest temperature observed for any fraction. This temperature coincides with the hardening temperature of mesophase [125], and consequently indicates the end of liquid phase polymerization. Consequently, Fr-3 should be the least reactive, in complete agreement with its composition (PAHs with molecular weight >300 amu and nonbasic PANCs).

Analysis of the coke derived from this fraction shows that the conversion to anisotropic material is still incomplete (Fig. 17). The predominant optical texture of the anisotropic material is flow domains, although isolated mesophase spheres embedded in the isotropic material are observed. This suggests that a longer soaking time and/or higher temperature is required to convert Fr-3 to an anisotropic coke. Although the optical texture of the coke from Fr-3 is similar to that of Fr-2, the mechanisms involved in carbonization must be different. Thus, in the coke from Fr-3 the isotropic material is less abundant and the spheres are larger and less distorted than in Fr-2. The thermal reactivity of Fr-3 is lower than that of Fr-2, as indicated by the DTA curve (Fig. 16). Moreover, the fact that spheres are practically undistorted indicates that the viscosity of the system and carbonization rate are low. The higher molecular weight of the PAHs along with the presence of nonbasic PANCs seem to be the factors responsible for this different behavior. Heavy PAHs might be expected to produce mesophase at lower temperatures than light PAHs. However, in Fr-3 heavy PAHs coexist with carbazole-type compounds, which are very stable [9,126] and in certain proportions cannot be accommodated by the liquid crystal structures. Consequently mesophase growth is restricted [100].

Fractions Fr-4 and Fr-5 have a similar composition (basic nitrogen PANCs), the main difference being the presence of monophenols in Fr-4 and azacarbonyls in Fr-5 and the greater condensation degree of the components that elute in Fr-5 [70]. It is not surprising, therefore, that these fractions exhibit similar DTA curves (Fig. 16). In the case of these curves, the endothermic effect associated with weight loss is almost totally compensated by the effect of polymerization. In Fr-5 a peak is observed at ~200–250°C that represents a break from the exothermic character of the curve. This peak might be due to the removal of CO and CO_2 from azacarbonyls. The polymerization in Fr-4 and Fr-5 begins at a lower temperature than in Fr-3 as a consequence of the greater reactivity of the nitrogen-rich fractions (Table 2). Common to both curves is the sharp endothermic peak at ~800–850°C, which might be attributed to the cleavage of C-N bonds. This temperature seems to be too low for the removal of nitrogen. However, Isaacs [92] found that during the carbonization of 7,8-benzoquinoline the nitrogen content began to decrease slightly at 800°C until it was completely removed at 1400°C.

Cokes obtained from Fr-4 and Fr-6 (basic PANCs/monophenols and highly polar compounds, respectively) are anisotropic (Fig. 17), showing an optical texture of elongated flow domains (Fr-4) and flow domains (Fr-5). Highly ordered cokes are associated with systems with low viscosity at this stage of mesophase initiation and mesophase development [127]. This is attributable to the high thermal stability of the system during mesophase development. However, these fractions begin to exhibit polymerization reactions at lower temperatures than do the PAH-rich fractions (Fig. 16). This higher reactivity could be due to the basic PANCs and also to the presence of oligoaryl structures, as indicated by the high quaternary carbon content in these fractions (Table 3). The former type of compounds may serve as a catalyst for thermally induced polymerization [101], while the latter type of compounds may undergo cyclodehydrogenation reactions to form large, planar, and more thermally stable aromatic structures that favor the formation of highly anisotropic carbons [21], such as those obtained from basic-nitrogen-rich fractions.

The highly polar fraction Fr-6 yields a DTA curve (Fig. 16) with a rapid sequence of sharp endo-exothermic peaks attributable to consecutive cracking and polymerization reactions. This reveals that a significant number of components of Fr-6 are unstable in the range 480–500°C, which is corroborated by an appreciable weight loss (Fig. 15). Finally, the Fr-7 curve exhibits a broad exothermic peak extending from 250°C to 500°C, possibly indicative of a continuous change in the molecular structure and size of its most representative components.

Fr-6 and Fr-7 produce cokes with an optical texture of small domains and small domains/mosaics (Fig. 17), respectively. The high polarity of these fractions leads, on carbonization, to high-viscosity and high-reactivity systems, where the solid-state carbonization process [102] might be of significance.

To sum up, the thermal effects evidenced by the DTA curves of the extrographic fractions of coal-tar pitches, and corroborated by the optical texture of the derived cokes, seem closely associated with the physicochemical phenomena expected during the pyrolysis of the main constituents. Initially, it seems evident that a certain interaction between the components of the different fractions occurs. However, the main thermal effects observed in the DTA curves of the fractions are not evident in the DTA curves of the parent coal-tar pitch. This may be due to the fact that such effects overlap in the pitch. Nevertheless, the information obtained from the extrographic fractions is very valuable because it allows the contribution of the different fractions to the pitch to be determined. Moreover, this study clearly reveals that it is virtually impossible to predict the pitch pyrolysis behavior and the optical texture of cokes in a simplistic way taking into account only the elemental composition of the pitch. Thus, Fr-3, Fr-4, and Fr-5 contain PANCs. However, the nonbasic PANCs in Fr-3 (benzologs of pyrrole) behave in a different way to the basic PANCs in Fr-4 and Fr-5 (benzologs of pyridine). Even the different condensation degree of the basic PANCs present in Fr-4 and Fr-5 gives rise to a different pyrolysis behavior, thereby yielding cokes with a different optical texture.

D. Carbonization of Coal-Tar Pitches

Understanding the mechanisms involved in carbonization and how the variables affect the process are of great interest for the optimization of pitch processing. Many factors influence the transformation of pitch into carbon. However, in this section we will mainly focus our attention on three of these: (1) experimental conditions, (2) role of solid particles and (3) composition of pitches.

1. Influence of Experimental Conditions

(a) Effect of Temperature and Soaking Time.   Temperature and time are probably the two most controlled experimental conditions and the easiest to modify. Selection of the final temperature is extremely important during pitch processing. The most dramatic changes in pitch occur during mesophase development. It is not surprising, therefore, that the monitoring of carbonization in the temperature range at which mesophase takes place is of special interest. As an example, in the preparation of pitch-based C/C composites, the temperature of pressing must be close to that of mesophase solidification. If mechanical pressure were applied at lower temperatures, the system would be too fluid and pitch would be exuded [128]. On the other hand, if temperatures were higher, the system would be excessively rigid (the pitch could already have been transformed into coke) and, consequently, the application of pressure could result in damage to the material [129].

Fig. 18 and Fig. 19 show the effect of temperature and soaking time on the mesophase development of an impregnating coal-tar pitch thermally treated in a stainless steel reactor with stirring (100 rpm) at 430°C for 3, 4, and 5 h. A generous nitrogen flow was used throughout the treatment to maintain an inert atmosphere and facilitate the removal of volatiles. As the temperature increases, pitches develop more mesophase, which becomes larger (Fig. 18). A similar effect is achieved by increasing the soaking time. However, the effect of temperature is more pronounced than that of soaking time (Fig. 19). At 440°C the impregnating coal-tar pitch develops 44.2 vol % of mesophase after 4 h. At 420°C, the same pitch requires 8 h to develop 41.6 vol %. The morphology of the mesophase is strongly affected by temperature and time. Thus, after 8 h at 420°C, 6.2 vol % of mesophase developed by the impregnating coal-tar pitch has coalesced, whereas the percentage of coalesced mesophase increases to 12.4 vol % at 440°C

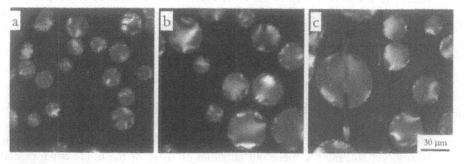

FIG. 18   Optical micrographs of an impregnating coal-tar pitch carbonized at 430°C for (a) 3, (b) 4, and (c) 5 h.

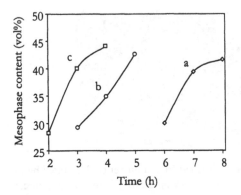

FIG. 19 Variation of mesophase content with time for an impregnating coal-tar pitch carbonized at (a) 420°C, (b) 430°C, and (c) 440°C.

after 4 h. This is of special interest for the preparation of mesocarbon microbeads with a narrow distribution of mesophase sphere diameters [130].

(b)   Effect of Heating Rate.   Carbonization heating rate is a parameter that is frequently associated with the evolution of volatiles, and consequently, with the porosity of the final material. Bermejo et al. [107] studied the effect of the heating rate on the pyrolysis behavior of a coal-tar pitch by means of thermal analysis techniques (TG/DTG and DTA). They noted that increasing heating rates displaced the initial and final temperatures of weight loss and the temperature of maximum rate of weight loss toward higher temperatures. However, the yield of carbonaceous residue at 960°C remained practically constant. At the same time, increasing the heating rate makes the DTA peaks more intense and modifies the general appearance of the curves. Consequently, by merely changing the heating rate the chemistry of the carbonization process can be altered.

Furthermore, the development of mesophase occurs at a range of temperatures, which starts with the formation of small spheres and finishes with solidification into coke. This range of temperatures, characteristic of each pitch, is covered in a shorter time when the heating rate is increased. Consequently, polymerization reactions, and initiation, growth, and coalescence of mesophase take place more rapidly. This modification in the kinetics of mesophase formation has repercussions on the structure and properties of the pitch coke. Figure 20 shows the optical micrographs of the pitch cokes obtained by the carbonization of a binder coal-tar pitch at different heating rates [131]. At 5°C min^{-1} the optical texture of the coke is flow domains coexisting with mosaics embedded by QI aggregates. At lower heating rate (0.1°C min^{-1}) the predominant optical texture of the coke is small domains/mosaics homogeneously distributed. This reduction in the size of the microcrystallites of the cokes improves certain mechanical

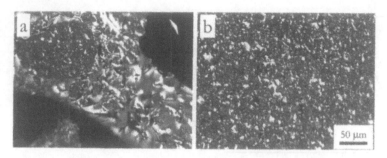

FIG. 20 Optical micrographs of the cokes obtained from a binder coal-tar pitch by carbonization to 1000°C at (a) 5°C min^{-1} and (b) following a 5-day carbonization process.

properties related to the optical texture, such as microstrength [131]. Moreover, the low heating rate makes it possible for a low-viscosity system to develop over a long period, allowing QI material to be distributed homogeneously within the coke.

(c)   Effect of Experimental Device.   The fact that different results could be obtained using similar raw materials may be due to the use of different experimental devices. Factors such as geometry, capacity, and load are important aspects to bear in mind. Research has been carried out to evaluate the effect of the experimental device on the kinetics, yield of carbonization, and properties of the intermediate reaction products and the final carbon.

Hüttinger & Wang studied the kinetics of mesophase formation from a petroleum pitch, using both a continuous [132] and a discontinuous stirred-tank reactor [133]. They found that the most significant difference between the continuous and discontinuous conditions was the broad residence time distribution in the former conditions. This caused more extensive polymerization/polycondensation of the pitch components, resulting in a product with a broader molecular mass distribution.

Table 5 shows the main characteristics of the reaction products obtained from a commercial impregnating coal-tar pitch by treatment at 430°C for 3, 4, and 5 h under a specific nitrogen flow of 15 L h^{-1}. The treatment was carried out using two different reactors: (1) reactor A, which was designed to facilitate the removal of volatiles, and (2) reactor B, in which the distillation of volatiles competed with the reflux. The differences between the properties of the reaction products obtained by both procedures may be related to the viscosity of the system. In the former reactor the removal of volatiles led to products with a higher softening point and carbon yield. Moreover, the type of reactor dramatically affects the development of mesophase. In reactor A single-mesophase spheres are formed, while in reactor B coalesced mesophase is mainly formed. This is because in

TABLE 5 Properties of a Thermally Treated Impregnating Coal-Tar Pitch Using a Reactor in Which Distillation is Favored (A) and a Reactor in Which Distillation Competes with Reflux (B)

Reactor	Treatment conditions			SP[d]	CY[e]	Mesophase content (vol %)	
	L^a	T^b	t^c			Spheres	Coalesced
A	1200	430	3	185	73.6	27.6	1.6
			4	202	76.9	32.6	2.2
			5	—	78.8	38.8	3.8
B	200	430	3	137	53.8	23.8	2.2
			4	152	57.8	24.4	9.6
			5	163	60.4	5.0	32.8

[a] Pitch load (g).
[b] Temperature (°C).
[c] Reaction time (h).
[d] Softening point (Mettler, °C).
[e] Carbon yield (Alcan, wt %).

reactor B the reflux of volatiles contributes to generating a system with low viscosity where the mobility and coalescence of spheres is favored. It is not surprising, therefore, that the design of the reactor, which may or may not allow the evacuation of volatiles, plays an imporatamt role in the development of mesophase and properties of the reaction products.

(d) Effect of Pressure. Another factor of special relevance in carbonization is the use of pressure. In pressurized carbonization, a closed system preventing the removal of volatiles is generated. As a result, the carbon yield increases [134,135]. In addition, the use of pressure modifies the chemistry of carbonization. Fig. 21 shows the optical micrographs of cokes obtained from a QI-free pitch, an impregnating (QI ~3 wt %) pitch and a binder coal-tar pitch (QI ~10 wt %), which were carbonized to 550°C, soaking time of 5 h, under nitrogen pressures of 1 and 10 bar. The most remarkable effect produced by pressure is the enhancement of the optical texture of the cokes (increase in the size of the microcrystallites). Hüttinger and Rosenblatt obtained similar results [135]. These authors reported that the volatiles, which in an open system would be removed, are now retained, forming a solvent phase of lower viscosity. This favors nucleation, growth, coalescence, and especially the reorientation of mesophase spheres after coalescence.

It is worth noting the effect of pressure on the distribution of the primary QI particles in the coke. When carbonization is carried out at atmospheric pressure, QI particles form aggregates of different sizes, which are relatively homoge-

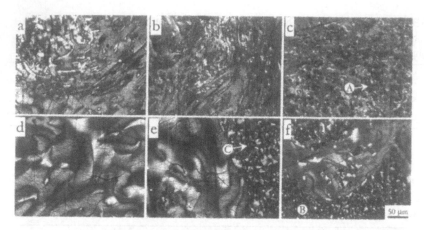

FIG. 21 Optical micrographs of cokes obtained from (a/d) QI-free coal tar pitch, (b/e) impregnating coal-tar pitch, and (c/f) binder coal-tar pitch, applying a nitrogen pressure of 1 bar (upper micrographs) and 10 bar (lower micrographs). A, B, and C represent QI aggregates of different sizes.

neously distributed in the coke (Fig. 21c, position A). However, when nitrogen pressure is applied, the aggregates of QI particles increase considerably in size (Fig. 21f, position B), even when the pitch contains a low amount of QI, as in the case of the impregnating coal-tar pitch (Fig. 21e, position C). It is not easy to explain this peculiar behavior of the primary QI material. A possible explanation might be that, under pressure, volatiles diffuse into the reaction mass. Consequently, the viscosity of the system is reduced [135,136], facilitating the segregation of the QI particles and the formation of large aggregates.

When high-pressure is used carbonization/graphitization is completed at much lower temperatures. At 600–800°C, Voronov and Kashevarova demonstrated by X-ray diffraction the formation of graphite from a coal-tar pitch by applying 8 GPa, using a toroid-type device [137]. It is not easy to explain the formation of graphite from coal-tar pitch at such low temperatures. They suggested that gases formed during carbonization/graphitization (CH_4, H_2, etc.) either partially intercalate into the layered carbon crystalline structure or displace to crystallite boundaries, preventing the formation of double and triple bonds between carbon atoms [137]. Other authors attribute this graphitization temperature to the generation of stresses that favor the formation of oriented regions [138–140]. These stresses supply the thermal energy needed to diminish structural defects, producing an increase in local preferred orientation [139]. Thus, the effect of pressure compensates for the effect of temperature, making it possible for graphitization to take place at lower temperatures than when pressure is not applied [140].

The use of pressure also prevents the exudation of pitch during its processing. Thus, densification of carbon materials by impregnation with melted coal-tar pitch significantly improves its efficiency by means of pressure, the result being a reduction in the number of densification cycles [128] and an improvement in the mechanical properties of the resultant materials [141].

2. Role of Solid Particles in Coal-Tar Pitch

(a) Effect of Primary QI Particles. Commercial coal-tar pitches, as mentioned above, contain solid particles embedded in other pitch components. These particles (primary QI) are present in coal-tar pitches in different proportions depending on their origin and determine the subsequent application of the pitch. In the densification of graphite electrodes, where low viscosity for an effective impregnation of the material is desired, coal-tar pitches with QI contents <2.5 wt % are required (Table 1). On the other hand, in the preparation of carbon anodes, coal-tar pitches with QI contents >10 wt % are frequently used (Table 1). In these pitches, QI particles play an important role in the wetting of the calcined petroleum coke grains, controlling penetration of the pitch into the coke and enhancing the density and mechanical properties of the anode [20]. These functions could erroneously lead to the idea that QI particles are inert constituents of coal-tar pitches with a role similar to that of a filler. This is categorically not the case. QI particles are very active components during the carbonization process [71,75,142–144]. QIs significantly affect the development of mesophase by restricting coalescence between spheres (Fig. 18). However, there is some uncertainty about the role of QI in the nucleation of mesogenic molecules. Taylor et al. [130] found that at very early stages of mesophase formation (<1 vol % of pitch ransformed into mesophase), mesophase spheres are not attached to QI particles, even when the concentration of QI particles is high. This shows that in this case QIs do not participate in the nucleation of the mesophase, with the association between mesophase and QI occurring in a postnucleation stage. At a later stage, when mesophase content is >1 vol %, QI particles begin to surround the mesophase spheres, until the spheres are completely covered by QI material (Fig. 18). The consumption of QI material as mesophase develops is illustrated schematically in Fig. 22 [143]. Different authors have tried to clarify the role of the QIs during mesophase formation and their results show that certain aspects remain obscure. Tilmanns et al. [145] stated that the presence of QIs accelerates the formation of mesophase, whereas Romovacek et al. [73] observed that QIs delay mesophase formation. On the other hand, Stadelhofer [146] showed that when QI content is <10 wt % no effects are observed in the initial stages of mesophase formation. These different points of view may be a consequence of

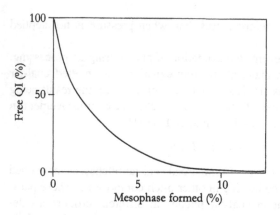

FIG. 22 Schematic relationship between "free" QI (not in contact with mesophase) and the amount of mesophase formed [143].

differences in the characteristics of the raw material and/or the experimental conditions used in the pyrolysis.

The effects of QI on mesophase development can be modeled by comparing the results obtained during the thermal treatment of an impregnating pitch in a stirring reactor and the same impregnating pitch freed of QIs by the removal of particles through filtration. QI material, even in small amounts as in an impregnating coal-tar pitch, is able to regulate the growth of the mesophase, restricting its coalescence. At the same time, QIs also regulate the mechanisms involved in mesophase development, as shown in Figs. 23 and 24. In the earliest stages of mesophase development, spheres are predominantly formed by the segregation of mesogens from the isotropic phase (kinetic rate constant, $k_1 > 0$) and, to a lesser extent, by the incorporation of mesogens to the existent spheres (kinetic rate constant, $k_2 \geq 0$, being $k_1 \gg k_2$). The presence or absence of primary QI particles does not affect the first step of mesophase formation as monitored by optical microscopy. Differences are first observed at more advanced stages of mesophase development. When the primary QIs are present, mesophase mainly grows as a result of the transfer of mesogens from the isotropic pitch ($k_2 > 0$) to the spheres. The process of formation or nucleation of new spheres stops ($k_1 = 0$) as no tiny spheres are observed. On the other hand, when primary QIs are removed, the picture is a different one (Fig. 24). Now mesophase spheres of all sizes can be seen in the pitch. Various mechanisms might be proposed for mesophase development and sphere growth. As in the previous case, mesogens might be incorporated into the mesophase. However new mesophase spheres are also formed ($k_1 > 0$) and, of course, the coalescence of spheres occurs without any hindrance ($k_3 > 0$). As the treatment continues these differences persist. When

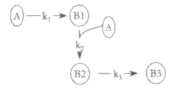

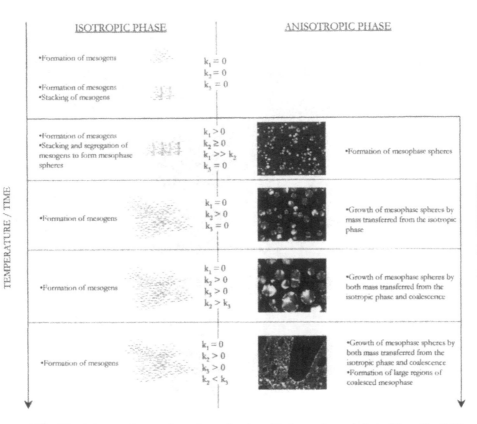

FIG. 23 Scheme of mesophase formation in a binder coal-tar pitch (~10 wt % of QI content). A = Mesogens in isotropic phase. B1 = New emerged mesophase spheres. B2 = Large mesophase spheres formed from B1 by incorporation of A. B3 = Coalesced mesophase.

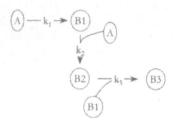

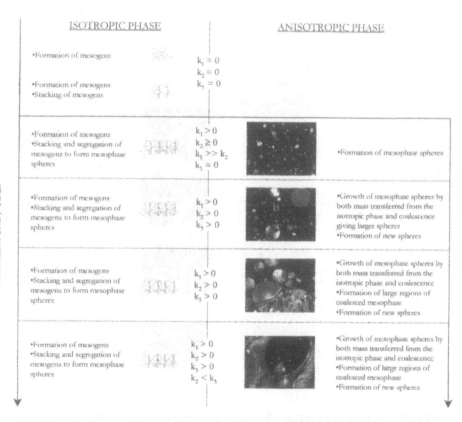

FIG. 24 Scheme of mesophase formation in a QI-free coal-tar pitch. A = Mesogens in isotropic phase. B1 = New emerged mesophase spheres. B2 = Large mesophase spheres formed from B1 by incorporation of A. B3 = Coalesced mesophase.

the mesophase content is high, coalescence also occurs in the pitch with the primary QI and then the mesophase becomes the continuous phase. But the most remarkable difference is that, even when only a small number of isotropic islands remain untransformed, the nucleation of new mesophase spheres still continues in the QI-free pitch. By contrast this process only occurs in the earliest stages of mesophase development when QI particles are present. More research is needed to achieve a better undersanding of the processes involved in these experiments and to clarify the role that primary QI particles play in mesophase formation and development.

The regulating effect of the QI particles on the mesophase is also observed in pitch blends prepared from coal-tar pitch and petroleum pitch [131]. The mechanisms have not yet been clarified. However, when a QI-free coal-tar pitch is carbonized, the mesophase develops in a similar way to that of a petroleum pitch (Fig. 24). In a QI-free coal-tar pitch, mesophase development is an overlapped process wherein coalesced mesophase coexists with large mesophase spheres and with newly emerged spheres.

The effects of QIs on mesophase development have a direct consequence on the optical texture of the coke and its properties. The result is that with the increase in QI content, the size of the microcrystallites decreases (Fig. 25). These variations in the optical texture affect the properties of the carbon materials derived from QI-containing pitches. As an example, the increase in QI content leads to pitch-based unidirectional C/C composites with higher flexural strength [75], interlaminar shear strength, and compressive strength [147]. This beneficial effect derives mainly from the reduction in the volumetric contraction of the matrix, which is due to the presence of QI, resulting in a smaller number of microcracks [75]. Moreover, when QIs are present fiber/matrix bonding is much stronger [75].

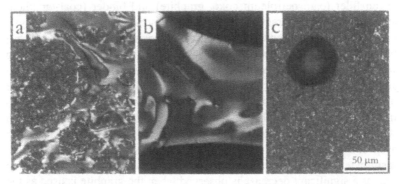

FIG. 25 Optical micrographs of cokes obtained from (a) a binder coal-tar pitch (11 wt % QI) and the same pitch (b) QI free and (c) QI enriched (47 wt %).

Another property affected by QIs is the reactivity of the carbon materials. As a result of the higher air reactivity of the QIs with respect to other pitch components, pitch-based unidirectional C/C composites are more rapidly oxidized as the QI content increases [83].

These results show that QI content must be controlled for the optimization of the mechanical properties and reactivity of the materials. Moreover, a high concentration of QI particles in the pitch may cause problems, related mainly to the processing of the pitch. Thus, in the preparation of particulate carbons and C/C composites, a surplus of QIs can lead to a lack of wettability, making the interaction between the filler (particulate carbons) or fibers (C/C composites) and the pitch more difficult. Nevertheless, Casal et al. [147] prepared unidirectional C/C composites using a pitch with 47 wt % of primary QIs.

(b)   Effect of Added Solid Particles.   The addition of solid particles to coal-tar pitches has been widely used as a procedure to modify the pyrolysis behavior of pitches. These particles include, for example, graphite [148], silica [144], and carbon black [144,148,149]. Carbon black has been successfully used to increase the carbon yield and to reduce pitch coke porosity [149] and hence to hinder the growth and coalescence of mesophase reducing pitch volumetric expansion and coke optical texture [144]. Menéndez et al. [149] reported that, depending on temperature, the specific carbon black used in their studies behaves in a different way. Above 425°C, it facilitates the polymerization of pitch molecules, whereas below 425°C hydrogenation of aromatics takes places, increasing pitch solubility in hexane. Differences in the structure, particle size, and surface activity of carbon blacks may cause these particles to exhibit different behaviors [150].

(c)   Effect of the Filler on Particulate Carbons.   Coal-tar pitches are widely used in the preparation of particulate carbons. In materials of this type, an interaction between filler (e.g., petroleum coke, graphite) and binder (coal-tar pitch) might be expected. This interaction might cause a modification of the pyrolysis behavior of the pitch compared to when pitch is carbonized alone. Méndez [151] has shown that this interaction occurs even at relatively low temperatures during the mixing stage. The toluene-insoluble content and glass transition temperature of the pitch are the parameters that are greatly altered by the presence of the filler. Table 6 shows the variations in these properties in a thermally treated impregnating coal-tar pitch (TT-CTP) and an air-blown impregnating coal-tar pitch (AB-CTP), before and after being mixed with petroleum coke and graphite. The filler may produce the opposite effect in the properties of the pitches. Thus, the glass transition temperatures increase when pitches are mixed with the petroleum coke, whereas a significant decrease is observed when the graphite is used as the filler. This behavior cannot be explained easily. One possible explanation could be that in the case of the graphite some components of the pitches, presumably

TABLE 6 Variation of Pitch Properties in the Presence of Different Fillers

Pitch	Filler	TI[a]	T_g[b]
TT-CTP	None	51.6	83.9
	Petroleum coke	66.4	92.8
	Amorphous graphite	65.4	76.3
AB-CTP	None	46.7	78.2
	Petroleum coke	66.3	92.4
	Amorphous graphite	65.1	71.2

[a] Toluene insolubles (wt %).
[b] Glass transition temperature (°C).

the largest and most aromatic, interact with graphite particles being adsorbed on the surface and consequently become insoluble. As a result, the solubility of the pitch decreases as does the glass transition temperature, only generated by the lower molecular weight molecules. In the case of the petroleum coke a different mechanism must be assumed. Moreover, it has also been observed that the temperature of coke formation may be affected by the characteristics of the filler (Fig. 26). Thus, in the presence of petroleum coke, at 500°C, TT-CTP has already been converted to coke (Fig. 26a), which is not the case in the presence of graphite, where TT-CTP is still partially isotropic (Fig. 26b). The total conversion of the pitch to coke does not occur until above 700°C (Fig. 26c).

However, it is not only the filler but also the pitch that may alter the nature of the interactions with the filler. Thus, during the pyrolysis of TT-CTP in the presence of graphite particles (Fig. 27a), mesophase develops initially on the surface of the particles, while in the case of AB-CTP, this does not occur (Fig. 27b) and the rate of mesophase formation is much higher.

FIG. 26 Optical micrographs of a thermally treated coal-tar pitch carbonized (a) in the presence of petroleum coke at 500°C and in the presence of graphite at (b) 500°C and (c) 700°C.

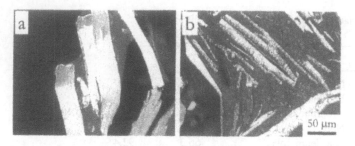

FIG. 27 Optical micrographs of graphite composites treated at 430°C with (a) a thermally treated pitch and (b) an air-blown pitch.

These results are evidence of the great influence that the filler may exert on the properties and pyrolysis behavior of a pitch and, consequently, on the processing of particulate carbons when coal-tar pitch is used as the binder.

3. Effect of Pitch Chemical Composition

Coal-tar pitches differ from each other not only in their primary QI content but also in their chemical composition, as observed in the extrographic distribution of fractions given in Fig. 10. QI particles control the development of mesophase, but ultimately it is the chemical composition of the pitch that determines its thermal reactivity and, consequently, its transformation into coke.

The different behavior of pitches with different composition during carbonization can be evidenced by thermal analysis. As an example, Fig. 28 shows the TG/DTG/DTA and DSC curves of an impregnating coal-tar pitch (CTP-I) and a binder coal-tar pitch (CTP-B). Weight loss occurs in both cases in a single step. CTP-I starts to lose weight at a lower temperature (187°C) and reaches its maximum rate ~20°C below CTP-B (Figs. 28a and 28b). CTP-B also shows a number of subsidiary peaks at >450°C. Nevertheless, both pitches complete their massive release of volatiles at the same temperature (~560°C), weight loss >600°C being practically negligible (Fig. 28a). At 100°C the carbonaceous residue in the impregnating coal-tar pitch is ~6 wt % lower than in the binder coal-tar pitch. This is a consequence of both the QI content and the chemical composition of the pitches. However, other features observed in the TG/DTG curves (onset temperature of weight loss, rate of maximum weight loss, or temperature of final weight loss) are mainly due to the different chemical compositions of the pitches (Fig. 10). As can be seen in Fig. 10, Fr-2 (light PAH) is more abundant in an impregnating coal-tar pitch than in a binder coal-tar pitch. On the other hand, a binder coal-tar pitch is richer in components that elute in Fr-3 (heavy PAH and nonbasic PANCs). This means that the PAH are lighter in an impregnating coal-tar pitch. Moreover, it explains why CTP-I starts its weight loss and

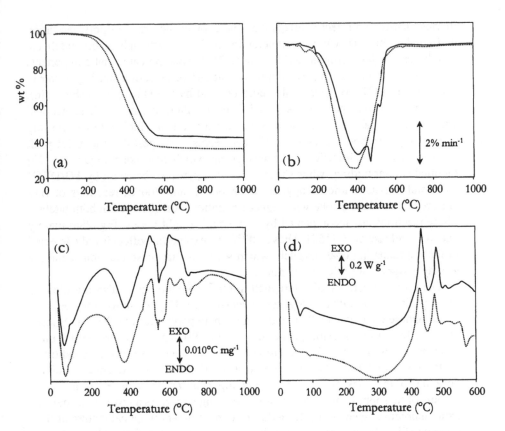

FIG. 28 (a) TG, (b) DTG, (c) DTA, and (d) DSC curves of an impregnating (dashed line) and a binder (continuous line) coal-tar pitch. TA instruments thermobalance, heating rate of 20°C min⁻¹.

reaches the rate of maximum weight loss at lower temperatures than CTP-B. In addition, the differences found in the magnitude of Fr-4 (basic PANCs and monophenols) could be responsible for the different thermal reactivity exhibited by both pitches (Figs. 28c and 28d). Monophenols, which are more abundant in CTP-I, infer a higher reactivity to pitch [40].

In general, the DTA and DSC curves show similar profiles for CTP-I and CTP-B. However, more careful observation reveals certain differences, mainly related to the chemical composition of the pitches. Thus, the endothermic effect below 300°C, ascribable to isomerizations and molecular rearrangement that produce more thermally stable structures [69], overlaps with distillation at a lower temperature in CTP-I than in CTP-B (241°C and 283°C, respectively). The endothermic distillation effect overlaps with the exothermic effect of polymerization and condensation at a lower temperature in CTP-I, although in this pitch the weight loss is more pronounced than in CTP-B. This indicates that components of CTP-I are more reactive, with polymerization/condensation starting earlier in CTP-I than in CTP-B. This is also corroborated by the DSC curves (Fig. 28d). After an exothermic maximum at ~525°C, the exothermic tendency is interrupted with an acute peak appearing at the same temperature at which both pitches stop losing weight (~560°C). This coincidence seems to suggest that this DTA peak is related to the solidification of the mesophase, which gives coke. Figueiras [125] reported similar findings when she studied pitch pyrolysis behavior by TG/DTG/DTA and hot-stage microscopy. Between 600°C and 700°C, reactions of ring closure that produce a coke with higher aromaticity are observed in both pitches. These reactions are accompanied by very small weight losses (Fig. 28a) mainly due to the release of H_2 [152]. Above 700°C the exothermic effect is only detected in the impregnating coal-tar pitch, which suggests that these reactions continue to play a significant role in this pitch.

The higher reactivity of impregnating pitch components is also revealed by the capacity of the pitches to generate mesophase. Fig. 29 shows the variation in mesophase content (Fig. 29a) and mean sphere perimeter (Fig. 29b) with soaking time for an impregnating coal-tar pitch (CTP-I) and a binder coal-tar pitch (CTP-B) carbonized in a glass test tube at 475°C. Under these specific conditions, CTP-I starts to form mesophase before CTP-B. Moreover, irrespective of the duration of the soaking time, impregnating coal-tar pitch develops a greater amount of mesophase, with this mesophase being larger. As a result, coke derived from impregnating coal-tar pitch exhibits a predominantly optical texture of flow domains (Fig. 30a, position A), coexisting with crystallites of a smaller size associated with QI aggregates (Fig. 30a, position B). Since a binder coal-tar pitch

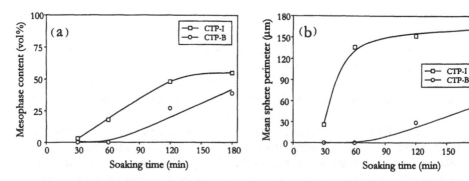

FIG. 29 Variation of (a) mesophase content and (b) mean sphere perimeter for an impregnating and a binder coal-tar pitch carbonized at 475°C.

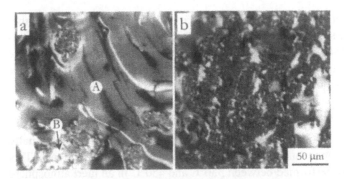

FIG. 30 Optical micrographs of cokes obtained at 900°C from (a) an impregnating and (b) a binder coal-tar pitch. A, domains and B, mosaics associated with QI aggregates.

has a higher QI content, the presence of mosaics associated with QI aggregates is more common in coke derived from this type of pitch (Fig. 30b).

V. MODIFICATION OF COAL-TAR PITCHES FOR HIGH-DENSITY CARBONS

In Section IV it was illustrated how the pyrolysis behavior of a pitch can be modified by changing the experimental conditions or adding solid particles. These modifications offer the possibility of improving the performance of carbon materials for specific applications. Moreover, it is also possible to manipulate the composition of a pitch before carbonizing in order to obtain a new pitch with improved properties (reduced swelling, high carbon yield, good wettability).

Several methodologies have been developed to modify pitch properties and so optimize pitch utilization. They can be grouped into two categories: (1) physical methods and (2) chemical methods. The most widely used physical methods are distillation and solvent extraction. The former produces a modification of the pitch by removing the lightest components [153,154], while solvent extraction leads to the removal and/or the concentration of some specific types of compounds [155].

The chemical methods modify the pitch by acting on the structure of the pitch components. During the treatment, chemical reactions take place, resulting in an irreversible change in pitch composition. Several additives and/or catalysts have been used to modify pitch composition. These include hydrogenation agents (e.g., Ni-Mo/Al$_2$O$_3$, Co-Mo/Al$_2$O$_3$) [156], AlCl$_3$ [157,158], FeCl$_3$ [159], sulfur [157,160,161], boron [157,162], and organic iron compounds [163]. The polymerization of coal-tar pitches using these types of catalysts and/or additives has hardly been exploited. This is because the use of these types of promoter is expensive and because coal-tar pitches can be successfully polymerized using other procedures that are simpler and more economic, e.g. thermal treatment in an inert atmosphere and air-blowing at moderate temperatures. These procedures are the most common because of the low investment cost required and the effectiveness of the treatments in achieving pitches with improved properties.

A. Thermally Treated Coal-Tar Pitches

Thermal treatment involves both the distillation of the lightest and thermally unreactive pitch components and the polymerization of the remaining ones. This treatment is usually performed between 350°C and 450°C in an inert atmosphere (e.g. nitrogen) to favor polymerization over distillation. Under these conditions several reactions take place producing free radicals by the homolytic cleavage of C-H bonds [88]. These free radicals react with each other in a hydrogen-transferable system, giving rise to dimers, trimers, and oligomers [84]. The result is the formation of large planar aromatic macromolecules that can form meso-phase. As previously mentioned, the molecular size of these macromolecules and the amount of mesophase generated are dependent on the experimental conditions (temperature, soaking time, etc.). These experimental conditions, therefore, must be determined with the utmost care in order to control the extent of the devolatil-ization process and thermal polymerization and, consequently, the properties of the final pitch.

There are two main reasons for thermally treating a pitch. The first is to adjust certain parameters of the pitch so that they are in accordance with its subsequent utilization. Carbon yield, softening point, or solubility parameters can be in-creased by thermal treatment without altering the binding capacity of the pitch (Table 7). At the same time, a general improvement in the structure and properties of the resulting pitch coke (porosity, density, optical texture) can be achieved [108,149,164]. These pitches are particularly appropriate for the preparation of matrices for different types of carbon composites [151,164,165]. The second rea-son derives from the great propensity of coal-tar pitches to generate mesophase. Coal-tar pitches are excellent sources of mesocarbon microbeads with a narrow size distribution [130], suitable for the preparation of polygranular carbons [166,167].

B. Air-Blown Coal-Tar Pitches

Similar effects to those obtained by the thermal treatment on the properties of coal-tar pitches can be achieved by air-blowing at moderate temperatures (Table 8) [44]. This is because the presence of oxygen enables polymerization of pitch components to take place at lower temperatures (<300°C). The global reaction is as follows [168]:

$$2Ar\text{-}H + \tfrac{1}{2}O_2 \rightarrow Ar\text{-}Ar + H_2O$$

TABLE 7 Properties of Pitches Obtained by Thermal Treatment of an Impregnating Coal-Tar Pitch at 430°C and Several Soaking Times, Using a Stirred Reactor

Sample	Treatment	C/H[a]	SP[b]	TI[c]	NMPI[d]	CY[e]	M[f]
A0	None	1.64	97	20.0	4.7	34.6	0
A1	430°C/2 h	1.82	149	43.9	20.7	54.0	10
A2	430°C/3 h	1.89	174	53.9	29.8	61.4	25
A3	430°C/4 h	1.95	190	57.5	34.7	65.6	37
A4	430°C/5 h	2.04	—	67.5	45.6	74.6	46
A5	430°C/6 h	2.06	—	69.0	49.6	79.4	65

[a] Carbon/hydrogen atomic ratio.
[b] Softening point (Mettler, °C).
[c] Toluene insolubles (wt %).
[d] N-Methyl-2-pyrrolidinone insolubles (wt %).
[e] Carbon yield (900°C, wt %).
[f] Mesophase content (vol %).

Oxygen induces polymerization without necessarily being incorporated in appreciable amounts into the reaction products (Table 8) and without affecting pitch graphitizability [169]. As a result of air-blowing, pitches increase their molecular weight [168]. The mechanisms involved in the formation of larger molecules are still not fully understood. It is generally believed that initially, the aliphatic carbon atoms in an α-position take up oxygen from aromatic rings [170,171] giving rise to epoxy radicals. These are intermediates of condensation reactions that are accompanied by the formation and release of water [172–175]. However, the

TABLE 8 Properties of Pitches Obtained from An Impregnating Coal-Tar Pitch by Air Blowing at 275°C and Several Soaking Times

Sample	Treatment	O[a]	C/H[b]	SP[c]	TI[d]	NMPI[e]	CY[f]
B0	None	1.80	1.64	97	20.0	4.7	34.6
B1	275°C/10 h	1.78	1.72	139	36.6	13.6	48.0
B2	275°C/18 h	1.81	1.83	168	44.6	18.1	57.6
B3	275°C/25 h	1.89	1.86	197	51.8	24.9	61.8
B4	275°C/30 h	1.86	1.87	210	52.0	27.1	62.7

[a] Oxygen content (wt %).
[b] Carbon/hydrogen atomic ratio.
[c] Softening point (Metter, °C).
[d] Toluene insolubles (wt %).
[e] N-Methyl-2-pyrrolidinone insolubles (wt %).
[f] Carbon yield (900°C).

mechanisms of the air-blowing process seem to be dependent on the experimental conditions [176,177] and chemical composition of the pitches [43,171]. In pitches bearing a small amount of aliphatic hydrogen, like a coal-tar pitch, air-blowing promotes the formation of oxy radicals [171], which lead to larger and more condensed molecules (Fig. 31a). On the other hand, when pitches are rich in aliphatic hydrogen and side chains, like petroleum pitch, oxidation takes place preferentially at the methyl or methylene sites (Fig. 31b) leading to intermolecular linking rather than the development of condensed aromatic rings [171]. The formation of such cross-linked structures may be involved in the formation of ether-type bonds. These bonds together with carbonyl groups are mainly responsible for the small incorporation of oxygen during the air-blowing process [43,170,172]. Fernández et al. [43] applied air-blowing to a binder and an impregnating coal-tar pitch, observing the coexistence of the two basic mechanisms mentioned above. In the impregnating coal-tar pitch, which is less aromatic, air-blowing

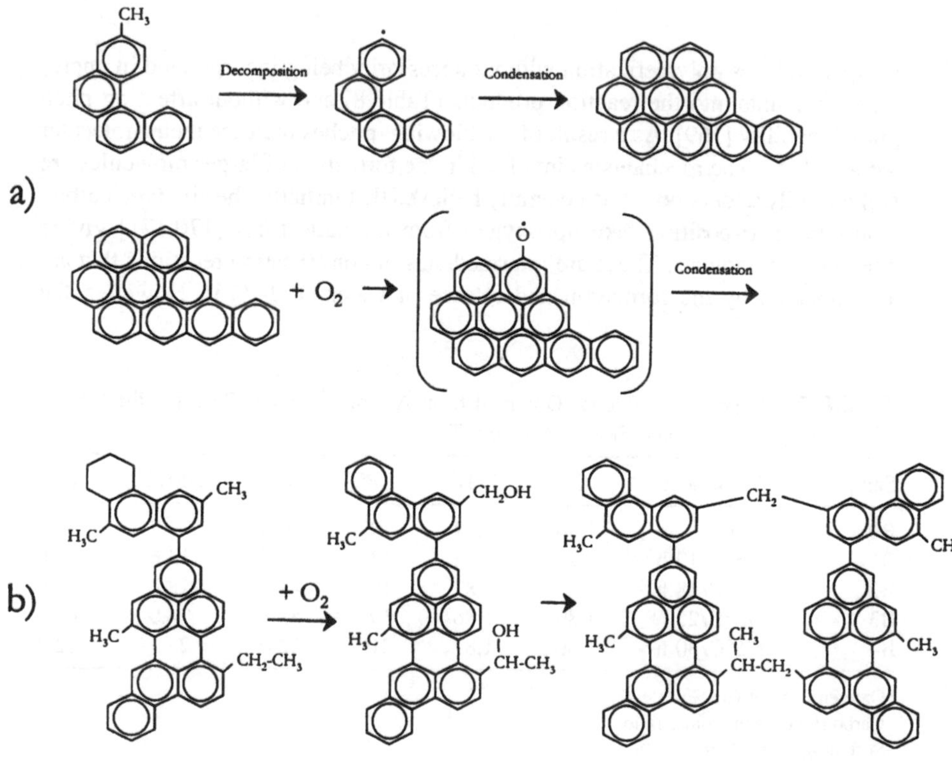

FIG. 31 Oxidation schemes proposed for (a) a coal-tar pitch and (b) a petroleum pitch [171].

produces a substantial delay in the development of the mesophase as a consequence of the formation of cross-linked structures (side bonds) that contribute to generating a system with high viscosity where the mobility and assembling of molecules are restricted. By contrast, in the binder coal-tar pitch the mechanism that leads to the formation of large planar macromolecules prevails.

Studies on the air-blowing of aromatic model compounds [178] and industrial mixtures of aromatic hydrocarbons, such as anthracene oil [176], showed alkyl and bridged alkyl aromatics to be the most reactive compounds. However, mutual influences of the reactivity of hydrocarbons were observed when standard mixtures were air-blown [179], making it more difficult to establish the mechanisms involved.

Pitch air-blowing is a process that continues to attract the attention of scientists who are trying to determine the molecular structure of resultant compounds because of its importance in industrial processing.

C. Comparative Study of Thermally Treated Pitches and Air-Blown Pitches

Pitches with a similar softening point and/or similar carbon yield can be obtained by thermal treatment and air-blowing. If one takes into account just these two parameters alone it might seem that the properties of thermally treated pitches are comparable to those of air-blown pitches. This is categorically not the case. When thermally treated and air-blown pitches are characterized in detail, significant differences emerge not only in their composition but also in their pyrolysis behavior, as has recently been demonstrated by Blanco et al. [44]. These authors thermally treated an impregnating coal-tar pitch at 430°C for 2, 3, 4, 5, and 6 h (Table 7), and also subjected the same pitch to air-blowing at 275°C for 10, 18, 25, and 30 h (Table 8). Although the pitch underwent polymerization in both treatments, the extent of the polymerization and the structure of the polymerized product were completely different depending on the treatment applied. In general terms, it can be said that for pitches with a similar softening point (e.g. A2/B2 or A3/B3) the air-blown pitches are more soluble and have a lower carbon yield than the thermally treated pitches.

The toluene-insoluble content is a widely used parameter for monitoring the extent of polymerization in coal-derived samples [43]. This parameter is effective for comparing samples submitted to the same type of treatment. However, when comparing thermally treated and air-blown pitches the different structures of the material polymerized by both treatments must be borne in mind. In any case, the toluene-insoluble content increases as a consequence of the distillation of the lightest components and the polymerization of the other pitch components. As an example, Fig. 32 shows the DTG curves of the parent pitch and the pitches with similar toluene-insoluble contents obtained by thermal treatment and air-

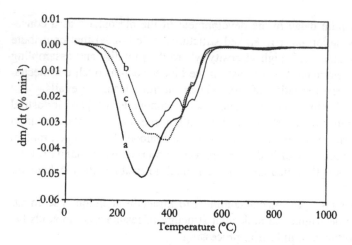

FIG. 32 DTG curves of (a) parent pitch, (b) pitch thermally treated at 430°C/2 h, and (c) pitch air-blown at 275°C/18 h.

blowing. These curves clearly reveal the different pyrolysis behaviors of the treated pitches. The large band centered at ~300°C in the parent pitch substantially decreases in the treated pitches, as a consequence of their different chemical composition and/or molecular structure. This band, which can be assigned to the distillation of the lightest compounds of the pitch, shifts in the treated pitches to higher temperatures. At the same time, pitch treatment also causes magnification of the small shoulders observed above 300°C in the parent pitch. This magnification is more pronounced in the case of air-blown pitch where the band centered at ~400°C is the most intense of the DTG curve (Fig. 32).

In order to find out more about the differences between the polymerized products obtained by thermal treatment and air-blowing, pitches were characterized in greater detail. Iodine adsorption is a technique that has been successfully used to characterize coal-tar pitches [127,180]. It is an established fact that iodine reacts reversibly with polyaromatic compounds, producing charge-transfer complexes in which several molecules of iodine are complexed with one polyaromatic compound [181]. This interaction is favored by the planarity and basicity of the polyaromatic compounds [180]. Figure 33 shows the variation of the iodine index with toluene insolubles for pitches obtained by thermal treatment and air-blowing. The decrease in the iodine uptake showed by the air-blown pitches with the severity of the treatment is related to the nonplanarity of the structures generated. Such treatment preferentially leads to three-dimensional molecules (cross-linked). In the case of the thermally treated pitches, the iodine uptake does not significantly change with the severity of the treatment, as a result of the

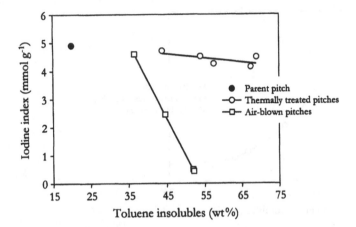

FIG. 33 Variation of iodine index with toluene insolubles for pitches obtained by thermal treatment and air blowing.

planarity of the macromolecules. The formation of these nonplanar large molecules in the air-blown pitches may be the reason for the high-intensity peaks shown in the exclusion region of the SEC chromatograms obtained with NMP as eluent (Fig. 34), while in the case of thermally treated pitches the planarity of the equivalent polymerized molecules makes them insoluble in NMP and, consequently, they are not observed in the SEC chromatogram (Fig. 34).

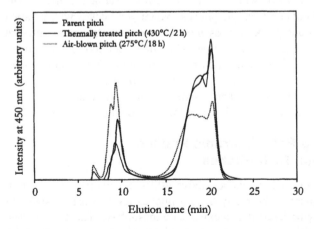

FIG. 34 Size exclusion chromatograms of (a) parent pitch, (b) pitch thermally treated at 430°C/2 h, and (c) pitch air-blown at 275°C/18 h.

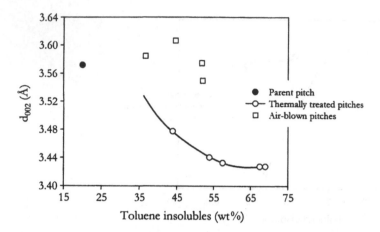

FIG. 35 Variation of interlaminar space (d_{002}) with toluene insolubles for pitches obtained by thermal treatment and air blowing.

Although pitches have no crystallographic order and their X-ray diffraction patterns are not well-resolved, the information obtained by this technique is further evidence of the different chemical structures of the components of thermally treated and air-blown pitches inferred by iodine adsorption. Crystallographic parameters, interlayer spacing (d_{002}), and crystallite size (L_c) progressively increase with soaking time in the thermally treated pitches (Figs. 35 and 36), whereas they remain unaltered in the case of air-blown pitches.

The advantage of pitch air-blowing over pitch thermal treatment is that by air-blowing it is possible to obtain isotropic pitches, which yield cokes with a varied range of optical textures and properties based on the experimental conditions. The disadvantages of air-blowing are the long period of time required, the sensitivity of the pitch properties to variations in experimental conditions (temperature, time, air flow rate, etc.), and the formation of three-dimensional structures that may adversely affect their wettability and their viscoelastic properties at low temperatures as the pitches become more elastic when the air blowing increases [182].

D. High Softening Point Isotropic Pitches and Mesophase-Rich Pitches

The thermal treatment of coal-tar pitches above 400°C results in the formation of mesophase. Consequently, thermally treated pitches are biphasic systems in which the isotropic polymerized pitch coexists with mesophase. There are at least two purposes for separating the isotropic and anisotropic phases of a partially anisotropic pitch. The first is to determine the contribution of the isotropic and

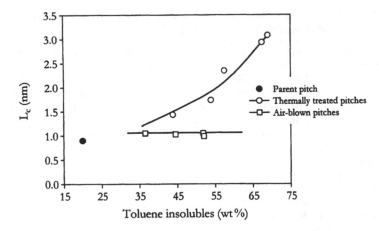

FIG. 36 Variation of crystallite size (L_c) with toluene insolubles for pitches obtained by thermal treatment and air blowing.

the anisotropic phases to the properties and behavior of the bulk pitch. The second is that the polymerized isotropic material and the mesophase are very attractive precursors for the preparation of a great variety of high-density carbon materials, such as isotropic carbon fibers or polygranular graphites as will be described below. For these reasons many investigations have been devoted to developing strategies for mesophase separation from the isotropic phase. High-temperature centrifugation [183,184] and solvent extraction [185] are perhaps the most frequently used techniques for achieving this separation. More recently, a procedure for the separation of mesophase in thermally treated coal-tar pitches by means of filtration under pressure, using moderate temperatures has been developed [186]. This procedure is based on the principle that in partially anisotropic coal-tar pitches mesophase spheres do not deform when submitted to shear stress at a certain temperature, behaving like solid particles in the liquid isotropic pitch [187]. This property allows mesophase to be effectively separated from the isotropic phase by heating. The application of pressure facilitates the filtration of the liquid (isotropic phase), with the solid (mesophase) remaining in the filter cake. By this procedure mesophase can be concentrated in percentages of up to ~90 vol % [130].

The thermally treated pitches described above have been successfully filtered, giving rise to two fractions (isotropic and anisotropic). Since all of the isotropic fractions and similarly all of the anisotropic fractions show a regular tendency in their properties and behavior, mainly attributed to the extent of polymerization, the results presented here focus only on one of the pitches (A2, 430°C/3 h). A complete study of the separation of the fractions in the other thermally treated pitches is reported by Blanco et al. [188].

318

Granda et al.

TABLE 9 Properties of a Pitch Thermally Treated at 430°C for 3 h (A2) and Its Isotropic Phase (12) and Coexisting Mesophase (M2)

Sample	M^a	C/H^b	SP^c	TI^d	$NMPI^e$	R_β^f	CY^g
A2	25	1.89	174	53.9	29.8	24.1	61.5
I2	0	1.85	169	45.8	16.3	29.5	56.2
M2	80	2.05	—	66.7	53.4	13.3	74.9
A2 (theo)	25	1.91	—	52.3	27.9	24.4	62.0

[a] Mesophase content (vol %).
[b] Carbon/hydrogen atomic ratio.
[c] Softening point (Mettler, °C).
[d] Toluene insolubles (wt %).
[e] N-Methyl-2-pyrrolidinone insolubles (wt %).
[f] Beta resin determined as TI–NMPI (wt%).
[g] Carbon yield (900 °C, wt %).

Table 9 shows the properties of the isotropic and anisotropic phases separated by filtration from a thermally treated coal-tar pitch (430°C/3 h) [188]. After filtration two fractions are obtained: the filtered fraction and the residue. The former is completely isotropic whereas the latter is enriched in mesophase with some remnants of isotropic material (Fig. 37). The mesophase spheres do not distort or coalesce during filtration. The characterization of these fractions clearly reveals their different nature. All of the QI particles, which were previously attached to the mesophase spheres, are in the anisotropic fraction. The isotropic fraction, therefore, is a primary QI-free pitch. The parameters summarized in Table 9 show that the properties of the whole pitch can be considered as an average of both fractions. The isotropic fraction is composed of less aromatic compounds that render it more soluble than the anisotropic fraction.

The differences between the two fractions are also patent from their pyrolysis

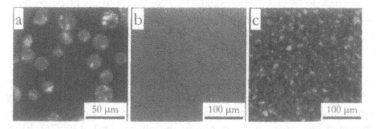

FIG. 37 Optical micrographs of (a) an impregnating coal-tar pitch thermally treated at 430°C/3 h and its (b) isotropic and (c) anisotropic fractions.

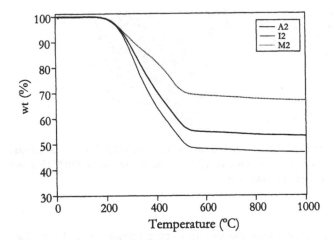

FIG. 38   TG curves of a thermally treated coal-tar pitch, A2, and its coexisting phases, I2 (isotropic fraction) and M2 (anisotropic fraction).

behavior. The thermogravimetry analyses show that the thermally treated pitch exhibits a behavior intermediate between its two coexisting phases (Figs. 38 and 39). The isotropic fraction loses weight in a completely different way than the anisotropic fraction. Most of its weight loss occurs below 400°C, while the anisotropic fraction undergoes a noticeable weight loss between 400°C and 600°C.

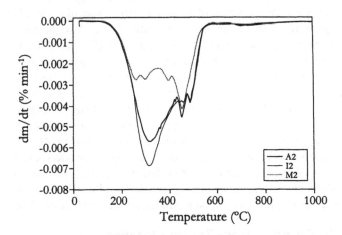

FIG. 39   DTG curves of a thermally treated coal-tar pitch, A2, and its coexisting phases, I2 (isotropic fraction) and M2 (anisotropic fraction).

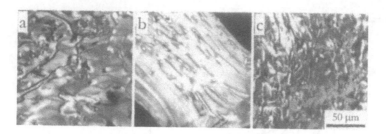

FIG. 40 Optical micrographs of cokes obtained by carbonization at 900°C in a horizontal tube furnace from (a) an impregnating coal-tar pitch thermally treated at 430°C/3 h and its (b) isotropic and (c) anisotropic fractions.

(Fig. 38). This is in agreement with the structures predominant in each fraction: lighter in the isotropic fraction and heavier, highly aromatic and more condensed in the anisotropic fraction. The fact that between 400°C and 600°C the isotropic and the anisotropic fractions lose comparable masses indicates that polymerization occurs in both coexisting phases. A comparison of the three DTG curves reveals that below 400°C the thermally treated pitch behaves in a similar way to the isotropic fraction, while above 400°C it exhibits an intermediate behavior between isotropic and anisotropic fractions.

Given the very different behavior of the two fractions, it is not surprising that cokes derived from these samples have very different microstructures (Fig. 40). The isotropic fraction gives rise to a coke with a needle-type microstructure (Fig. 40b). The formation of this type of microstructure occurs in a system of low viscosity to which not only the presence of large aromatic molecules but also the total absence of QI particles in this fraction contributes. In contrast, the anisotropic fraction produces a coke made up of small crystallites (Fig. 40c). In this case, the decrease in microstructure size may be due to the fact that this fraction generates high-viscosity systems upon carbonization, restricting the coalescence and formation of large anisotropic regions. In addition, this fraction is enriched in QI particles (10.2 wt %), which also restrict the development of large orientated microstructures. Judging from the optical texture, the coke obtained from the anisotropic fractions resembles that of the thermally treated pitch (Fig. 40a).

As precursors for carbon materials, the isotropic phase has been successfully used for the preparation of general-purpose carbon fibers [189,190] whereas the anisotropic part is adequate for high-density polygranular graphites because this fraction has the characteristics of mesocarbon microbead material [166,191,192].

E. High-Density Carbons from Treated Pitches

Treated pitches, obtained from commercial coal-tar pitches, have been successfully used in the preparation of a great diversity of high-density materials (Fig. 41).

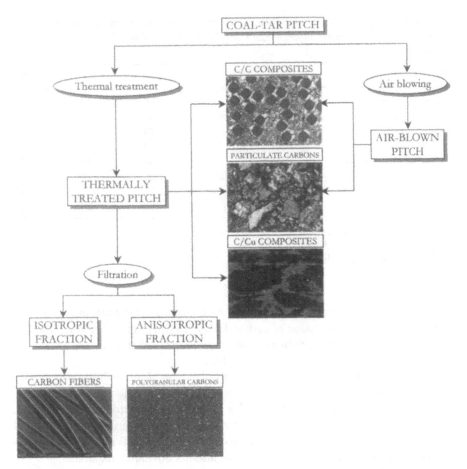

FIG. 41 Current applications of treated coal-tar pitches and their separated fractions.

Unidirectional C/C composites and fine-particulate carbons have been prepared from commercial coal-tar pitches [193,194]. However, the resultant materials show structural imperfections (e.g., swelling, large pores and cracks) generated during carbonization because of the massive release of volatiles originating from the pitch. The use of thermally treated and air-blown pitches reduces the massive release of volatiles, and consequently, the resultant materials exhibit an improved structure and superior mechanical properties [164,194]. As an example, Table 10 illustrates the improvement achieved using treated coal-tar pitches in the preparation of fine-particulate carbons.

Additional important uses of the thermally treated pitches derive from the possibility of separating the two coexisting phases. Since the isotropic fraction

TABLE 10 Properties of Fine-Particulate Carbons Prepared from Several Treated Pitches Derived from an Impregnating Coal-Tar Pitch, and Using Different Reinforcements

Reinforcement	Pitch treatment	d_{ap}[a]	P[b]	CS[c]
Foundry coke	None	1.09	42	18
	275°C/18 h/air	1.44	23	43
	430°C/3 h/N_2	1.44	21	82
	430°C/5 h/N_2	1.43	21	62
Amorphous graphite	None	1.49	29	27
	275°C/18 h/air	1.67	20	49
	430°C/3 h/N_2	1.57	22	47
	430°C/5 h/N_2	1.57	20	27

[a] Apparent density (water, g cm^{-3}).
[b] Open porosity (water, vol %).
[c] Compressive strength (MPa).

yields needle-type microstructures upon carbonization, this fraction is used as a precursor for the preparation of carbon fibers [189]. On the other hand, the isolated mesophase spheres, with a narrow distribution of diameters, are used in the preparation of polygranular carbons [166].

The applications mentioned above are only some of the multiple possibilities for treated coal-tar pitches and their fractions. The preparation of pitch/graphite/copper systems for electric applications [195] or the insertion of lithium in the mesophase spheres for lithium-ion batteries [196] are just a few of the potential applications of these versatile carbon precursors. However, for the optimal development of these materials further investigation is obviously required.

VI. COAL-TAR PITCH ENVIRONMENTAL AND TOXICOLOGICAL ISSUES

The toxicological aspects of coal-tar pitches have attracted great interest, and, consequently, much work has been carried out with the aim of evaluating the toxicity of the different PAHs present in pitches, and the effects caused by these compounds in humans [197-202]. Based on these studies, the U.S. Environmental Protection Agency, in a conscientious report [3], described 16 PAHs as priority pollutants, potentially harmful for public health (Table 11). However, not all the PAHs have the same propensity to produce cancer. A relative potency factor has, therefore, been defined [153,154,203,204] establishing benzo[a]pyrene as the reference PAH and assigning it a carcinogenic factor of 1. The carcinogenic potency of the other PAHs is then expressed relative to that of benzo[a]pyrene.

TABLE 11 Molecular Weights and Relative Potential Factors of the 16 PAHs Designed as Priority Pollutants by the U.S. Environmental Protection Agency

PAH	$M_w{}^a$	RPF[b]	PAH	$M_w{}^a$	RPF[b]
Naphthalene	128	0.000	Benzo[k]fluoranthene	252	0.010
Acenaphthene	154	0.000	Benzo[a]anthracene	228	0.033
Acenaphthylene	152	0.000	Fluoranthene	202	0.034
Fluorene	166	0.000	Indeno[1,2,3-c,d]pyrene	276	0.100
Phenanthrene	178	0.000	Benzo[b]fluoranthene	252	0.100
Anthracene	178	0.000	Chrysene	228	0.260
Pyrene	202	0.000	Benzo[a]pyrene	252	1.000
			Benzo[g,h,i]perylene	276	1.000
			Dibenzo[a,h]anthracene	278	1.400

[a] Molecular weight (amu).
[b] Relative potency factor, benzo[a]pyrene = 1 by definition.

Since these 16 PAHs are found in coal-tar pitches in substantial concentrations [153,154,204], pitch suppliers and consumers have developed strategies to mitigate their emission. These strategies include (1) the use of coal-tar pitches with a higher softening point [153,154,203–205]; (2) in the case of Söderberg carbon anodes, a reduction in the percentage of pitch in the paste (dry paste [153]); and (3) the development of blended pitches based on coal-tar and petroleum derivatives [131,206,207]. The use of blended pitches in the Söderberg plants may bring about a significant reduction in the emissions of PAHs. Koppers Industries has developed a coal-tar/petroleum pitch (~40 wt % of petroleum material) suitable for use in the aluminum industry, which has shown an average reduction of ~40 wt % on the benzo[a]pyrene equivalent index when compared with a typical binder coal-tar pitch [154].

The development of blended pitches offers the possibility not only of incorporating new pitches to the binder market but also of introducing a new raw material for the preparation of high-density materials based on the positive characteristics of the two components of the blend.

VII. SUMMARY AND CONCLUSIONS

As stated in the introduction, despite the huge amount of research performed over the years and the great advances achieved in understanding coal-tar pitch composition and carbonization behavior, there are still some aspects that need further development. The need for this is even greater when dealing with new pitches with high softening points. Topics to be considered for future research include (1) the optimization of current solid-state techniques for the characterization of the whole pitch, such as ^{13}C NMR; (2) the search for new techniques for the characterization of the largest components present in the coal-tar pitch and its fractions and/or further developments of current ones, such as matrix-assisted laser desorption ionization mass spectrometry and high-performance size exclusion chromatography; (3) the establishment of adequate processing conditions, which allow maximum benefit from inherent pitch composition to be obtained while reducing the environmental impact; (4) the development of data bases with up-to-date information, to facilitate new initiatives.

Despite all of the current problems associated with the production and use of coal-tar pitch, its role in the field of carbon materials cannot be ignored because of its unique composition. The possibility of tailoring the structure and properties of carbon materials through the processing of coal-tar pitch and the possibility of blending different pitches or pitch fractions will create a wider market for both conventional and high-technology applications.

REFERENCES

1. NR Turner. JOM—The Minerals, Metals and Materials Society 45(11):39–42, 1993.
2. RH Wombles, MDK Kiser. Light Metals 537–541, 2000.
3. U.S. Environmental Protection Agency: Locating and estimating air emissions from sources of polycyclic organic matter. Office of Air Quality, Research Triangle Park, NC, EPA-454/R-98-014, July 1998.
4. Council Directive 90/394/EEC of 28 June 1990 on the protection of workers from the risks related to exposure to carcinogens at work.
5. Council Directive 1999/38/EV of 29 April 1999 amendment for the second time Directive 90/394/ECC on the protection of workers from the risks related to exposure to carcinogens at work and extending it to mutagens.
6. International Agency for Research on Cancer. Overall evaluations of carcinogenicity to humans. In IARC Monographs Database on Carcenogenic Risks to Human (http://www.iarc.fr).
7. RA Greinke. In: PA Thrower, ed. Chemistry and Physics of Carbon, Vol. 24. New York: Marcel Dekker, 1994, pp. 1–43.
8. JD Brooks, GH Taylor. In: PL Walker, Jr, ed. Chemistry and Physics of Carbon, Vol. 4. New York: Marcel Dekker, 1968, pp. 243–286.
9. E Fitzer, K Mueller, W Chaefer. In: P.L. Walker, Jr, ed. Chemistry and Physics of Carbon, Vol. 7. New York: Marcel Dekker, 1971, pp. 237–383.
10. H Marsh, PL Walker, Jr. In: PA Thrower, ed. Chemistry and Physics of Carbon, Vol. 15. New York: Marcel Dekker, 1979, pp. 229–286.
11. RA Forrest, H Marsh, C Cornford, BT Kelly. In: PA Thrower, ed. Chemistry and Physics of Carbon, Vol. 19. New York: Marcel Dekker, 1984, pp. 211–330.
12. A Oberlin, S Bonnamy, P Rouxhet. In: PA Thrower and LR Radovic, eds. Chemistry and Physics of Carbon, Vol. 26. New York: Marcel Dekker, 1999, pp. 1–148.
13. I Mochida, K Shimizu, Y Korai, H Otsuke, S Fujiyama. Carbon 26:843–852, 1988.

14. Y Korai, M Nakamura, I Mochida, Y Sakai, S Fujiyama. Carbon 29:561–567, 1991.
15. I Mochida, SH Yoon, Y Korai, K Kanno, Y Sakai, M Komatsu. In: H. Marsh, F. Rodríguez-Reinoso, eds. Sciences of Carbon Materials. Alicante, Spain: Publicaciones de la Universidad de Alicante, 2000, pp. 259–285.
16. S Ōtani. Carbon 3:31–38, 1965.
17. I. Mochida, SH Ion, Y Korai, K Kanno, Y Sakai, M Komatsu. ChemTech February: 29–37, 1995.
18. I Mochida, R Fujiura, T Kojima, H Sakamoto, K Kanno. Carbon 32:961–969, 1994.
19. YC Chang, HJ Sohn, CH Ku, YG Wang, Y Korai, I Mochida. Carbon 37:1285–1297, 1999.
20. RJ Gray, KC Krupinski. In: H Marsh, EA Heintz, F Rodríguez-Reinoso, eds. Introduction to Carbon Technologies. Alicante, Spain: Publicaciones de la Universidad de Alicante: 1997, pp. 329–423.
21. M Zander, G Collin. Fuel 72:1281–1285, 1993.
22. MD Guillén, MJ Iglesias, A Domínguez, CG Blanco. J Chromatogr 591:287–295, 1992.
23. MD Guillén, MJ Iglesias, A Domínguez, CG Blanco. Energy and Fuels 6:518–525, 1992.
24. R. Álvarez, MA Díez, R García, AI González de Andrés, CE Snape, SR Moinelo. Energy and Fuels 7:953–959, 1993.
25. JR Kershaw, KJT Black. Energy and Fuels 7:420–425, 1993.
26. CG Blanco, J Blanco, P Bernad, MD Guillén. J Chromatogr 539:157–167, 1991.
27. MJ Lázaro, CA Islas, AA Herod, R Kandiyoti. Energy and Fuels 13:1212–1222, 1999.
28. R Menéndez, C Blanco, R Santamaría, J Bermejo, I Suelves, AA Herod, R Kandiyoti. Energy and Fuels 15:214–223, 2001.
29. R Menéndez, J Bermejo, A Figueiras. In: H Marsh, F Rodríguez-Reinoso eds. Sciences of Carbon Materials. Alicante, Spain: Publicaciones de la Universidad de Alicante, 2000, pp. 173–204.
30. M Zander. In: H Marsh, F Rodríguez-Reinoso, eds. Sciences of Carbon Materials. Alicante, Spain: Publicaciones de la Universidad de Alicante, 2000, pp. 205–258.
31. AL Lafleur, Y Nakagawa. Fuel 68:741–752, 1989.
32. AA Herod, S.F Zhang, BR Johnson, KD Bartle, R Kandiyoti. Energy and Fuels 10:743–750, 1996.
33. MJ Lázaro, AA Herod, M Domin, Y Zhou, CA Islas, R Kandiyoti. Rapid Commun Mass Spectrom 13:1401–1412, 1999.
34. AA Herod, BJ Stokes, HR Shulten. Fuel 72:31–43, 1993.
35. AA Herod, R Kandiyoti. J Chromatogr. A 708:143–160, 1995.
36. W Boenigk, MW Haenel, M Zander. Fuel 69: 1226–1232, 1990.
37. M Zander. Fuel 66:1536–1539, 1987.
38. AA Herod. In: AH Neilson ed. The Handbook of Environmental Chemistry, Vol. 3, Part I, PAHs and Related Compounds. Berlin: Springer-Verlag, 1998, pp. 271–323.
39. J Černý. Fuel 68:402–405, 1989.
40. R Menéndez, M Granda, J Bermejo. Carbon 35:555–562, 1997.
41. M Zander. Fuel 66:1459–1466, 1987.

42. Y Martín, R García, RA Solé, SR Moinelo. Chromatographia 47:373–382, 1998.
43. JJ Fernández, A Figueiras, M Granda, J Bermejo, R Menéndez. Carbon 33:295–307, 1995.
44. C Blanco, R Santamaría, J Bermejo, R Menéndez. Carbon 38:517–523, 2000.
45. YQ Fei, K Sakanishi, YN Sun, R Yamashita, I Mochida. Fuel 69:261–262, 1990.
46. NA Kekin, VG Gordienko. Coke and Chemistry 6:60–61, 1986.
47. MD Guillén, J Blanco, JS Canga, CG Blanco. Energy and Fuels 5:188-192, 1991.
48. JM Mulligan, KM Thomas, AP Tytko. Fuel 66:1472–1480, 1987.
49. M Alula, M Diack, R Gruber, G Kirsch, JC Wilhem, D Cagniant. Fuel 68:1330–1335, 1989.
50. Y Martín, R García, RA Solé, SR Moinelo. Energy and Fuels 10:436–442, 1996.
51. M Farcasiu. Fuel 56:9–14, 1977.
52. RA Greinke, LM O'Connor. Anal Chem 52:1877–1881, 1980.
53. KD Bartle, DG Mills, MJ Mulligan, IO Amaechina, N Taylor. Anal Chem 58:2403–2408, 1986.
54. N Evans, TM Haley, JC Mulligan, KM Thomas. Fuel 65:694–703, 1989.
55. KD Bartle, M Zander. Erdöl und Kohle 36:15–22, 1983.
56. M Sasaki, T Yokono, M Satou, Y Sanada. Energy and Fuels 5:122–125, 1991.
57. M Zander. Fuel 70:563–565, 1991.
58. I. Halász. Erdöl und Kohle 32:571, 1979.
59. GP Blümer, HW Kleffner, W Lücke, M Zander. Fuel 59:600–602, 1980.
60. GP Blümer, HW Kleffner, W Lücke, J Palm, M Zander. Erdöl und Kohle 34:81, 1981.
61. M Granda, J Bermejo, SR Moinelo, R Menéndez. Fuel 69:702–705, 1990.
62. NY Beilina, EN Kozhueva, OE Golubkov. Solid Fuel Chem 24 (5):124–127, 1990.
63. VL Cebolla, JV Weber, M Swistek, A Krzton, J Wolszczak. Fuel 73:950–956, 1994.
64. J Machnikowski, H Machnikowska, MA Díez, R Álvarez, J Bermejo. Chromatogr 778:403–413, 1997.
65. SR Moinelo, R Menéndez, J Bermejo. Fuel 67: 682–687, 1988.
66. M Granda, R Menéndez, P Bernad, J Bermejo. Fuel 72:397–403, 1993.
67. J Černý, G Šebor, J Mitera. Fuel 70:857–860, 1991.
68. B Bogdoll, I Halász. Erdöl und Kohle 34:549–556, 1981.
69. M Granda. The influence of chemical composition of pitches on their pyrolysis behavior. PhD dissertation, University of Oviedo, Oviedo, Spain, 1992.
70. M Granda, R Menéndez, SR Moinelo, J Bermejo, CE Snape. Fuel 72:19–23, 1993.
71. R Menéndez, M Granda, J Bermejo, H Marsh. Fuel 73:25–34, 1994.
72. H Marsh, CS Latham, EM Gray. Carbon 23:555–570, 1985.
73. GR Romovacek, JP McCullough, AJ Perrotta. Fuel 62: 1236–1238, 1983.
74. R Álvarez, CS Canga, MA Díez, E Fuente, R García, AI González de Andrés. Fuel Proc Technol 47:281–293, 1996.
75. A Figueiras, M Granda, E Casal, J Bermejo, J Bonhomme, R Menéndez. Carbon 36:883–891, 1998.
76. DR Ball. Carbon 16:205–209, 1978.
77. HA Kremer. Chem and Ind 18:702–713, 1982.
78. DL Belitskus, KC Krupinski. Light Metals, 583–588, 1995.

79. M Forrest, H Marsh. In: EL Fuller, Jr., ed. Coal and Coal Products; Analytical Characterization Techniques. ACS Symposium Series 205, Divisions of Analytical, Fuel, and Colloid and Surface Chemistry, Washington, DC, 1982, pp. 1–25.

80. R Menéndez, M Granda, J Bermejo. In: H Marsh, EA Heintz, F Rodríguez-Reinoso, eds. Introduction to Carbon Technologies. Alicante, Spain: Publicaciones de la Universidad de Alicante, 1997, pp. 461–490.

81. JW Patrick, A Walker. Fuel 64:136–138, 1985.

82. R Loison, P Foch, A Boyer (Eds.). In: Coke Quality and Production. Cambridge, UK: Butterworths, 1989, pp. 158–200.

83. M Granda, E Casal, J Bermejo, R Menéndez. Carbon 39:483–492, 2001.

84. IC Lewis. Carbon 18:191–196, 1980.

85. M Zander, J Haase, H Dreeskamp. Erdöl und Kohle 35:65–69, 1982.

86. RH Hurt, Y Hu. Carbon 37:281–292, 1999.

87. I Mochida, Y Korai, CH Ku, F Watanabe, Y Sakai. Carbon 38:305–328, 2000.

88. IC Lewis. Carbon 20:519–529, 1982.

89. S Stein. Carbon 19:421–429, 1981.

90. T Edstrom, IC Lewis. Carbon 7:85–91, 1969.

91. IC Lewis. Fuel 66:1527–1531, 1987.

92. LG Isaacs. Carbon 8:1–5, 1970.

93. A Martínez-Alonso, J Bermejo, JMD Tascón. J Thermal Anal 38:811–820, 1992.

94. AW Scaroni, RG Jenkins, PL Walker Jr. Carbon 29:969–980, 1991.

95. T Ida, K Akada, T Okura, M Miyake, M Nomura. Carbon 33:625–631, 1995.

96. RA Greinke, IC Lewis. Carbon 22:305–314, 1984.

97. RA Greinke. Carbon 28:701–706, 1990.

98. I Mochida, SI Inoue, K Maeda, K Takeshita. Carbon 15:9–16, 1977.

99. I Mochida, T Ando, K Maeda, K Takeshita. Carbon 16:453–458, 1978.

100. H Marsh, JM Foster, G Hermon, M Iley, JN Melvin. Fuel 52:243–252, 1973.

101. M Sato, Y Matsui, K Fujimoto. Characterization of charge-transfer adducts of coal-tar pitches. Extended Abstracts of 17th Biennial Conference on Carbon. Lexington, KY, 1985, pp. 326–327.

102. W Boenigk, MW Haenel, M Zander. Fuel 74:305–306, 1995.

103. JC Fetzer, JR Kershaw. Fuel 74:1533–1536, 1995.

104. M Inagaki, K Kuroda, M Sakai, E Yasuda, S Kimura. Carbon 22:335–339, 1984.

105. J Bermejo, M Granda, R Menéndez, R García, JMD Tascón. Fuel 76:179–187, 1997.

106. A Martínez-Alonso, J Bermejo, M Granda, JMD Tascón. Fuel 71: 611–617, 1992.

107. J Bermejo, M Granda, R Menéndez, JMD Tascón. Carbon 32:1001–1010, 1994.

108. J Bermejo, R Menéndez, A Figueiras, M Granda. Fuel 74:1792–1799, 1995.

109. MG Raspapov, VP Balykin, GD Kharlampovich. Solid Fuel Chem 20:108–113, 1986.

110. TV Dobroserdova, TM Kulikova, EK Smetanina, VI Golenkov, NI Andronova. Coke Chem 12:42–45, 1981.

111. VI Denisenko, AN Chistyakov, MV Vinogradov, ML Itskov. Solid Fuel Chem 18: 118–124, 1984.

112. AI Demidova, VI Boronina, TS Bykova, IV Chizhov. Coke Chem 3:55–57, 1989.

113. DD Rustschev. Application of DTA- and TG- analysis to investigate the carboniza-

tion process. Preprints of 1st Cokemaking Congress, Essen, Germany, 1987, D4 1–12.

114. RW Wallouch, NN Murty, EA Heintz. Carbon 10:729–735, 1972.
115. B Rand. In: JD Bacha, JW Newman, JL White, eds. Petroleum-Derived Carbons. Washington, DC: ACS Symposium Series 303, 1986, pp. 45–61.
116. P Ehrburger, C Martin, J Lahaye, JL Saint-Romain, P Couderc. Fuel Proc Technol 20:61–67, 1988.
117. M Alula, D Cagniant, JC Lauer. Fuel 69:177–182, 1990.
118. E Fitzer, K Müller. Ber Dt Keram Ges 48:269–275, 1971.
119. IC Lewis, T Edstrom. J Org Chem 28:2050–2057, 1963.
120. JM Guet, DT Choubar. Carbon 23:273–280, 1985.
121. I Katime, LC Cesteros, JR Ochoa. Thermochim Acta 59:25–30, 1982.
122. NA Bacon, WA Barton, LJ Lynch, DS Webster. Carbon 25:669–678, 1987.
123. S Evans, H Marsh. Carbon 9:733–746, 1971.
124. I Mochida, H Marsh. Fuel 58:626–632, 1979.
125. A Figueiras. Influence of chemical composition of pitches on structure and properties of unidirectional C/C composites. PhD dissertation, University of Oviedo, Oviedo, Spain, 1997.
126. JR Pels, F Kapteijn, JA Moulijn, Q Zhu, KM Thomas. Carbon 33:1641–1653, 1995.
127. R Menéndez, H Marsh, C Calvert, T Takekawa. Fuel Proc Technol 20:197–205, 1988.
128. M Granda, JW Patrick, A Walker, E Casal, J Bermejo, R Menéndez. Carbon 36: 943–952, 1998.
129. E Casal, M Granda, J Bermejo, R Menéndez. J Microsc 201:324–332, 2001.
130. F Fanjul, M Granda, J Bermejo, R Menéndez. Preparation of precursors of graphites from coal-tar pitches (in Spanish). Proceedings of the 5th Conference of the Spanish Carbon Group. Oviedo, Spain, 1999, pp. 145–148.
131. M Pérez, M Granda, R García, E Romero, R Menéndez. Light Metals 573–579, 2001.
132. KJ Hüttinger, JP Wang. Carbon 30:9–15, 1992.
133. KJ Hüttinger, JP Wang. Carbon 30:1–8, 1992.
134. E Fitzer, B Terwiesch. Carbon 11:570–574, 1973.
135. KJ Hüttinger, U Rosenblatt. Carbon 15:69–74, 1977.
136. V Krebs, M Elalaoui, JF Mareche, G Furdin, R Bertau. Carbon 33:645–651, 1995.
137. OA Voronov, LS Kashevarova. Inorg Mater 29:304–306, 1993.
138. M Inagaki, RA Meyer. In: PA Thrower, LR Radovic, eds. Chemistry and Physics of Carbon, Vol. 26. New York: Marcel Dekker, 1999, pp. 149–244.
139. DB Fischbach. In: PL Walker, Jr., ed. Chemistry and Physics of Carbon, Vol. 7. New York: Marcel Dekker, 1971, pp. 1–105.
140. RM Bustin, JN Rouzaud, JV Ross. Carbon 33:679–691, 1995.
141. T Hosomura, H Okamoto. Mater Sci and Eng A143:223–229, 1991.
142. R Menéndez, EM Gray, H Marsh, RW Pysz, EA Heintz. Carbon 29:107–118, 1991.
143. GH Taylor, GM Pennock, JD FitzGerald, LF Brunckhorst. Carbon 31:341–354, 1993.
144. K Kuo, H Marsh, D Broughton. Fuel 66:1544–1551, 1987.
145. H Tilmanns, G Pietzka, H Pauls. Fuel 57:171–173, 1978.
146. JW Stadelhofer. Fuel 59:360–361, 1980.

147. E Casal, J Viña, J Bonhomme, M Granda, R Menéndez. J Mater Chem 10:2637–2641, 2000.
148. H Marsh, F Dachille, M Iley, PL Walker Jr., PW Whang. Fuel 52:253–261, 1973.
149. R Menéndez, JJ Fernández, J Bermejo, V Cebolla, I Mochida, Y Korai. Carbon 34:895–902, 1996.
150. JJ Fernández. Preparation of C/C composites from coal-tar pitch. Chemical modification of pitches by different treatments. PhD dissertation, University of Oviedo, Oviedo, Spain, 1996.
151. A Méndez. Carbon composite materials for break systems. PhD dissertation, University of Oviedo, Oviedo, Spain, 2001.
152. J Bermejo, AL Fernández, M Granda, F Rubiera, I Suelves, R Menéndez. Fuel 80: 1229–1238, 2001.
153. AA Mirtchi, AL Proulx, L Castonguay. Light Metals, 601–607, 1995.
154. ER McHenry, WE Saver. Light Metals, 463–468, 1996.
155. DM Riggs. In: JD Bacha, JW Newman, JL White, eds. Division of Petroleum Chemistry. Washington DC: ACS Vol. 29(2), 1984, pp. 480–494.
156. A Ishihara, X Wang, H Shono, T Kabe. Energy Fuels 7: 334–336, 1993.
157. VV Kulakov, EI Neproshin, EN Fedeneva. Solid Fuel Chem 18(5):123–125, 1984.
158. I Mochida, Y Sone, Y Korai. Carbon 23:175–178, 1985.
159. S Otani, A Oya. Bull Chem Soc Jpn 45:623–624, 1972.
160. F Kolář, J Svítilová, K Balík. Ceramics—Silikáty 39:81–120, 1995.
161. E Fitzer, W Hüttner, LM Manocha. Carbon 18:291–295, 1980.
162. A Becker, J Gremmels, KJ Hüttinger. Carbon 37:953–960, 1999.
163. M Braun, J Kramer, KJ Hüttinger. Carbon 33:1359–1367, 1995.
164. R Menéndez, M Granda, JJ Fernández, A Figueiras, J Bermejo, J Bonhomme, J Belzunce. J Microsc 185:146–156, 1997.
165. C Blanco, R Santamaría, J Bermejo, J Bonhomme, R. Menéndez. J Microsc 196: 213–223, 1999
166. F Fanjul, M Granda, J Bermejo, R Menéndez. Preparation of fine-grained isotropic carbons by the sintering of powdered mesophase. Abstracts and Programme of the Eurocarbon'2000, Deutsche Keramische Gesellschaft e.V., Berlin, 2000, Vol. 2, pp. 501–502.
167. WR Hoffmann, KJ Hüttinger. Carbon 32:1087–1103, 1994.
168. JB Barr, IC Lewis. Carbon 16, 439–444, 1978.
169. JJ Fernández, A Figueiras, M Granda, J Bermejo, JB Parra, R Menéndez. Carbon 33:1235–1245, 1995.
170. T Metzinger, KJ Hüttinger. Carbon 35:885–892, 1997.
171. S Zeng, T Maeda, K Tokumitsu, J Mondori, I Mochida. Carbon 31:413–419, 1993.
172. M Hein. Erdöl und Kohle 43:354–358, 1990.
173. IC Lewis, LS Singer. J Phys Chem 85:354–360, 1981.
174. T Matsumoto, I Mochida. Carbon 30:1041–1046, 1992.
175. CQ Yang, JR Simms. Carbon 31:451–459, 1993.
176. AL Fernández, M Granda, J Bermejo, R Menéndez. Carbon 38:1315–1322, 2000.
177. I Mochida, T Inaba, Y Korai, H Fujitsu, K Takeshita. Carbon 21:543–552, 1983.
178. C. Yamaguchi, J Mondori, A Matsumoto, H Honma, H Kumagai, Y Sanada. Carbon 33:193–201, 1995.

179. JG Lavin. Carbon 30:351–357, 1992.
180. H López, T Yokono, K Murakami, Y Sanada, H Marsh. Fuel 66:866–867, 1987.
181. NM Rodríguez, H Marsh. Fuel 66:1727–1732, 1987.
182. R Menéndez, O Fleurot, C Blanco, R Santamaría, J Bermejo, D Edie. Carbon 36: 973–979, 1998.
183. LS Singer, DM Riffle, AR Cherry. Carbon 25:249–257, 1987.
184. M Kodama, T Fjirua, K Esumi, K Meguro, H Honda. Carbon 26:595–598, 1988.
185. A Gschwindt, KJ Hüttinger. Carbon 32:1105–1118, 1994.
186. C Blanco, R Santamaría, J Bermejo, R Menéndez. Carbon 35:1191–1193, 1997.
187. C Blanco, O Fleurot, R Menéndez, R Santamaría, J Bermejo, D Edie. Carbon 37: 1059–1064, 1999.
188. C Blanco, R Santamaría, J Bermejo, R Menéndez. Carbon 38:1169–11176, 2000.
189. V Prada, C Blanco, R Santamaría, J Bermejo, R Menéndez. Preparation of carbon fibers from the isolated isotropic phase of a thermally treated coal-tar pitch. Extended Abstracts and Program of the 24th Biennial Conference on Carbon, American Carbon Society, Charleston, SC, 1999, Vol. 1, pp. 374–375.
190. E Mora. Development of new precursors for carbon fibres from pitches. Msc dissertation, University of Oviedo, Oviedo, Spain, 2001.
191. F Fanjul. Preparation of precursors of polygranular graphite from pitches. Msc dissertation, University of Oviedo, Oviedo, Spain, 2000.
192. F Fanjul, M Granda, R Santamaría, J Bermejo, R Menéndez. J Anal Appl Pyrol 58–59: 911–926, 2001
193. A Figueiras, JJ Fernández, M Granda, J Bermejo, E Casal, R Menéndez. J Microsc 177:218–229, 1995.
194. C Blanco, R Santamaría, J Bermejo, R Menéndez. Carbon 38:1043–1051, 2000.
195. M Pérez, M Granda, J Bermejo, R Menéndez. C/Cu composites. Effect of composition on the structure of the material (in Spanish). Proceedings of the 5th Conference of the Spanish Carbon Group. Oviedo, Spain, 1999, pp. 141–144.
196. JS Kim. J Power Sources 97:70–72, 2001.
197. C Tremblay, B Armstrong, G Thériault, J Brodeur. Am J Ind Med 27:335–348, 1995.
198. A Rønneberg, A Andersen. Occup Environ Med 52:250–254, 1995.
199. A Rønneberg. Occup Environ Med 52:255–261, 1995.
200. M De Méo, C Genevois, H Brandt, M Laget, H Bartsch, M Castegnaro. Chem-Biol Interact 101:73–88, 1996.
201. MR Cullen, H Checkoway, BH Alexander. Occup Environ Med 53:782–786, 1996.
202. G Benke, M Abramson, M Sim. Ann Occup Hyg 42:173–189, 1998.
203. AA Mirtchi, L Noël. Polycyclic aromatic hydrocarbons (PAH's) in pitches used in the aluminium industry. Extended Abstracts of the Carbon '94, Spanish Carbon Group, Granada, Spain, 1994, pp. 794–795.
204. M Eie, M Sørlie, HA Øye. Light Metals 469–475, 1996.
205. M Eie, M Sørlie. Environmental aspects of selecting pitches for Søderberg anode operation. Extended Abstracts of the Carbon '94, Spanish Carbon Group, Granada, Spain, 1994, pp. 824–825.
206. ER McHenry. Light Metals 543–548, 1995.
207. NR Turner, SH Alsop, O Malmros, D Whittle, BE Hansen, EH Stenby, SI Andersen. Light Metals 565–572, 2001.

Index

9 780367 395278